A first course in

# Quantum Mechanics

## THE MODERN UNIVERSITY PHYSICS SERIES

This series is intended for readers whose main interest is in physics, or who need the methods of physics in the study of science and technology. Some of the books will provide a sound treatment of topics essential in any physics training, while other, more advanced, volumes will be suitable as preliminary reading for research in the field covered. New titles will be added from time to time.

Clark: *A First Course in Quantum Mechanics (revised edition)*
Lothian: *Optics and Its Uses*
Lovell, Avery and Vernon: *Physical Properties of Materials*
Tritton: *Physical Fluid Dynamics*

# A first course in Quantum Mechanics

## revised edition

H. CLARK
*Principal Lecturer in mathematics at Sunderland Polytechnic*

**Van Nostrand Reinhold (UK) Co. Ltd.**

*First published 1974*
*Revised edition 1982*
*Reprinted 1983*



**Published by Van Nostrand Reinhold (UK) Co. Ltd.**
**Molly Millars Lane, Wokingham, Berkshire, England**

**Library of Congress Cataloging in Publication Data**
Clark, H. (Hylton)
A first course in quantum mechanics.

(The Modern university physics series)
Includes bibliographies and index.
1. Quantum theory. I. Title II. Series
QC174. 12. C5 1982 530.1′2 81-12959
AACR2

ISBN 0 442-30173-1 (pbk.)

Printed and bound in Hong Kong

# Preface

This book is intended to be a realistic first course in quantum mechanics and as such no claim is made that the subject is covered comprehensively. However, it is hoped that the material included will not only give a good basic grounding in the subject but will introduce the student to the more advanced techniques essential for a serious study.

After a brief survey of classical mechanics and of the historical background of Quantum Theory, a postulate statement is given for Quantum Mechanics. One chapter is devoted to matrix representation and two chapters to approximate methods of solution. Group theory is introduced in chapter nine. This is a topic of growing importance. The final chapter is an introduction to Dirac's relativistic theory.

This book should be useful to first degree students in physics, chemistry and mathematics and also to research workers in industry, polytechnics and universities.

Problems are included at the end of each chapter and a few of these have been used to extend the theory slightly. SI units are used and the charge on the electron is taken as $-e = -1{\cdot}602 \times 10^{-19}$ C. A list of references and further reading is given at the end of each chapter.

I would like to thank my first teacher of the subject Professor H. C. Bolton whose influence I gratefully acknowledge.

*Tynemouth,*
*June 1973.*

H.C.

# Preface to the Revised Edition

The most important change in this revised edition is the addition of a new chapter entitled 'Electrons in Solids'. Solid State Theory has been a major growth area in physics during the last few decades and it is hoped this new chapter will give the basic elements of the theory.

The opportunity had also been taken to correct some minor misprints that occurred in the first edition.

The author gratefully acknowledges permission from the Macmillan Press to use part of his book *Solid State Physics – an introduction to its theory*.

*Tynemouth,*
*May 1981*

*Hylton Clark*

# Contents

**Chapter 12–The Dirac Equation**



**Chapter 13–Electrons in Solids**

CHAPTER 1

# Classical Dynamics

## 1.1 Introduction

Although classical mechanics does not apply to the dynamics of atomic systems it certainly affords an excellent approximation to a wide range of phenomena. Quantum mechanics is a more general theory which contains classical mechanics as a limiting case and in fact historically quantum mechanics was developed by analogy with classical theory although this approach is not followed in this text.

It is important that the reader understands something of the more general formulations of classical mechanics and is familiar with the terms Lagrangian, conjugate momentum, Hamiltonian and Poisson bracket. A brief review is given in this chapter but for a more complete treatment the reader is referred to *Principles of Mechanics*, by Synge and Griffith, or *Classical Mechanics,* by Goldstein.

## 1.2 The Lagrange Equations

The configuration of a mechanical system is determined when the values of a set of 'generalized co-ordinates' are given. The set of co-ordinates chosen may be determined by the symmetry of the system. Some or all of these generalized co-ordinates can be arbitrarily varied independently without violating the constraints of the system and the number of such co-ordinates is called the number of degrees of freedom of the system. In a holonomic system the number of degrees of freedom is equal to the number of co-ordinates. A smooth sphere in contact with a fixed plane constitutes a holonomic system, five co-ordinates being necessary to define the configuration, each of which may be varied independently.

On the other hand a rough sphere rolling on a fixed plane forms a non-holonomic system as there are now only three degrees of freedom whereas the number of co-ordinates is still five. The rest of this chapter is concerned with holonomic systems only.

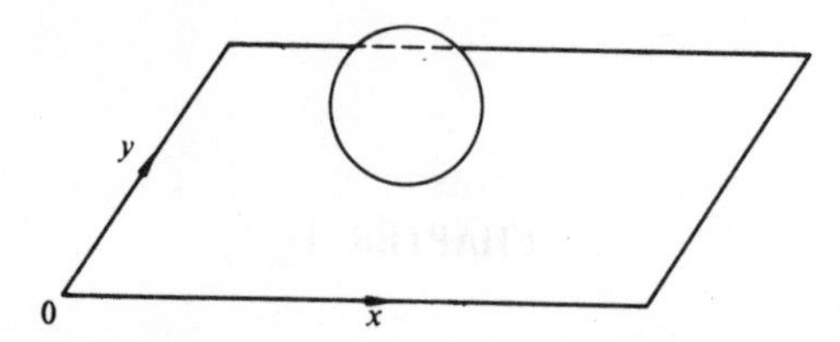


Figure 1.1 A smooth sphere in contact with a fixed plane. The mass centre is specified by two co-ordinate $(x, y)$. Three angles (Eulerian), specify the orientation of the sphere. There are five degrees of freedom.

Lagrange's equation can be derived from Newton's laws of motion. Consider a system composed of $N$ particles of masses $m_i$ $(i = 1, \ldots, N)$. At time $t$ the position vector $\mathbf{r}_i$ of the $i$th particle is determined by the set of $n$ generalized co-ordinates $q_j$ $(j = 1, \ldots, n)$, i.e.

$$\mathbf{r}_i = \mathbf{r}_i(q_1, \ldots, q_n, t) \tag{1.1}$$

(for $N$ free point particles $n = 3N$). If it is assumed that the system has fixed constraints (i.e. scleronomic), then $\mathbf{r}_i$ does not explicitly involve the time.

The velocity of the particle is then

$$\dot{\mathbf{r}}_i \equiv \frac{d\mathbf{r}_i}{dt} = \sum_{j=1}^{n} \frac{\partial \mathbf{r}_i}{\partial q_j} \frac{dq_j}{dt}. \tag{1.2}$$

The $n$ derivatives $\dot{q}_j$ are called the generalized velocities of the system. The particle-velocity $\dot{\mathbf{r}}_i$ is a function of all the variables $q_j$, $\dot{q}_j$ and from (1.2)

$$\frac{\partial \dot{\mathbf{r}}_i}{\partial \dot{q}_j} = \frac{\partial \mathbf{r}_i}{\partial q_j}, \tag{1.3}$$

and

$$\frac{d}{dt}\left(\frac{\partial \mathbf{r}_i}{\partial q_j}\right) = \frac{\partial \dot{\mathbf{r}}_i}{\partial q_j}. \tag{1.4}$$

Suppose the particles each undergo a small displacement consistent with the constraints of the system such that all the forces remain constant during the displacement. In this virtual displacement the total work done is

$$\delta W = \sum_i \mathbf{F}_i \, . \, \delta \mathbf{r}_i \tag{1.5}$$

where $\mathbf{F}_i$ is the force on the $i$th particle and $\delta\mathbf{r}_i$ is its displacement. From Newton's laws of motion

$$\mathbf{F}_i = m_i\ddot{\mathbf{r}}_i \tag{1.6}$$

and so

$$\delta W = \sum_i m_i\ddot{\mathbf{r}}_i \,.\, \delta\mathbf{r}_i \,. \tag{1.7}$$

If it is assumed that the internal forces do no work then $\mathbf{F}_i$ may be identified with the externally applied force.

Since

$$\delta\mathbf{r}_i = \sum_j \frac{\partial \mathbf{r}_i}{\partial q_j}\,\delta q_j$$

then from 1.3 and 1.7

$$\delta W = \sum_j \left(\sum_i m_i\ddot{\mathbf{r}}_i . \frac{\partial \dot{\mathbf{r}}_i}{\partial \dot{q}_j}\right)\delta q_j \,. \tag{1.8}$$

The total kinetic energy of the system is

$$T = \tfrac{1}{2}\sum_i m_i\dot{\mathbf{r}}_i^2 \tag{1.9}$$

$$\frac{\partial T}{\partial \dot{q}_j} = \sum_i m_i\dot{\mathbf{r}}_i \,.\, \frac{\partial \dot{\mathbf{r}}_i}{\partial \dot{q}_j} \tag{1.10}$$

and

$$\frac{\mathrm{d}}{\mathrm{d}t}\left(\frac{\partial T}{\partial \dot{q}_j}\right) = \sum_i m_i\left(\ddot{\mathbf{r}}_i \,.\, \frac{\partial \dot{\mathbf{r}}_i}{\partial \dot{q}_j}\right) + \sum_i m_i\left(\dot{\mathbf{r}}_i \,.\, \frac{\mathrm{d}}{\mathrm{d}t}\frac{\partial \dot{\mathbf{r}}_i}{\partial \dot{q}_j}\right). \tag{1.11}$$

But

$$\frac{\mathrm{d}}{\mathrm{d}t}\left(\frac{\partial \dot{\mathbf{r}}_i}{\partial \dot{q}_j}\right) = \frac{\mathrm{d}}{\mathrm{d}t}\left(\frac{\partial \mathbf{r}_i}{\partial q_j}\right) = \frac{\partial \dot{\mathbf{r}}_i}{\partial q_j}$$

and so

$$\frac{\mathrm{d}}{\mathrm{d}t}\left(\frac{\partial T}{\partial \dot{q}_j}\right) = \sum_j m_i\ddot{\mathbf{r}}_i \,.\, \frac{\partial \dot{\mathbf{r}}}{\partial \dot{q}_j} + \frac{\partial T}{\partial q_j} \,. \tag{1.12}$$

From (1.8) and (1.12) the work done may be expressed as

$$\delta W = \sum_j \left(\frac{\mathrm{d}}{\mathrm{d}t}\frac{\partial T}{\partial \dot{q}_j} - \frac{\partial T}{\partial q_j}\right)\delta q_j \,. \tag{1.13}$$

Generalized forces $Q_j$ may be introduced so that (c.f. (1.5))

$$\delta W = \sum_j Q_j \delta q_j \tag{1.14}$$

and these generalized forces are given in terms of the applied external forces.

As the system is holonomic it is possible to consider a displacement involving one degree of freedom only, e.g.

$$\delta q_i \neq 0, \qquad \delta q_j = 0 \qquad j \neq i.$$

From (1.13) and (1.14)

$$\frac{\mathrm{d}}{\mathrm{d}t}\frac{\partial T}{\partial \dot{q}_i} - \frac{\partial T}{\partial q_i} = Q_i \qquad i = 1, \ldots, n. \tag{1.15}$$

Only conservative forces will be considered and they can be derived from a potential function $V(q_1, \ldots, q_n)$ where $\delta W = -\delta V$, i.e.

$$-\delta V = \sum_j Q_j \delta q_j$$

and so

$$Q_i = -\frac{\partial V}{\partial q_i}. \tag{1.16}$$

Equation (1.15) becomes

$$\frac{\mathrm{d}}{\mathrm{d}t}\frac{\partial T}{\partial \dot{q}_j} - \frac{\partial T}{\partial q_i} = -\frac{\partial V}{\partial q_i}. \tag{1.17}$$

As the potential function does not depend on the generalized velocities it is possible to define a Lagrangian

$$L(q_1, \ldots, q_n, \dot{q}_1, \ldots, \dot{q}_n) = T(q_1, \ldots, q_n, \dot{q}_1, \ldots, q_n) - V(q_1, \ldots, q_n) \tag{1.18}$$

such that

$$\frac{\mathrm{d}}{\mathrm{d}t}\frac{\partial L}{\partial \dot{q}_i} - \frac{\partial L}{\partial q_i} = 0 \qquad i = 1, \ldots, n. \tag{1.19}$$

This final result is Lagrange's equations for a conservative holonomic system.

As an almost trivial example consider the motion of a compound pendulum. A uniform rod of a length $2a$ and mass $m$ is free to rotate about a horizontal axis through one end. This is a holonomic system with one generalized co-ordinate, the angle $\theta$ (see Fig. 1.2).

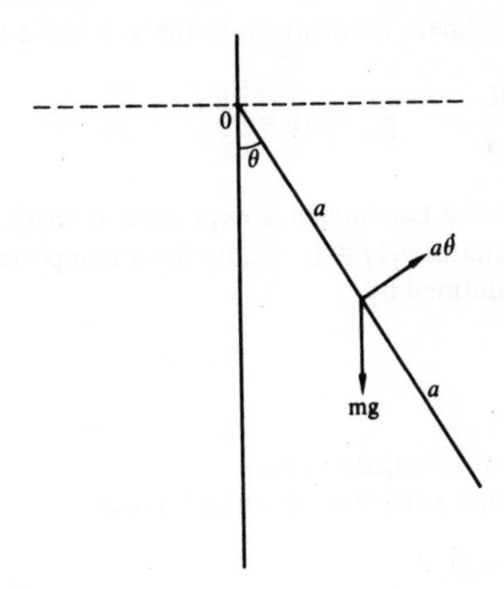


Figure 1.2 Compound pendulum.

The kinetic energy of the system is $T = \frac{1}{2}I\dot{\theta}^2$ where $I$ is the moment of inertia about the pivot, i.e.

$$T = \tfrac{2}{3}ma^2\dot{\theta}^2.$$

With a suitable choice for zero the potential energy is $V = -mga \cos\theta$. The Lagrangian function is then

$$L = \tfrac{2}{3}ma^2\dot{\theta}^2 + mga \cos\theta$$

and Lagrange's equation becomes

$$\tfrac{4}{3}a\ddot{\theta} + g \sin\theta = 0.$$

For small oscillations the motion is simple harmonic with frequency

$$\nu = \frac{1}{2\pi}\sqrt{\frac{3g}{4a}}.$$

## 1.3 Hamilton's Equations

To derive Hamilton's equations it is necessary to generalize the concept of momentum. The Lagrangian of a point particle moving in a conservative field is

$$L = \frac{m}{2}(\dot{x}^2 + \dot{y}^2 + \dot{z}^2) - V(x, y, z).$$

The components of linear momentum in the $x$, $y$ and $z$ directions are

$$p_x = m\dot{x} = \frac{\partial L}{\partial \dot{x}}, \qquad p_y = m\dot{y} = \frac{\partial L}{\partial \dot{y}}, \qquad p_z = m\dot{z} = \frac{\partial L}{\partial \dot{z}}.$$

Accordingly, when the Lagrangian is expressed in terms of a set of generalized co-ordinates $q_j (j = 1, \ldots, n)$ the $n$ components of generalized momenta may be defined by

$$p_i = \frac{\partial L}{\partial \dot{q}_i}. \tag{1.20}$$

$p_i$ is the momentum conjugate to $q_i$.

For the compound pendulum discussed above

$$p_\theta = \tfrac{4}{3} ma^2 \dot{\theta} = I\dot{\theta}.$$

A Hamiltonian function is defined by

$$H = \sum_i p_i \dot{q}_i - L. \tag{1.21}$$

From (1.20) the components of momentum are expressed in terms of the $q$'s and $\dot{q}$'s. This set of equations can be solved for the $\dot{q}$'s in terms of the $q$'s and $p$'s and in this way the Hamiltonian is expressed as a function of the $q$'s and $p$'s,

$$H = H(q_1, \ldots, q_n, p_1, \ldots, p_n). \tag{1.22}$$

It is clear from (1.2) and (1.9) that the kinetic energy is a homogeneous quadratic function of the $\dot{q}$'s and so from Euler's theorem

$$\sum_j \dot{q}_j \frac{\partial T}{\partial \dot{q}_j} = 2T. \tag{1.23}$$

From (1.18) and (1.20) $p_j = \partial T / \partial \dot{q}_j$ and so for a conservative field (1.21) becomes

$$\begin{aligned} H &= 2T - (T - V) \\ &= T + V. \end{aligned} \tag{1.24}$$

The Hamiltonian is equal to the sum of the potential and kinetic energies.

To obtain Hamilton's equations consider a small change in the generalized co-ordinates. In terms of differentials

$$\mathrm{d}H = \sum_i \mathrm{d}p_i \dot{q}_i + \sum_i p_i \,\mathrm{d}\dot{q}_i - \sum_i \frac{\partial L}{\partial q_i} \mathrm{d}q_i - \sum_i \frac{\partial L}{\partial \dot{q}_i} \mathrm{d}\dot{q}_i. \tag{1.25}$$

By definition of the generalized momenta $p_i = \partial L/\partial \dot{q}_i$ and from Lagrange's equations (1.19)

$$\frac{\partial L}{\partial q_i} = \dot{p}_i.$$

Thus

$$\mathrm{d}H = \sum_i \dot{q}_i \,\mathrm{d}p_i - \sum_i \dot{p}_i \,\mathrm{d}q_i.$$

As

$$H = H(q_1, \ldots, q_n, p_1, \ldots, p_n)$$

then

$$\frac{\partial H}{\partial q_i} = -\dot{p}_i \qquad \text{and} \qquad \frac{\partial H}{\partial p_i} = \dot{q}_i. \tag{1.26}$$

These final relations are Hamilton's equations. (Canonical equations.)

Hamilton's equations form a set of $2n$ first order equations in the $2n$ variables $q_i$ and $p_i$ and the co-ordinates and momenta enter the theory on equal basis. The complete solution of Hamilton's equations involves $2n$ arbitrary constants. If the values of all the co-ordinates and momenta are known at some instant then the future state of the system is uniquely determined by Hamilton's equations.

As a simple illustration consider a point mass moving under gravity. The kinetic energy is

$$T = \frac{m}{2}(\dot{x}^2 + \dot{y}^2 + \dot{z}^2)$$

and the potential energy is $V = mgz$ where $z$ is the perpendicular height from some chosen zero and $g$ is the acceleration due to gravity.

The Lagrangian is

$$L = \frac{m}{2}(\dot{x}^2 + \dot{y}^2 + \dot{z}^2) - mgz$$

and the momenta conjugate to $x$, $y$, $z$ are

$$p_x = \frac{\partial L}{\partial \dot{x}} = m\dot{x}, \qquad p_y = m\dot{y}, \qquad p_z = m\dot{z}.$$

The Hamiltonian is

$$\begin{aligned} H &= T + V \\ &= \frac{1}{2m}(p_x^2 + p_y^2 + p_z^2) + mgz. \end{aligned}$$

Note that the Lagrangian is used to relate $\dot{q}_i$ and $p_i$ and hence to obtain the Hamiltonian.

A more interesting example is afforded by the rigid rotator. This system is composed of two point masses $m_1$ and $m_2$, a fixed distance

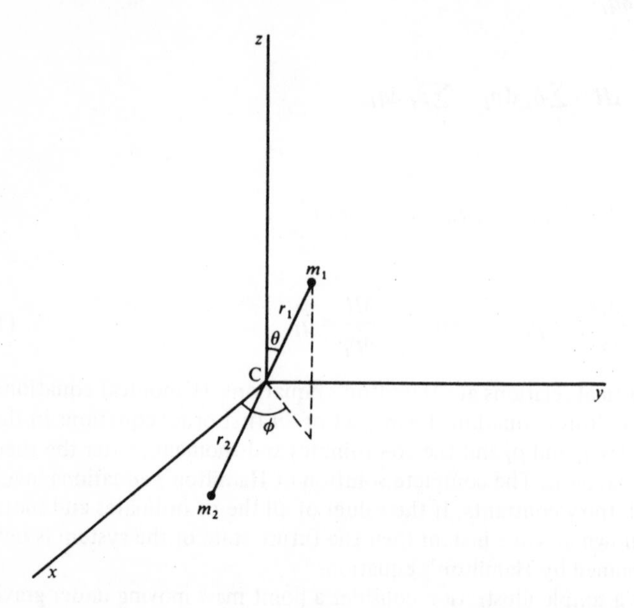


Figure 1.3 Rigid rotator. $C$ is mass centre.

apart, rotating about the mass centre $C$, and arises in discussing the energy levels of diatomic molecules. Let $C$ coincide with the origin of the co-ordinate system and let $m_1, m_2$ be at the distances $r_1, r_2$ respectively from $C$. This is a holonomic system with two degrees of freedom, the polar angles $\theta$ and $\phi$. The potential energy of the system is zero and the kinetic energy is

$$T = \tfrac{1}{2}m_1V_1^2 + \tfrac{1}{2}m_2V_2^2$$

where $V_1$, $V_2$ are the speeds of $m_1, m_2$ respectively. In terms of the polar angles

$$T = \tfrac{1}{2}m_1[r_1^2\dot{\theta}^2 + r_1^2 \sin^2\theta\dot{\phi}^2] + \tfrac{1}{2}m_2[r_2^2\dot{\theta}^2 + r_2^2 \sin^2\theta\dot{\phi}^2].$$

The moment of inertia about the mass centre is

$$I = m_1r_1^2 + m_2r_2^2$$

and the Lagrangian is

$$L = T = \tfrac{1}{2}I(\dot{\theta}^2 + \sin^2\theta\dot{\phi}^2).$$

The momenta conjugate to $\theta$ and $\phi$ are

$$p_\theta = \frac{\partial L}{\partial \dot{\theta}} = I\dot{\theta}, \qquad p_\phi = \frac{\partial L}{\partial \dot{\phi}} = I \sin^2\theta\dot{\phi}$$

and the Hamiltonian of the system is

$$H(\theta, \phi, p_\theta, p_\phi) = T = \frac{1}{2I}\left(p_\theta^2 + \frac{1}{\sin^2\theta} p_\phi^2\right).$$

*Conservation of energy*

The total energy of the system is

$$E = T + V = H(q_1, \ldots, q_n, p_1, \ldots, p_n)$$

and the rate of change of the energy is

$$\begin{aligned}\frac{\mathrm{d}E}{\mathrm{d}t} &= \sum_j \left(\frac{\partial H}{\partial q_j}\dot{q}_j + \frac{\partial H}{\partial p_j}\dot{p}_j\right)\\ &= \sum_j \left(\frac{\partial H}{\partial q_j}\frac{\partial H}{\partial p_j} - \frac{\partial H}{\partial p_j}\frac{\partial H}{\partial q_j}\right)\\ &= 0.\end{aligned} \qquad (1.27)$$

Throughout the dynamical motion the energy remains constant. This result of course was to be expected as only conservative forces have been considered (1.16).

*Poisson brackets*

Suppose $F$ and $G$ are two functions of the $2n$ variables $q_i, p_i$. The Poisson bracket of $F$ and $G$, denoted $\{F, G\}$ is defined by

$$\{F, G\} = \sum_j \left(\frac{\partial F}{\partial q_j}\frac{\partial G}{\partial p_j} - \frac{\partial F}{\partial p_j}\frac{\partial G}{\partial q_j}\right). \qquad (1.28)$$

It is immediately obvious that

$$\{F, G\} = -\{G, F\}$$

and

$$\{F, F\} = 0. \qquad (1.29)$$

Simple important cases arise when $F$ and $G$ are $q_i$ and $p_i$.

$$\{q_i, q_j\} = 0, \qquad \{p_i, p_j\} = 0, \qquad \{q_i, p_j\} = \delta_{ij}, \qquad (1.30)$$

where $\delta_{ij}$ is the Kronecker delta defined by

$$\begin{aligned}\delta_{ij} &= 0 \qquad i \neq j\\ &= 1 \qquad i = j.\end{aligned}$$

In classical mechanics Poisson brackets occur when discussing the rate of change of functions of the $2n$ variables $q_i$ and $p_i$. $F$ is such a function and

$$\frac{\mathrm{d}F}{\mathrm{d}t} = \sum_j \left( \frac{\partial F}{\partial q_j} \dot{q}_j + \frac{\partial F}{\partial p_j} \dot{p}_j \right) = \sum_j \left( \frac{\partial F}{\partial q_j} \frac{\partial H}{\partial p_j} - \frac{\partial F}{\partial p_j} \frac{\partial H}{\partial q_j} \right)$$

i.e.

$$\frac{\mathrm{d}F}{\mathrm{d}t} = \{F, H\}. \tag{1.31}$$

If $\{F, H\} = 0$ then $F$ is a constant of the motion.

There is a correspondence between Poisson brackets and the commutators of quantum mechanics.

## 1.4 Time Involved Explicitly

In the more general case it may be possible to describe a system by a Lagrangian satisfying (1.19) but which explicitly involves the time. The conjugate momenta are defined in the usual manner (1.20) and so a time-dependent Hamiltonian may be found (1.21). Hamilton's equations (1.26) still follow from the analogue of (1.25) but in addition there is the result

$$\frac{\partial H}{\partial t} = -\frac{\partial L}{\partial t}. \tag{1.32}$$

The appropriate generalization of (1.27) is

$$\frac{\mathrm{d}H}{\mathrm{d}t} = \frac{\partial H}{\partial t}. \tag{1.33}$$

If $F$ explicitly involves the time then (1.31) becomes

$$\frac{\mathrm{d}F}{\mathrm{d}t} = \{F, H\} + \frac{\partial F}{\partial t}. \tag{1.34}$$

A system is conservative if either of the equivalent statements

$$\frac{\partial L}{\partial t} = 0, \qquad \frac{\partial H}{\partial t} = 0 \tag{1.35}$$

is satisfied.

*An electron in an electromagnetic field*

For some systems it is not always obvious what the precise form of the Lagrangian should be. Consider an electron with charge $-e$ in an electromagnetic field with electric vector $\mathscr{E}$ and magnetic induction vector $\mathbf{B}$ (S.I. units). The force exerted on the electron moving with velocity $\dot{\mathbf{r}}$ is

$$\mathbf{F} = -e[\mathscr{E} + \dot{\mathbf{r}} \wedge \mathbf{B}]. \tag{1.36}$$

The field may be described by the scalar and vector potentials $\phi$ and $\mathbf{A}$ with

$$\mathbf{B} = \text{curl } \mathbf{A} \qquad \text{and} \qquad \mathscr{E} = -\text{grad } \phi - \frac{\partial \mathbf{A}}{\partial t}. \tag{1.37}$$

In general these potentials are functions of position and time. These equations do not specify $\phi$ and $\mathbf{A}$ uniquely and it is convenient to impose the condition

$$\text{div } \mathbf{A} + \mu\epsilon \frac{\partial \phi}{\partial t} = 0. \tag{1.38}$$

$\mu$ and $\epsilon$ are the permeability and permittivity of the medium. The equation of motion of the electron (mass $m$) is then

$$m\ddot{\mathbf{r}} = \mathbf{F} = e \text{ grad } \phi + e \frac{\partial \mathbf{A}}{\partial t} - e\dot{\mathbf{r}} \wedge \text{curl } \mathbf{A}. \tag{1.39}$$

It will now be shown that the Lagrangian that describes the system is

$$L(\mathbf{r}, \dot{\mathbf{r}}, t) = \tfrac{1}{2} m\dot{r}^2 + e\phi - e\dot{\mathbf{r}} \,.\, \mathbf{A}. \tag{1.40}$$

The generalized co-ordinates will be taken to be the cartesian co-ordinates of the electron and so,

$$\frac{\partial L}{\partial x_i} = e \frac{\partial \phi}{\partial x_i} - e \sum_j \dot{x}_j \frac{\partial A_j}{\partial x_i}. \tag{1.41}$$

Also

$$\frac{\partial L}{\partial \dot{x}_i} = m\dot{x}_i - eA_i. \tag{1.42}$$

Lagrange's equations (1.19) become

$$m\ddot{x}_1 = e \frac{\partial \phi}{\partial x_1} + e \frac{\partial A_1}{\partial t} - e \left[ \dot{x}_2 \left( \frac{\partial A_2}{\partial x_1} - \frac{\partial A_1}{\partial x_2} \right) + \dot{x}_3 \left( \frac{\partial A_3}{\partial x_1} - \frac{\partial A_1}{\partial x_3} \right) \right]$$

and similarly for $x_2$ and $x_3$. These equations are simply the components of (1.39) so verifying that (1.40) is the correct Lagrangian.

The momentum conjugate to $x_i$ is from (1.42)

$$p_i = m\dot{x}_i - eA_i. \tag{1.43}$$

The conjugate momentum is not equal to the linear momentum unless $\mathbf{A} = 0$, i.e. unless there is no magnetic field. The Hamiltonian is

$$\begin{aligned} H &= \sum_j p_j \dot{q}_j - L = \mathbf{p} \cdot \dot{\mathbf{r}} - L \\ &= (m\dot{\mathbf{r}} - e\mathbf{A}) \cdot \dot{\mathbf{r}} - \tfrac{1}{2} m\dot{r}^2 - e\phi + e\dot{\mathbf{r}} \cdot \mathbf{A} \\ &= \tfrac{1}{2} m\dot{r}^2 - e\phi. \end{aligned} \tag{1.44}$$

In terms of conjugate momenta

$$H(\mathbf{r}, \mathbf{p}, t) = \frac{1}{2m} (\mathbf{p} + e\mathbf{A})^2 - e\phi. \tag{1.45}$$

It is a simple matter to confirm that

$$\frac{\partial H}{\partial t} = -\frac{\partial L}{\partial t}.$$

If the fields are static then $\phi$ and $\mathbf{A}$ do not depend explicitly upon the time and consequently neither do $L$ nor $H$ and so the energy is conserved. In a purely magnetostatic field $\phi = 0$ and the speed of the electron is a constant of the motion.

## 1.5 Hamilton's Principle

*The calculus of variations*

Let A and B be two fixed points in the $xy$-plane with co-ordinates $(x_1, y_1)$, $(x_2, y_2)$ and let $F$ be a functional form of the independent variable $x$, a function $y(x)$ and the derivative $y' \equiv \mathrm{d}y/\mathrm{d}x$. For a given curve $y = y(x)$ through A and B the integral

$$I = \int_{\mathrm{A}}^{\mathrm{B}} F(x, y, y')\, \mathrm{d}x \tag{1.46}$$

has a definite value. This integral will in general take different values for different curves through the fixed points A and B. It is desired to find the curve $y = y_s(x)$ which makes $I$ take a stationary value, if this exists.

Let $g(x)$ be an essentially arbitrary differentiable function such that

$$g(x_1) = g(x_2) = 0. \tag{1.47}$$

Then $y(x) = y_s(x) + \beta g(x)$ where $\beta$ is small is the equation of a neighbouring curve through the points A and B.

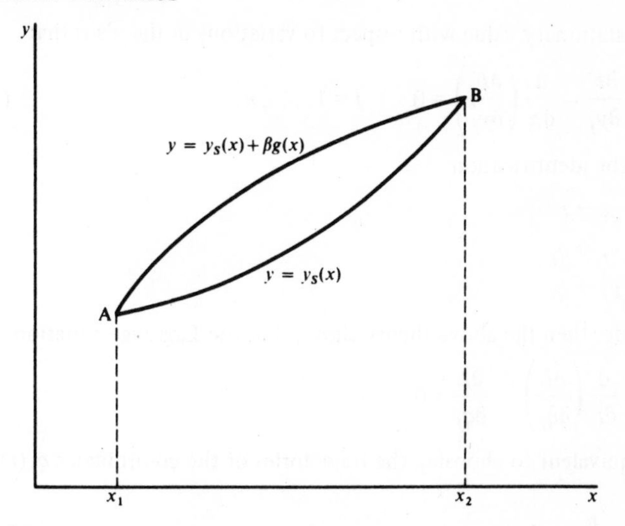


Figure 1.4

The integral (1.46) may be regarded as a differentiable function of $\beta$.

$$I(\beta) = \int_A^B F(x, y_s(x) + \beta g(x), y_s'(x) + \beta g'(x))\,\mathrm{d}x.$$

Since this integral is stationary when $\beta = 0$ then $I'(0) = 0$. Differentiating under the integral sign gives

$$\int_A^B [g(x)F_y(x, y_s, y_s') + g'(x)F_{y'}(x, y_s, y_s')]\,\mathrm{d}x = 0. \tag{1.48}$$

The second term can be integrated by parts and using (1.47) equation (1.48) becomes

$$\int_A^B g(x)\left[F_y(x, y_s, y_s') - \frac{\mathrm{d}}{\mathrm{d}x}F_{y'}(x, y_s, y_s')\right]\mathrm{d}x = 0.$$

Since $g(x)$ is arbitrary, the condition for $I$ to take a stationary value is

$$\frac{\partial F}{\partial y} - \frac{\mathrm{d}}{\mathrm{d}x}\left(\frac{\partial F}{\partial y'}\right) = 0. \tag{1.49}$$

This differential equation for $y$ is called the Euler–Lagrange equation.

The extension of the theory to include $n$ variables $y_j\,(j = 1, \ldots, n)$ is straightforward and the condition that the integral

$$I = \int_A^B F(x, y_1, \ldots, y_n, y_1', \ldots, y_n')\,\mathrm{d}x \tag{1.50}$$

has a stationary value with respect to variations in the $y$'s is that

$$\frac{\partial F}{\partial y_j} - \frac{\mathrm{d}}{\mathrm{d}x}\left(\frac{\partial F}{\partial y_j'}\right) = 0 \qquad j = 1, \ldots, n. \tag{1.51}$$

If the identification

$$x \to t$$
$$y_j \to q_j$$
$$y_j' \to \dot{q}_j$$

is made, then the above theory shows that the Lagrange equations

$$\frac{\mathrm{d}}{\mathrm{d}t}\left(\frac{\partial L}{\partial \dot{q}_j}\right) - \frac{\partial L}{\partial q_j} = 0$$

are equivalent to choosing the trajectories of the co-ordinates $q_i(t)$ so that

$$\delta \int_{\mathrm{A}}^{\mathrm{B}} L(t, q_1, \ldots, q_n, \dot{q}_1, \ldots, \dot{q}_n)\,\mathrm{d}t = 0. \tag{1.52}$$

A represents the fixed initial configuration of the system at time $t_1$, and B the fixed final configuration at $t_2$. This is Hamilton's principle. Of course for conservative systems the Lagrangian does not depend explicitly on the time and the variational principle determining the equations of motion (1.19) is

$$\delta \int_{\mathrm{A}}^{\mathrm{B}} (T - V)\,\mathrm{d}t = 0. \tag{1.53}$$

Generally the stationary value of the integral is a minimum.

## 1.6 The Hamilton–Jacobi Equation

The Hamilton–Jacobi equation is of special interest since the quantum mechanical Schrödinger equation includes the classical Hamilton–Jacobi equation as a limiting case.

Define the action integral

$$S = \int_{\mathrm{A}}^{\mathrm{B}} L(t, q_1, \ldots, q_n, \dot{q}_1, \ldots, \dot{q}_n)\,\mathrm{d}t \tag{1.54}$$

Suppose the integral (1.54) is evaluated along the path in the $(n + 1)$ dimensional space defined by $q_1, \ldots, q_n, t$ representing the natural motion of the system from a point A to a point B. This is the unique stationary path and the value of $S$ must only depend upon the values of the co-ordinates and time at A and B.

Consider now a neighbouring stationary path from the same initial point A to a new end point $B_1$. This is possible since the velocity components have not been defined at A. Let $\delta q_{iB}, i = 1, \ldots, n$ and $\delta t_B$ be the displacement of $B_1$ from B. The change in $S$ can be expressed in terms of these increments. Only an outline of the argument is given. (For further details see the references at the end of the chapter.)

With the aid of (1.21), the integral (1.54) becomes

$$S = \int_A^B \left( \sum_i p_i \, dq_i - H(q_1, \ldots, q_n, \dot{q}_1, \ldots, \dot{q}_n, t) \, dt \right). \tag{1.55}$$

After some manipulation, the variation in $S$ is

$$\delta S = \sum_i p_{iB} \delta q_{iB} - H_B \delta t_B$$

$$+ \int_A^B \left( \sum_i \delta p_i \, dq_i - \sum \delta q_i \, dp_i - \delta H \, dt + \delta t \, dH \right). \tag{1.56}$$

Note that there is no reason to take $\delta S = 0$ (c.f. equation 1.52) since $\delta S$ is the variation in $S$ between two stationary paths. The variation in $H$ is given by

$$\delta H = \sum_i \left( \frac{\partial H}{\partial q_i} \delta q_i + \frac{\partial H}{\partial p_i} \delta p_i \right) + \frac{\partial H}{\partial t} \delta t \tag{1.57}$$

and so

$$\delta S = \sum_i p_{iB} \delta q_{iB} - H_B \delta t_B$$

$$+ \int_A^B \left\{ \sum_i \left[ \left( \dot{q} - \frac{\partial H}{\partial p_i} \right) \delta p_i - \left( \dot{p}_i + \frac{\partial H}{\partial q_i} \right) \delta q_i \right] + \left( \dot{H} - \frac{\partial H}{\partial t} \right) \delta t \right\} dt. \tag{1.58}$$

As the integral is taken along a stationary path representing a natural motion of the system, Hamilton's equations (1.26) are satisfied (also (1.33)) and so the integral is zero.

$$\therefore \quad \delta S = \sum_i p_{iB} \delta q_{iB} - H_B \delta t_B. \tag{1.59}$$

Replace $q_{iB}$ and $t_B$ by $q_i$ and $t$ to represent any point on the path. As the increments are independent

$$S = S(q_1, \ldots, q_n, t) \tag{1.60}$$

with

$$\frac{\partial S}{\partial q_i} = p_i$$

and

$$\frac{\partial S}{\partial t} = -H(q_1, \ldots, q_n, p_1, \ldots, p_n, t). \tag{1.61}$$

Combining these two equations

$$\frac{\partial S}{\partial t} + H\left(q_1, \ldots, q_n, \frac{\partial S}{\partial q_1}, \ldots, \frac{\partial S}{\partial q_n}, t\right) = 0. \tag{1.62}$$

This partial differential equation satisfied by $S$ is the Hamilton–Jacobi equation. For a conservative system the Hamiltonian does not explicitly involve the time and the equation reduces to

$$H\left(q_1, \ldots, q_n, \frac{\partial S}{\partial q_1}, \ldots, \frac{\partial S}{\partial q_n}\right) = -\frac{\partial S}{\partial t} = E \tag{1.63}$$

where $E$ is the energy of the system.

For a rigid rotator with moment of inertia $I$ the Hamiltonian is

$$H = \frac{1}{2I}\left(p_\theta^2 + \frac{1}{\sin^2\theta} p_\phi^2\right)$$

and the Hamilton–Jacobi equation is

$$\frac{1}{2I}\left[\left(\frac{\partial S}{\partial \theta}\right)^2 + \frac{1}{\sin^2\theta}\left(\frac{\partial S}{\partial \phi}\right)^2\right] = -\frac{\partial S}{\partial t}.$$

The solution of a dynamical problem can be reduced to the solution of the Hamilton–Jacobi equation but this will not be pursued here. The interesting relationships between the Hamilton–Jacobi equation and the Schrödinger equation of quantum mechanics is mentioned in chapter four. Schrödinger used this equation in the development of wave mechanics. By comparing Hamilton's principle with Fermat's principle an analogy is obtained relating particle trajectories to rays and Hamilton's function $S$ to the phase of a wave.

## PROBLEMS

1 A particle mass of $m$ distance $r$ from the origin, is moving in a plane under an attractive force $\mu m/r^2$ directed towards the origin. Write down the kinetic and potential energies and verify that Lagrange's equations may be written

$$m\ddot{r} - mr\dot{\theta}^2 + \frac{\mu m}{r^2} = 0, \qquad m\frac{\mathrm{d}}{\mathrm{d}t}(r^2\dot{\theta}) = 0.$$

2 A double pendulum consists of two uniform rods AB and BC each of mass $m$ and length $l$ hinged together at B. The end A is hinged at a fixed

point O so that the rods can oscillate in a vertical plane. The configuration of the system is specified by the angles $\theta$ and $\phi$ between the rods AB, BC and the vertical. Write down the kinetic and potential energies and obtain the two Lagrange equations. Assuming the oscillations are small find the solutions of Lagrange's equations of the form

$$\theta = A_1 e^{i\omega t} \qquad \phi = A_2 e^{i\omega t}$$

where $A_1, A_2$ are constants. Confirm that the two normal frequencies are

$$\omega_1^2 = \frac{\sqrt{2}g}{l(\sqrt{2}-1)} \quad \text{and} \quad \omega_2^2 = \frac{\sqrt{2}g}{l(\sqrt{2}+1)}$$

where $g$ is the acceleration due to gravity.

3 Consider a system which moves according to equation (1.15)

$$\frac{d}{dt}\left(\frac{\partial T}{\partial \dot{q}_i}\right) - \frac{\partial T}{\partial q_i} = Q_i.$$

Integrate this equation with respect to time from $t = 0$ to $t = \tau$. Proceed to the limit $\tau \to 0$ and assume that the generalized force $Q_i \to \infty$ in such a manner that the impulsive force

$$I_i = \lim_{\tau \to 0} \int_0^\tau Q_i \, dt$$

is finite. Using $L = T - V$ show that

$$I_i = \delta \left(\frac{\partial L}{\partial \dot{q}_i}\right) = \delta p_i.$$

That is, the change in the generalized momentum is equal to the generalized impulsive force.

4 A particle of mass $m$ undergoes simple harmonic oscillations along the $x$-axis. The controlling force is directed towards the origin and has value $-kx$. Write down the kinetic and potential energies and hence the Lagrangian. Find the momentum conjugate to $x$ and show that the Hamiltonian of the system is

$$H = \frac{1}{2m}p^2 + \frac{1}{2}kx^2.$$

5 A two dimensional harmonic oscillator has kinetic energy $T$ and potential energy $V$ given by

$$T = \frac{m}{2}(\dot{q}_1^2 + \dot{q}_2^2) \qquad V = \frac{m}{2}\omega^2(q_1^2 + q_2^2).$$

Derive Lagrange's equations of motion and find the conjugate momenta $p_1, p_2$. Show that the Hamiltonian of the system is

$$H(p, q) = \frac{1}{2m}[p_1^2 + p_2^2 + m^2\omega^2(q_1^2 + q_2^2)].$$

New co-ordinates $Q_1, Q_2$ with momenta $P_1, P_2$ are defined by the transformation

$$q_1 = \sqrt{\frac{2}{m\omega}}\sqrt{P_1}\sin Q_1 \qquad q_2 = \sqrt{\frac{2}{m\omega}}\sqrt{P_2}\sin Q_2$$

$$p_1 = \sqrt{2m\omega}\sqrt{P_1}\cos Q_1 \qquad p_2 = \sqrt{2m\omega}\sqrt{P_2}\cos Q_2.$$

Confirm that the transformed Hamiltonian is

$$H' = \omega(P_1 + P_2).$$

A necessary condition for this transformation to be canonical is that Hamilton's equations (1.26) are satisfied for this new Hamiltonian. Show that these equations are

$$\dot{P}_i = 0 \quad \text{and} \quad \dot{Q}_i = \omega \quad i = 1, 2$$

and verify that they are indeed satisfied.

6 Prove the following properties of Poisson brackets

(i) $\{E + F, G\} = \{E, G\} + \{F, G\}$

(ii) $\{F, \{G, K\}\} + \{G, \{K, F\}\} + \{K, \{F, G\}\} = 0$

where $E, F, G, K$ are all functions of $q_i$ and $p_i$.

7 Suppose A, B are two fixed points in space, A higher than B and choose the two co-ordinate axes so that both points lie in the $(x, y)$ plane. Let the co-ordinates of A and B be $(x_1, y_1)$, $(x_2, y_2)$ respectively. $y = y(x)$ is the equation of a curve joining A and B. A smooth particle starts from rest at A and falls under gravity down the curve.

Show that the time taken to reach B is

$$T = \frac{1}{\sqrt{2g}}\int_{x_1}^{x_2}\sqrt{\frac{1 + y'^2}{(y_1 - y)}}\,dx \qquad \text{where} \qquad y' = \frac{dy}{dx}.$$

Show that $T$ is a minimum when the curve joining A and B satisfies

$$(1 + y'^2)(y_1 - y) = \text{constant}.$$

Introduce the parameter $\psi$ with $y' = \tan\psi$ and derive the solution

$$y - y_1 = A(1 + \cos 2\psi)$$
$$x = B - A(2\psi + \sin 2\psi).$$

These are the parametric equations of a cycloid. (This is the famous brachistochrone problem).

8 The Hamiltonian for a particle of mass $m$ moving vertically under gravity is

$$H(x, p) = \frac{1}{2m} p^2 + mgx$$

where $x$ is the height of the particle above the zero for the potential energy.

Obtain the Hamiltonian-Jacobi equation. Verify that

$$S = -a_1 t + \sqrt{\frac{2}{m}} \frac{2}{3g} (a_1 - mgx)^{3/2} + a_2$$

is a solution where $a_1$ and $a_2$ are constants. Using equation (1.61) explain the physical significance of the constant $a_1$ and find the momentum as a function of $x$.

## References

Goldstein, H. *Classical mechanics.* Addison-Wesley, Cambridge, Mass. (1950).

Synge, J. L. and Griffith, B. A. *Principles of mechanics.* McGraw-Hill, New York (1959).

CHAPTER 2

# Origins of the Quantum Theory

## 2.1 Failure of Classical Mechanics

The classical mechanics formulated by Newton and developed by Lagrange, Hamilton and others, is very successful in explaining the dynamical motion of macroscopic objects and before the beginning of this century was universally accepted as being the ultimate theory of dynamics.

A similarly satisfactory state of affairs existed concerning the theory of light. The work of the Frenchman A. J. Fresnel had led to the acceptance of the wave nature of light. Later the Scotsman Clerk Maxwell developed the electromagnetic theory of light and in 1887 the German physicist Hertz confirmed experimentally that electromagnetic waves are emitted by oscillatory electric charges.

However, by the end of the nineteenth century these classical theories were unable adequately to explain a growing number of small-scale (atomic) phenomena. Physicists were beginning to explore the structure of matter and the nature of radiation and the interaction of radiation with matter. As a consequence, the foundations of physics were radically re-examined and the first quarter of the present century was to see the development of the theories of relativity and quantum mechanics. This chapter attempts to trace the history of the more obvious failures of classical mechanics and how they were overcome, firstly by Planck, Einstein and Bohr, and later by Schrödinger and others.

*Black-body radiation*

An obvious break-down in classical theory occurred in the attempts to account for the continuous spectrum emitted by a black-body. A black-body is defined as one that absorbs all incident electromagnetic radiation.

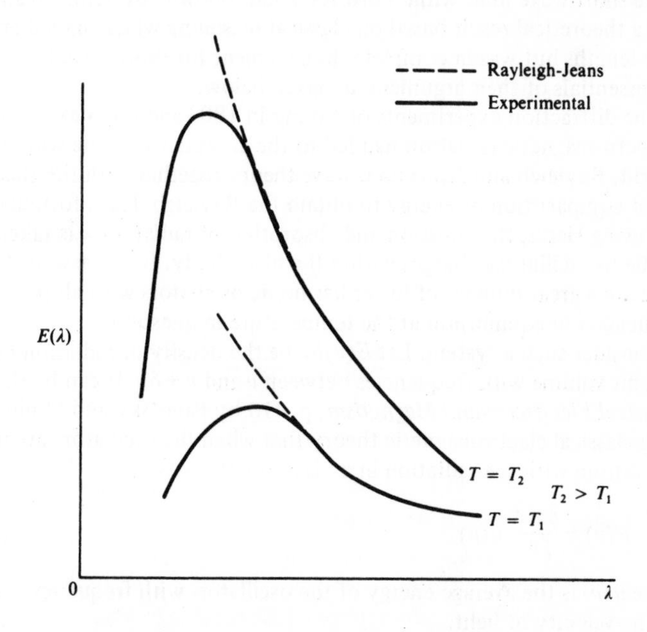


Figure 2.1 Radiation density from a black-body radiator.

It follows from thermodynamics that such a body is a better radiator at every frequency that any other body at the same temperature. The frequency distribution of the spectrum depends only on the temperature and not on the material of the black-body. A true black-body does not exist but it is clear that the radiation emitted from a small opening of a hollow enclosure is equivalent to black-body radiation.

In 1879 Stefan suggested that the rate of heat loss from a hot body is proportional to $T^4$ where $T$ is the absolute temperature and Boltzmann, using classical thermodynamics was able to derive the result

$$E = \frac{4\sigma}{c} T^4$$

where $E$ is the total radiation density from all frequencies and $\sigma$ is a constant; $c$ is the velocity of light. Wien in 1893 also used a classical argument to show that the wave-length, which is emitted with maximum intensity from a black-body, is proportional to $1/T$. This is the Wien displacement law and is in agreement with experiment.

So far classical theory was adequate. However, there was no theory to explain the observed frequency distribution of the black-body radiator (Fig. 2.1). Wien proposed a semi-empirical theory which agreed in the short-wave limit while Lord Rayleigh followed by James Jeans gave a theoretical result based on classical reasoning which agreed at long wave-lengths but was in complete disagreement for short wave-lengths. The essentials of their argument are given below.

The diffraction experiments of Young in 1803 and the Maxwell theory of electromagnetic radiation had led to the acceptance of the wave nature of light. Rayleigh and Jeans used wave theory together with the classical law of equipartition of energy to obtain the Rayleigh–Jeans formula. Following Hertz, the emission and absorption of radiation was taken to be due to oscillatory charges within the black-body. It was assumed that there are a great number of linear harmonic oscillators with all possible frequencies in equilibrium at the temperature in question.

Consider such a system. Let $E(\nu)\delta\nu$ be the density of radiation energy per unit volume with frequencies between $\nu$ and $\nu + \delta\nu$. It can be shown (*Classical Electricity and Magnetism*, p. 410 by Panofsky and Phillips) using classical electromagnetic theory that when the oscillators are in equilibrium with the radiation in a black-body cavity

$$E(\nu) = \frac{8\pi\nu^2}{c^3}\, u(\nu) \tag{2.1}$$

where $u(\nu)$ is the average energy of the oscillators with frequency $\nu$ and $c$ is the velocity of light.

The classical 'Law of Equipartition of Energy' states that the mean energy of the oscillators with frequency $\nu$ is

$$u(\nu) = kT \tag{2.2}$$

where $k$ is Boltzmann's constant and $T$ is the absolute temperature. From (2.1) and (2.2)

$$E(\nu)\delta\nu = \frac{8\pi\nu^2}{c^3}\, kT\,\delta\nu$$

or

$$E(\lambda)\delta\lambda = \frac{8\pi}{\lambda^4}\, kT\,\delta\lambda. \tag{2.3}$$

This is the Rayleigh–Jeans law and it implies that the energy radiated in a given wave-length range $\delta\lambda$ increases without bound as $\lambda$ becomes smaller. This non-physical result is of course experimentally incorrect (Fig. 2.1) and the Rayleigh–Jeans law is in agreement with experiment only for long wave-lengths. Also, equation (2.3) implies that the total energy emission is

$$E = \int_0^\infty E(\nu)\, d\nu = \frac{8\pi}{c^3} kT \int_0^\infty \nu^2\, d\nu \tag{2.4}$$

and clearly $E$ is infinite. The Rayleigh–Jeans law demands that the total energy radiated per unit time is infinite at all temperatures except $T = 0$. This obviously false result is often referred to as the ultra-violet catastrophe.

Clearly there was a fundamental error in classical theory.

## 2.2 Origin of the Quantum Theory

*Max Planck*

In 1900, before the publication of the Rayleigh–Jeans law, the German physicist Max Planck discovered an empirical relation that agreed with the entire continuous spectrum emitted by a black-body radiator. He then worked to find a physical explanation for his formula and succeeded in doing this after several weeks work. His results, published in 1901, were the origin of the quantum theory. The first quarter of the present century saw the subsequent development of the 'Old Quantum Theory' which was to be superceded in 1925 by the 'wave mechanics'.

Planck also assumed that a black-body is composed of oscillators in equilibrium with the radiation field and that equation (2.1) holds true. He realized that some drastic change was necessary to obtain a fit to the experimental data and his basic assumption was that the material oscillators can only have discrete energy levels rather than a continuous range of energies that had been assumed before. In particular Planck assumed that an oscillator with frequency $\nu$ can only take the values

$$\epsilon_n = nh\nu \qquad n = 0, 1, 2, \ldots \tag{2.5}$$

where $h$ is a universal constant, later to be known as the Planck constant. Planck chose the value $h = 6{\cdot}55 \times 10^{-34}$ joule sec (units of action) to agree with the experimental results but the value accepted today is

$$h = 6{\cdot}624 \times 10^{-34} \text{ joule sec.} \tag{2.6}$$

Note that the difference between consecutive energy levels of a Planck oscillator is an extremely small 'quantum of energy'.

Planck did not introduce the concept of the photon. He merely suggested that matter absorbed or emitted radiation energy in discrete amounts. (i.e. multiples of $h\nu$).

The argument below is not in the form given by Planck who used a thermodynamic derivation. Consider the $N_\nu$ oscillators with frequency $\nu$.

In equilibrium at temperature $T$ the number of oscillators with energy $\epsilon_n$ is, from the classical Boltzmann expression, equal to

$$N_\nu e^{-\epsilon_n/kT} \Big/ \sum_{n=0}^{\infty} e^{-\epsilon_n/kT}. \tag{2.7}$$

The mean energy per oscillator of frequency $\nu$ is

$$u(\nu) = \sum_{n=0}^{\infty} \epsilon_n e^{-\epsilon_n/kT} \Big/ \sum_{n=0}^{\infty} e^{-\epsilon_n/kT}. \tag{2.8}$$

Substitution of equation (2.5) gives

$$u(\nu) = h\nu/(e^{h\nu/kT} - 1). \tag{2.9}$$

The Planck expression for the density of radiation is obtained from (2.1) and (2.9) to be

$$E(\nu) = \frac{8\pi\nu^2}{c^3} \frac{h\nu}{(e^{h\nu/kT} - 1)}. \tag{2.10}$$

Planck's law (above) agrees very closely with the observed radiation distribution when $h$ is chosen correctly (see (2.6)). In the long wavelength limit

$$e^{h\nu/kT} \sim 1 + \frac{h\nu}{kT}$$

and Planck's expression reduces to the Rayleigh-Jeans law. The total radiation density arising from all frequencies can be obtained using (2.10) and

$$E = \int_0^\infty E(\nu)\, d\nu = \frac{8}{15} \frac{\pi^5 k^4}{c^3 h^3} T^4. \tag{2.11}$$

The energy density is proportional to the fourth power of the absolute temperature. This result was first suggested by Stefan.

*Albert Einstein—The photo-electric effect*

In 1905, Albert Einstein an employee of the Swiss patent office, took the next important step in the development of the quantum theory. Planck had suggested that oscillators absorb or emit radiation in energy bundles but that in free space the radiation obeys the continuous laws of Maxwell. Einstein pointed out that these ideas are not compatible and suggested that electromagnetic radiation only exists in discrete energy bundles later to be called photons. This was a step that Planck had hesitated to take as it opened the old wave-particle dialogue about the nature of light. The first major success in the new theory was Einstein's explanation of the photo-electric effect.

In 1887, Hertz in his 'hoop' experiments confirmed the existence of the electromagnetic waves predicted by Maxwell. During the same experiment he noticed that when light flashes from the transmitting circuit shone on the ends of his detector 'hoop', the weak sparks in the gap passed more easily. The Hertz effect was explained by Lenard who showed that a negatively-charged metal loses its charge when exposed to ultra-violet light because the electrons gain energy from the radiation and are emitted. This is now known as the photo-electric effect. Further experimental investigation of this effect showed that the number of electrons emitted increases when the intensity of light is increased but that the maximum observed speed of the electrons only increases if the frequency of the light is increased. This final result cannot be explained by the classical Maxwell theory which clearly demands that the electron speed should depend upon the light intensity. Einstein gives the explanation in terms of photons.

Einstein looked upon the photo-electric problem as one of particle collisions. He looked upon a monochromatic beam of light as being composed of 'particles' called photons, each of which has an energy $h\nu$ where $\nu$ is the frequency of the light.

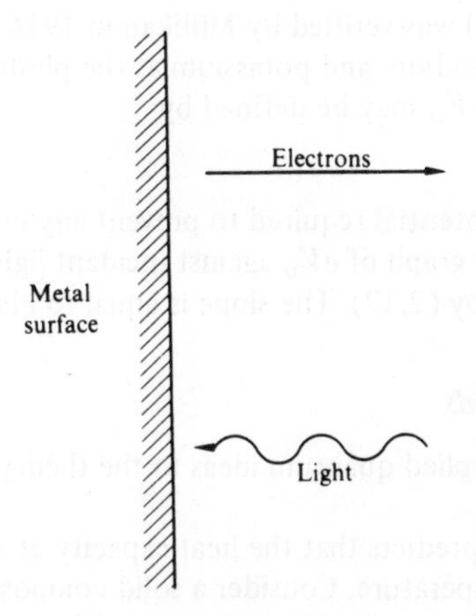


Figure 2.2

Essentially, Einstein's argument was as follows. When an incident photon collides with an electron in a metal surface, the energy of the photon is transferred to the electron and when the electron is emitted from the metal surface it has a kinetic energy given by

$$\tfrac{1}{2}mv^2 = h\nu - P \qquad (2.12)$$

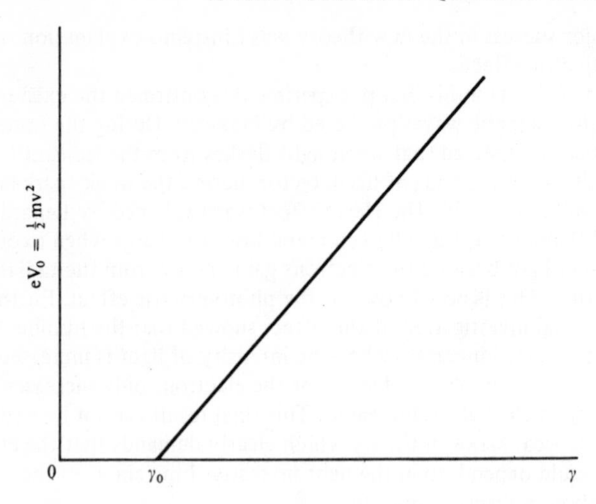


Figure 2.3 Dependence of stopping potential on incident light frequency.

where $\frac{1}{2}mv^2$ is the electron kinetic energy and $P$ is the work required to extract the electron from the metal. $P$ varies from metal to metal and is of the order of a few electron volts.

The result (2.12) was verified by Millikan in 1916 in a series of experiments using sodium and potassium as the photo-electric surface. A stopping particle $V_0$ may be defined by

$$V_0 e = \frac{1}{2}mv^2.$$

$V_0$ is the applied potential required to prevent any emission of electrons and the experiment graph of $eV_0$ against incident light frequency is linear as predicted by (2.12). The slope is equal to Planck's constant.

*Specific heat of solids*

In 1906 Einstein applied quantum ideas to the theory of the specific heat of solids.

Classical theory predicts that the heat capacity at constant volume is independent of temperature. Consider a solid composed of $N$ identical atoms each of which can vibrate about its equilibrium position. The classical 'Law of Equipartition of Energy' demands that the mean energy of such a three-dimensional oscillator in equilibrium at absolute temperature $T$ is $3kT$ and the mean energy of the solid is

$$u = 3NkT. \tag{2.13}$$

For one mole, $N = 6{\cdot}025 \times 10^{23}$ (Avogadro's number) and

$$u = 3RT \tag{2.14}$$

where $R$ is the gas constant. The lattice contribution to the molar heat capacity is

$$C_v = \left(\frac{\partial u}{\partial T}\right)_v = 3R \sim 6 \text{ cal/deg mole.} \tag{2.15}$$

Classical theory predicts that the heat capacity per mole is the same for all substances at all temperatures. At room temperatures and above many solids obey this law but agreement fails entirely at low temperatures. Experimentally it is found that the heat capacity approaches zero as $T \to 0$.

Einstein, following Planck, assumed a solid can be represented by a collection of harmonic oscillators which can only take discrete energy values $nh\nu$, $n$ integer, where $\nu$ is the oscillator frequency. Einstein considered the simplest model and assumed all the oscillators have the same frequency $\nu_0$.

The internal energy of the solid is then

$$u = N \frac{3h\nu_0}{(e^{h\nu_0/kT} - 1)}. \tag{2.16}$$

The lattice heat capacity at constant volume is

$$C_v = \frac{\partial u}{\partial T} = 3Nk \left(\frac{\theta_E}{T}\right)^2 \frac{e^{\theta_E/T}}{[e^{\theta_E/T} - 1]^2} \tag{2.17}$$

where the Einstein temperature $\theta_E$ is defined by $h\nu_0 = k\theta_E$. The Einstein temperature for a material may be chosen to make equation (2.17) give

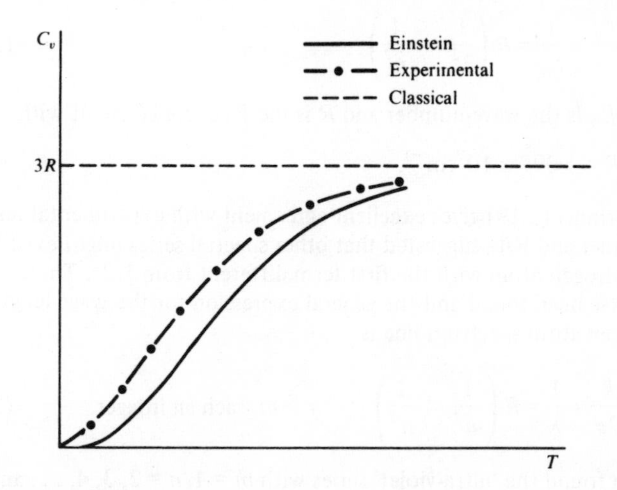


Figure 2.4 Molar heat capacity.

a reasonable fit to the experimental results and is of the order of a few hundred degrees absolute.

Einstein's formula is in fair agreement with experiment for $T \gg \theta_E$ but at low temperatures when $T \ll \theta_E$ the agreement is poor. For non-metals, the heat capacity approaches zero as $T^3$ whereas Einstein's formula predicts

$$c_v = 3Nk\left(\frac{\theta_E}{T}\right)^2 e^{-\theta_E/T} \qquad T \ll \theta_E. \tag{2.18}$$

Debye (1912) successfully explained the $T^3$ variation at low temperatures by considering a spectrum of frequencies to be present.

## 2.3 The Hydrogen Atom

Experimental spectroscopy developed rapidly during the second half of the nineteenth century. Because the emission of light was thought to be the result of vibrations, harmonic relations between the spectrum lines were looked for without success. However in 1885 Johann Balmer discovered that the wavelengths of all the spectrum lines of the hydrogen atom then known could be expressed by the formula

$$\lambda_n = \frac{n^2}{n^2 - 4} b \qquad n = 3, 4, 5, \ldots$$

where $b$ is a constant. This equation is generally written

$$\frac{k_n}{2\pi} = \frac{1}{\lambda_n} = R\left(\frac{1}{2^2} - \frac{1}{n^2}\right) \tag{2.19}$$

where $k_n$ is the wave-number and $R$ is the Rydberg constant with

$$R = 1{\cdot}097 \times 10^7 \text{ m}^{-1}.$$

The formula (2.19) gives excellent agreement with experimental results.

Balmer and Ritz suggested that other spectral series might exist for the hydrogen atom with the first term different from $1/2^2$. These have of course been found and the general expression for the wave-length of a hydrogen atom spectrum line is

$$\frac{k}{2\pi} = \frac{1}{\lambda} = R\left(\frac{1}{m^2} - \frac{1}{n^2}\right) \qquad n > m, \text{ each an integer.} \tag{2.20}$$

Lyman found the 'ultra-violet' series with $m = 1, n = 2, 3, 4, \ldots$ and Paschen discovered the 'infra-red' series with $m = 3, n = 4, 5, 6, \ldots$.

From (2.20), the wave number of any line of the hydrogen atom spectrum is the difference between members of the series of 'terms'

$$t_n = \frac{R}{n^2} \qquad n = 1, 2, \ldots \tag{2.21}$$

The spectrum lines of other elements can also be obtained as the difference of two terms, i.e.

$$\frac{k}{2\pi} = \frac{1}{\lambda} = t_i - t_j. \tag{2.22}$$

However the terms $t_i$, $t_j$ generally are more complicated than those used for the hydrogen spectrum. This is the Ritz Combination principle. Although these empirical formulae were very useful they had no theoretical explanation.

In 1911 Ernest Rutherford suggested the 'nuclear atom'. From his experiments on the scattering of α-particles Rutherford was lead to suggest that an atom consists of a small positively charged nucleus surrounded by outer electrons. (The electron had been established by J. J. Thomson in 1897.)

Unfortunately classical deductions from Rutherford's 'nuclear atom' do not explain the observed atomic spectra. No stationary configuration of charges can be in stable equilibrium under their own electrostatic forces (Earnshaw's theorem) and so the electrons must be in motion relative to the nucleus. Such electrons must orbit in some way and the corresponding acceleration will result in energy loss by electromagnetic radiation. Hence electrons will gradually spiral into the nucleus. Moreover the frequency of the radiation emitted will gradually increase as this spiralling continues and a continuous spectrum will be produced instead of the observed discrete spectral lines.

*Niels Bohr*

In 1913 the Danish physicist Niels Bohr combined the 'nuclear atom' concept of Rutherford with Planck's quantum hypothesis to give a theory that accounted for the observed spectrum of hydrogen. Bohr made two basic postulates.

(i) He assumed that the electron is a particle that can revolve about the nucleus only in certain allowed orbits which may be circular or elliptical. Each orbit represents a stationary energy state in which electromagnetic radiation is not emitted, in conflict with the predictions of classical theory.

The allowed orbits or states are those in which the angular momentum of the electron about the nucleus is an integral multiple of $\hbar = h/2\pi$.

(ii) Emission or absorption of radiation occurs when the electron makes a quantum jump from one stationary state to another. When the atom makes a transition from a higher energy state $E_i$ to a lower energy state $E_f$ Bohr suggested, (following Planck), that the energy difference is radiated as a photon of frequency $\nu$ with

$$E_i - E_f = h\nu. \tag{2.23}$$

An outline of the Bohr theory is given below.

Bohr used classical methods to calculate the energy of his allowed stationary states. The force of attraction between the hydrogen nucleus and the electron is

$$\frac{e^2}{4\pi\epsilon_0 r^2} \tag{2.24}$$

where $r$ is the distance of the electron from the nucleus and $\epsilon_0$ is the permittivity of free space. The electron must move around the nucleus in an orbit that is a conic section. The simplest case is that of a circle and the inward acceleration is

$$\frac{mv^2}{r} \tag{2.25}$$

where $m$ is the electron mass and $v$ is the speed. From Newton's second law and equations (2.24) and (2.25) above, the kinetic energy is seen to be

$$T = \frac{1}{2} mv^2 = \frac{e^2}{8\pi\epsilon_0 r}. \tag{2.26}$$

The potential energy of the electron is given by

$$V = -\frac{e^2}{4\pi\epsilon_0 r}$$

and so the total energy of the electron in a circular orbit of radius $r$ is

$$E = T + V = \frac{-e^2}{8\pi\epsilon_0 r}. \tag{2.27}$$

The zero of energy is taken to be the ionized atom where $r = \infty$.

From Bohr's first postulate the radii of the allowed orbits are given by

$$mvr = n\hbar \qquad n = 1, 2, 3, \ldots. \tag{2.28}$$

From (2.26) and (2.28) it is possible to eliminate $v$ to give

$$r = \frac{n^2\hbar^2 4\pi\epsilon_0}{me^2} \tag{2.29}$$

and the energies of the allowed states are

$$E_n = -\frac{me^4}{2(4\pi\epsilon_0)^2\hbar^2}\frac{1}{n^2} \qquad n = 1, 2, \ldots. \tag{2.30}$$

When an electron jumps from the $n_i$ orbit to the $n_f$ orbit the frequency of the radiation emitted is given by

$$\nu = \frac{E_i - E_f}{h}$$

i.e.

$$\nu = \frac{2\pi^2 me^4}{(4\pi\epsilon_0)^2 h^3}\left[\frac{1}{n_f^2} - \frac{1}{n_i^2}\right]. \tag{2.31}$$

The wave number of the radiation is given by

$$\frac{k}{2\pi} = \frac{1}{\lambda} = \frac{\nu}{c}$$

and so

$$\frac{k}{2\pi} = \frac{2\pi^2 me^4}{(4\pi\epsilon_0)^2 h^3 c}\left[\frac{1}{n_f^2} - \frac{1}{n_i^2}\right].$$

This result agrees with the formula (2.20) if the Rydberg constant

$$R = \frac{2\pi^2 me^4}{(4\pi\epsilon_0)^2 h^3 c}. \tag{2.32}$$

The value of $R$ calculated from (2.32) using the accepted values of $m$, $e$, $c$ and $h$ agrees excellently with the experimental value. This theory was a mixture of classical and quantum concepts. Bohr assumed that classical theory could be used to calculate the energies of the states but used his own quantum argument to determine the states allowed.

Bohr's theory dealt only with circular orbits. A. Sommerfeld (1916) (see also Wilson 1915), generalized the Bohr theory to include elliptic orbits. He said that the allowed states of a periodic system are determined by the quantum conditions

$$\oint p_i \, dq_i = n_i h \qquad n_i \text{ integer} \tag{2.33}$$

where $p_i$ is the momentum conjugate to the co-ordinate $q_i$ and the integration is over a complete period of the motion.

Consider an electron revolving about the hydrogen nucleus in a plane elliptic orbit, with the nucleus at one focus. The single Bohr condition (2.28) is replaced by the two conditions

$$\oint p_\phi \, d\phi = n_\phi h \qquad \oint p_r \, dr = n_r h \tag{2.34}$$

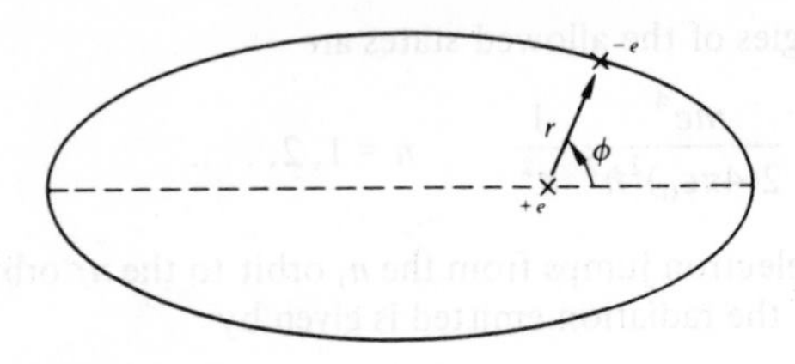


Figure 2.5 Elliptic electron orbit.

where $n_\phi$, $n_r$ are both integers. The momentum $p_\phi$ is simply the angular momentum and from Kepler's law is a constant.

$$\therefore \quad p_\phi 2\pi = n_\phi h \qquad \text{or} \qquad p_\phi = n_\phi \hbar$$

as assumed by Bohr. $n_\phi$ is called the angular or azimuthal quantum number. $n_r$ is the radial quantum number and the principle quantum number is defined by

$$n = n_r + n_\phi. \tag{2.35}$$

It can be shown that Bohr's energy expression (2.30) holds for elliptic orbits where the $n$ in the formula is now the principle quantum number. The introduction of elliptic orbits does not produce any new levels for the hydrogen atom although more than one state (orbit) may have the same energy. However, when the relativistic mass variation of the electron is included it is found that this 'degeneracy' is removed and that the energy depends upon both $n$ and $n_\phi$. Of course, a particle electron possesses three degrees of freedom and it is to be expected that a third quantum number should be necessary to define a state completely. This is the magnetic quantum number.

*Simple harmonic oscillator*

When the Sommerfeld rule is applied to the linear harmonic oscillator the energy levels used by Planck are obtained. Consider a particle of mass $m$ oscillating about the origin in the $x$-direction. The energy $E$ of the system is given by

$$E = \frac{p^2}{2m} + 2\pi^2\nu^2 m x^2$$

where $p$ is the momentum conjugate to $x$ and $\nu$ is the frequency. The Sommerfeld 'action' integral (2.33) becomes

$$\oint p(x)\,\mathrm{d}x = 2m \int_{-a}^{a} \left[\frac{2E}{m} - 4\pi^2\nu^2 x^2\right]^{1/2} \mathrm{d}x = nh \qquad n \text{ integer}$$

where

$$a = \sqrt{\frac{2E}{m}} \Big/ 2\pi\nu.$$

This integral can be evaluated by substituting

$$x = \sqrt{\frac{2E}{m}} \sin\theta / 2\pi\nu$$

to give

$$\frac{E}{\nu} = nh$$

i.e.

$$E = nh\nu.$$

This is equation (2.5).

*Bohr correspondence principle*

The Bohr theory of hydrogen atom even when developed using the Sommerfeld condition was clearly not complete. It did give an explanation of the Zeeman and Stark effects, but unfortunately the Bohr theory predicted too many spectrum lines. Some type of 'selection rules' had to be imposed. Also, although it was obvious that different spectral lines had different intensities the Bohr theory had no explanation of this. In 1918 Bohr added another rule to his theory which was to prove very successful in overcoming the difficulties mentioned above. This was the correspondence principle.

In his original paper (1913) Bohr reasoned that in the larger atomic orbits, where the energy jumps are smoothed out, the quantum theory must give the same results as classical theory. Consider an electron jump from a hydrogen state with $n_i = N + 1$ to a final state $n_f = N$. From (2.30) the frequency of the emitted radiation is

$$\nu = \frac{2\pi^2 me^4}{(4\pi\epsilon_0)^2 h^3} \left[ \frac{1}{N^2} - \frac{1}{(N+1)^2} \right].$$

In the large quantum number limit $N \to \infty$ this frequency becomes

$$\nu = \frac{4\pi^2 me^4}{(4\pi\epsilon_0)^2 h^3} \; \frac{1}{N^3} \qquad (2.36)$$

Now consider the frequency calculated from classical electromagnetic theory. Classically the emission frequency is equal to the frequency of revolution of the electron in the circular orbit. (Electrons in elliptic orbits also emit frequencies which are integral multiples of the fundamental frequency.) That is, the classical frequency is

$$\nu = v/2\pi r$$

where $v$ is the electron velocity and $r$ is the orbit radius. On substituting for $v$ and $r$ using (2.28) and (2.29) it is easily seen that the classical frequency is equal to the limiting quantum frequency (2.36). In the case where the energy change associated with a transition is small compared with the energies of the initial and final states, the quantum description approaches the classical description.

In 1918 Bohr extended the original form of the correspondence princple and applied it to large energy jumps and managed to obtain rules for calculating intensities. Although it proved quite successful it was an unsatisfactory mixture of quantum and classical theory and will not be considered further.

## 2.4 Failure of the Old Quantum Theory

There were obvious weaknesses in the 'old quantum theory' of Bohr and Sommerfeld. It was seen from the beginning to be a mixture of quantum and classical ideas. The electron was regarded as a point charge and the Sommerfeld rules picked out the allowed orbits which had energies calculated using the classical theory. However, even though classical theory demanded that the accelerating electrons must emit radiation, Bohr postulated that this did not happen for electrons in a stationary orbit. Also, non-periodic phenomena were not included. There was no quantum theory of $\alpha$-particle emission.

Although the Bohr–Sommerfeld theory allowed only integral quantum numbers it was discovered that certain systems demanded half-integral quantum numbers. Such an example was the rotation spectra of the hydrogen halides molecules. The explanation of this requires a spin structure for the 'particle electron'. In some cases quantitative disagreement occurred between the old quantum theory and experiment (e.g. the normal state of the helium atom) and in others even qualitative agreement was not even achieved. Disatisfaction with the old quantum theory eventually lead to the discovery of the matrix mechanics of Heisenberg and the equivalent wave mechanics of Schrödinger (1926).

## 2.5 Wave-Particle Dualism

### *Dual nature of light*

By the end of the nineteenth century, the wave theory of light was firmly established on the basis of interference and diffraction experiments and the success of the classical electromagnetic theory. This view was challenged by Einstein in his theory of the photoelectric effect in which energy is transferred to the electrons in quanta of energy $h\nu$ called photons. Further evidence concerning the dual nature of light was given by the production of X-rays.

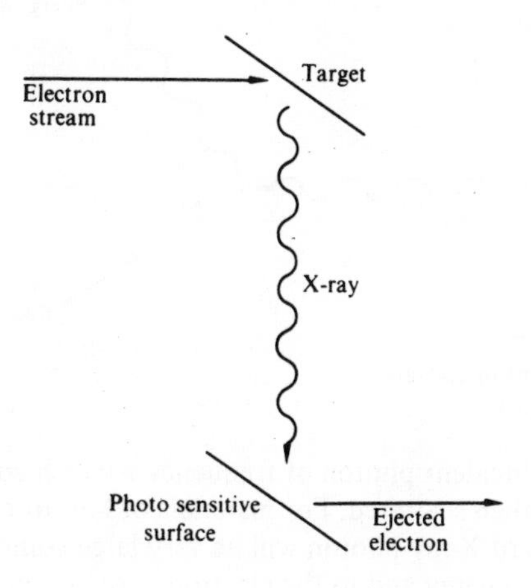


Figure 2.6

Roentgen discovered X-rays in 1895 and the wave nature of X-rays was established by von Laue in 1912 by his diffraction experiments. X-ray production may be regarded as the inverse of the photoelectric effect and is brought about by directing a stream of electrons onto the anti-cathode in an X-ray tube. If the X-rays produced are then used in a photo-electric experiment to produce electrons it is found that the maximum energy of a photo-electron is equal to the energy of one of the electrons in the stream which originally produced the X-rays. This implies that the X-rays transport their energy across space in photon bundles without any dissipation.

In 1922–23 Compton took the photon concept a step further when he associated a momentum with the photon in his explanation of the scattering of hard X-rays.

*Compton effect*

The American physicist Arthur Compton discovered that when hard (short wave-length) X-rays are scattered by atoms of low atomic number (graphite) the scattered radiation includes not only the original wavelength but also softer X-rays of longer wave-length. This is contrary to the classical theory which demands that the scattering electrons vibrate with the frequency of the incident radiation and so emit radiation of exactly this frequency. However Compton was able to explain this phenomenon by assuming that the X-rays consist of a collection of photon particles each having an energy $h\nu$ and also a momentum $h\nu/c$ where $c$ is the velocity of light.

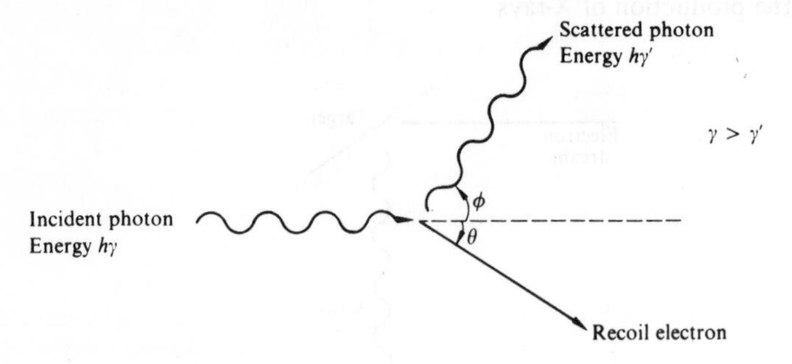


Figure 2.7 Compton scattering.

Consider an incident photon of frequency $\nu$ which collides with an electron and is then scattered. For materials of low atomic number the energy of the hard X-ray photon will be very large compared with the electron binding energy and so the electron may be considered to be free. During the collision, energy is conserved and so

$$h\nu = h\nu' + T \tag{2.37}$$

where $T$ is the kinetic energy of recoil of the electron and $\nu'$ is the frequency of the scattered photon. From special relativity theory

$$T = m_0c^2\left[\frac{1}{\sqrt{1-\left(\frac{v}{c}\right)^2}} - 1\right] \tag{2.38}$$

where $m_0$ is the electron rest mass and $v$ is the speed of recoil. Applying the low of conservation of momentum

$$\frac{h\nu}{c} = \frac{h\nu'}{c}\cos\phi + \frac{m_0}{\sqrt{1-\left(\frac{v}{c}\right)}}v\cos\theta$$

and

$$0 = \frac{h\nu'}{c}\sin\phi - \frac{m_0}{\sqrt{1-\left(\frac{v}{c}\right)^2}}v\sin\theta. \tag{2.39}$$

There are three equations involving the three unknowns $\nu'$, $v$ and $\theta$ and on solving it is found that the increase in wave-length is

$$\delta\lambda = \lambda' - \lambda = \frac{2h}{m_0 c}\sin^2\frac{\phi}{2} \tag{2.40}$$

where $\lambda$ and $\lambda'$ are the wave-lengths of the incident and scattered X-rays respectively. Note that $\delta\lambda$ is constant for a given scattering angle $\phi$ and so the percentage increase in wave-length is greater for small wave-lengths. This formula is in quite good agreement with experiment.

When the frequency of the incident radiation is reduced, the binding energy of the electron cannot be ignored and instead of Compton scattering the photoelectric effect takes place or the X-ray undergoes Laue scattering.

The work of Einstein and Compton when taken in conjugation with the work on interference and diffraction leads to the paradoxical result that electromagnetic radiation has a dual nature. Sometimes it behaves with wave-like properties and sometimes with particle-like properties. The attempts made to understand this paradox have resulted in the new quantum theory.

*Dual nature of matter*

In 1897 the English physicist J. J. Thomson succeeded in deflecting cathode rays by an electrostatic field and proved conclusively that the rays consist of material particles with a negative charge. For the next quarter century no doubt was expressed as to the particle nature of the electron.

During 1922–23 the Frenchman Louis de Broglie suggested that not only does radiation have a dual nature but material particles also require a wave-particle description. He was primarily interested in the nature of light and on the basis of Einstein's relativity theory he was convinced of the dual wave-particle nature of light.

From special relatively, the relation between the total energy $E$ of a particle in field free space and its momentum $p$ and rest mass $m_0$ is

$$\left(\frac{E}{c}\right)^2 = p^2 + m_0^2 c^2. \tag{2.41}$$

Taking the rest mass of a photon as zero

$$\left(\frac{E}{c}\right)^2 = p^2 \qquad \text{or} \qquad E = pc.$$

This final result is consistent with the Einstein formula

$$E = h\nu$$

and the Compton result

$$p = h/\lambda.$$

In relativity both radiation and matter appear as forms of energy. L. de Boglie suggested that any moving particle has associated with it a group of waves whose wave-length $\lambda$ depends on the momentum of the particle according to the equation

$$p = h/\lambda. \tag{2.42}$$

This is the equation used by Compton for photons. The wave-length associated with large bodies (e.g. a billiard ball) moving at low speeds is far tòo small to be detected experimentally. However, the wave-length becomes appreciable for atomic particles. An electron with a 10 eV kinetic energy has a wave-length of $5{\cdot}3 \times 10^{-8}$ cm (5 Å).

De Broglie noticed correspondences between the classical theory of light and the classical theory of mechanics which help to suggest equation (2.42). The French mathematician Fermat had reduced the laws of geometrical optics to the principles of 'least-time'. That is, a light ray follows the path requiring the least time. The time taken for a light ray to travel from a point A to a point B is given by the line integral

$$\int_A^B \frac{1}{V_p(x, y, z)}\, ds$$

where $V_p$ is the phase velocity in the medium and is a function of position in general. As

$$\frac{1}{V_p} = \frac{1}{\nu} \times \frac{1}{\lambda}$$

where $\nu$ is the constant frequency, Fermat's principle may be written

$$\delta \int_A^B \frac{1}{\lambda(x, y, z)}\, ds = 0. \tag{2.43}$$

There is a corresponding principle in classical mechanics. This is Hamilton's principle (1.52) which states

$$\delta \int_A^B L\, dt = 0.$$

For a single particle of constant total energy $E$ and kinetic energy $T$ then

$$L = 2T - E \qquad \text{as} \qquad E = T + V$$

and so

$$\delta \int_A^B 2T \, dt = 0.$$

By changing the variable of integration from the time to distance it is not difficult to show that this final result can be written

$$\delta \int_A^B p(x, y, z) \, ds = 0 \tag{2.44}$$

and the similarity between (2.43) and (2.44) is striking. The relation

$$p = \text{constant}/\lambda$$

is immediately suggested.

This analogy between the variational principles of geometrical optics and particle mechanics had of course been noticed much earlier and had resulted in the Hamilton–Jacobi equation discussed in Chapter 1. Schrödinger used this equation in his development of the Schrödinger wave equation for quantum mechanics but the discussion of the relation between the Hamilton–Jacobi and Schrödinger equations is left till Chapter 4.

The first experimental verification of the de Broglie matter waves was obtained in 1925 by Davison and Germer who were conducting experiments involving the scattering of slow electrons by nickel in a vacuum. De Broglie's formula was completely verified by G. P. Thomson using a thin film of gold for his scattering material.

Another aspect of the relation between particles and waves is illustrated by considering the concepts of phase and group velocity.

The phase velocity $V_p$ of a de Broglie wave is given by

$$V_p = \nu\lambda \tag{2.45}$$

where $\nu$ and $\lambda$ are the frequency and wave-length associated with the particle. Writing, with the usual notation

$$E = h\nu \qquad \text{and} \qquad p = h/\lambda$$

then

$$V_p = E/p.$$

In free space, relativity gives

$$E = \frac{m_0}{\sqrt{1 - \left(\frac{v}{c}\right)^2}} c^2 \qquad p = \frac{m_0}{\sqrt{1 - \left(\frac{v}{c}\right)^2}} v \tag{2.46}$$

where $m_0$ is the particle rest mass and $v$ is the particle velocity.

$$V_p = c^2/v. \tag{2.47}$$

As energy cannot be transmitted faster than light, $v < c$ and so

$$V_p > c.$$

The de Broglie waves for a free particle travel with a phase velocity faster than light. However there is no conflict with relativity as the energy of the waves (or particle) is transmitted with the group velocity not the phase velocity. This is explained below.

*Group and phase velocities*

The equation of motion of a plane wave of frequency $\nu$ and wavelength $\lambda$ moving in the $x$-direction is

$$\psi(x, t) = A \sin 2\pi \left( \nu t - \frac{x}{\lambda} \right)$$

where $A$ is the amplitude. This wave moves with a phase velocity $V_p = \nu\lambda$.

It is possible to form a concentrated wave packet by taking a sum of a large number of plane waves with slightly different wave-lengths and frequencies. If the medium is dispersive so that plane waves with different frequencies move with different phase velocities, the wave packet constructed moves with a velocity different from the wave velocities. This new velocity is called the group velocity.

The essentials of the argument are obtained by considering the interference of two waves of similar but different frequencies and wave-lengths. Take

$$\psi(x, t) = A [\sin(\omega t - kx) + \sin(\omega' t - k' x)]$$

where $k$, $k'$ denote the wave numbers such that

$$k = 2\pi/\lambda \qquad \text{and} \qquad k' = 2\pi/\lambda'$$

and $\omega$, $\omega'$ denote the angular frequencies given by

$$\omega = 2\pi\nu \qquad \text{and} \qquad \omega' = 2\pi\nu'.$$

It follows that

$$\psi(x, t) = 2A \cos\left\{ \left( \frac{\omega' - \omega}{2} \right) t - \left( \frac{k' - k}{2} \right) x \right\}$$

$$\times \sin\left\{ \left( \frac{\omega' + \omega}{2} \right) t - \left( \frac{k' + k}{2} \right) x \right\}.$$

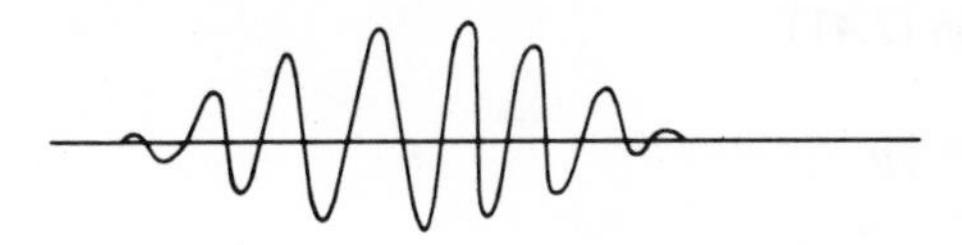

Figure 2.8 Wave packet.

Put $\omega' = \omega + \delta\omega$ and $k' = k + \delta k$ and then approximately

$$\psi(x, t) = \left[2A \cos\left\{\frac{\delta\omega}{2}t - \frac{\delta k}{2}x\right\}\right] \sin(\omega t - kx). \tag{2.48}$$

The combined disturbance represents a carrier sine wave with frequency $\nu$ and wave number $k$ with an amplitude that slowly fluctuates so that

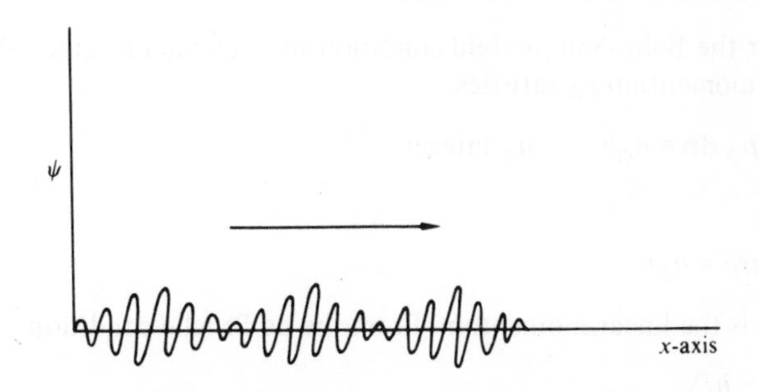


Figure 2.9 Groups of waves or beats.

there are groups of waves with gaps in between. The groups of waves move with a group velocity

$$V_g = \frac{\delta\omega}{\delta k} \quad \text{where} \quad \omega = 2\pi\nu.$$

In the limit as $\delta k \to 0$ the group velocity becomes

$$V_g = \frac{\partial\omega}{\partial k}. \tag{2.49}$$

For the de Broglie waves corresponding to a particle, the total energy $E$ and linear momentum $p$ are given by

$$E = h\nu \qquad p = \hbar k$$

and so the group velocity is

$$V_g = \frac{\partial E}{\partial p}. \tag{2.50}$$

From equation (2.41)

$$\frac{2E}{c^2}\frac{\partial E}{\partial p} = 2p$$

and so the group velocity is (from 2.46)

$$V_g = \frac{pc^2}{E} = \frac{vc^2}{c^2} = v.$$

The group velocity of the de Broglie waves is equal to the corresponding particle velocity and so from (2.47)

$$V_g V_p = c^2. \tag{2.51}$$

*De Broglie waves and the Bohr atom*

Consider the Bohr–Sommerfeld condition for a circular atomic orbit. The angular momentum $p_\phi$ satisfies

$$\oint p_\phi \, \mathrm{d}\phi = n_\phi h \qquad n_\phi \text{ integer}$$

i.e.

$$2\pi r p = n_\phi h$$

where $p$ is the linear momentum. From the de Broglie condition

$$p = h/\lambda$$

and so the circumference of the circular orbit is

$$2\pi r = n_\phi \lambda.$$

The Bohr–Sommerfeld quantization rule is equivalent to demanding that the length of the circumference of a stationary Bohr orbit is equal to an integral number of de Broglie wave-lengths. This is additional evidence that particles have wavelike properties.

## 2.6 The New Quantum Mechanics

*Werner Heisenberg*

In 1925 the young German physicist Werner Heisenberg introduced the new matrix mechanics. His work was based upon the inadequacies of the Bohr correspondence principle rather than the hypothesis of de Broglie waves. He found it necessary to introduce non-commutative algebra into physics. If $q$ and $p$ represent the position and conjugate momentum for a particle then Heisenberg found that in his atomic theory $qp$ is not equal to $pq$. In 1925 he applied his theory successfully to the simple harmonic oscillator and explained the phenomena of zero-point energy.

Heisenberg found it necessary to introduce a new calculus. Max Born, a professor in the university of Göttingen where Heisenberg was working, pointed out that this 'new' calculus was precisely the calculus of matrices previously discovered by the English mathematician A. Cayley. Born and a colleague Jordan sought to extend Heisenberg's work to obtain a new theory of the atom. They obtained the important result

$$pq - qp \equiv -i\hbar, \tag{2.52}$$

where $i$ is the square root of minus one. The significance of this will become apparent in a later chapter (Chapter 5).

In January 1926 the Austrian Wolfgang Pauli successfully applied the Heisenberg theory to the hydrogen atom problem and in the same month the young Englishman Paul Dirac produced a more elegant, though more abstract, application of the Heisenberg theory to the hydrogen atom. He also obtained the relation between the Poisson brackets of classical mechanics and the commutators of quantum mechanics and showed that there is a very close relationship between classical and quantum mechanics.

*Erwin Schrödinger*

At about the same time the young Austrian Erwin Schrödinger was developing his wave mechanics which seemed at the time to be completely distinct from Heisenberg's work. He solved the hydrogen problem and published his results in the same year as Pauli and Dirac, 1926. In February he gave the solution for a harmonic oscillator. Bohr had found it necessary to postulate the existence of quantum numbers; Schrödinger was able to derive them.

Schrödinger knew that in many classical wave problems, discrete frequencies arise naturally when boundary conditions are imposed. Consider the vibrations of a string of length $l$ with both ends fixed. This transverse displacement $\phi(x, t)$ of an elastic string lying along the $x$-axis obeys the wave equation

$$\frac{\partial^2 \phi}{\partial x^2} - \frac{1}{V_p^2}\frac{\partial^2 \phi}{\partial t^2} = 0 \tag{2.53}$$

where $V_p$ is the phase velocity of the wave and $t$ is the time. Variable separable solutions that satisfy the boundary conditions

$$\phi(0, t) = \phi(l, t) = 0$$

exist and have the form

$$\phi(x, t) = A \sin \frac{n\pi x}{l} \sin\left(\frac{n\pi V_p t}{l} + \epsilon\right), \tag{2.54}$$

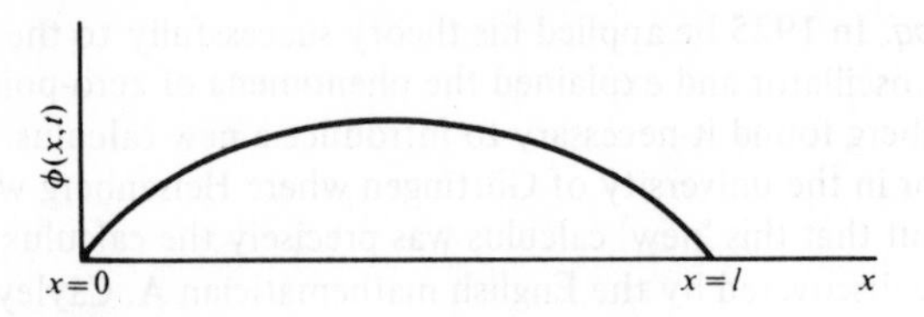


Figure 2.10 Vibrating string.

where $A, \epsilon$ are constants and $n$ can take the integer values 1, 2, . . .. The solutions (2.54) are called 'eigenfunctions' with corresponding discrete 'eigenfrequencies'.

$$\nu_n = nV_p/2l.$$

Schrödinger was able to set up the wave equations governing the motion of a particle. His equation is obtained below although the argument given is in no sense a proof. Wave mechanics cannot be derived from classical mechanics but must be postulated as in Chapter 3 of this text. The truth of the postulates depends upon the accuracy of the predicted results.

From classical mechanics, the total energy of a particle of mass $m$ in a conservative field is given by

$$E = \frac{p^2}{2m} + V \tag{2.55}$$

where $p$ is the particle momentum and $V$ is the potential energy. The particle momentum can be written

$$p = +\sqrt{2m(E-V)}.$$

The phase velocity of the de Broglie matter waves is

$$V_p = \frac{E}{p} = \frac{E}{\sqrt{2m(E-V)}}. \tag{2.56}$$

By combining the wave equation (2.53) and this phase velocity expression the wave equation for matter can be written

$$\nabla^2\Psi - \frac{2m(E-V)}{E^2}\frac{\partial^2\Psi}{\partial t^2} = 0 \tag{2.57}$$

where $\Psi(x, y, z, t)$ denotes the amplitude of the matter wave at a point $(x, y, z)$ at time $t$.

States of constant energy correspond to states of constant frequency and for these the solutions can be written

$$\Psi(x, y, z, t) = \psi(x, y, z) \exp\left[-2\pi i \nu t\right]. \tag{2.58}$$

Equation (2.57) becomes

$$\nabla^2 \Psi + \frac{2m[E - V]}{E^2} 4\pi^2 \nu^2 \Psi = 0.$$

Using the Einstein relation $E = h\nu$ then finally

$$-\frac{\hbar^2}{2m} \nabla^2 \Psi + V\Psi = E\Psi \qquad \text{or} \qquad -\frac{\hbar^2}{2m} \nabla^2 \psi + V\psi = E\psi. \tag{2.59}$$

This final equation is known as Schrödinger's time-independent wave equation and gives the states of constant energy.

*Linear box*

Consider the idealized problem of a particle restrained to lie in a linear box of length $l$ by walls of infinite potential. Suppose

$$V(x) = 0 \qquad 0 < x < l$$
$$= \infty \qquad \text{otherwise.}$$

The Schrödinger equation is

$$-\frac{\hbar^2}{2m} \frac{d^2 \psi}{dx^2} = E\psi \qquad 0 < x < l.$$

The boundary conditions to impose are

$$\psi(0) = \psi(l) = 0$$

as the particle must lie within the box. The eigenfunctions (allowed solutions) are

$$\psi_n(x) = A \sin \frac{n\pi x}{l} \qquad n = 1, 2, \ldots$$

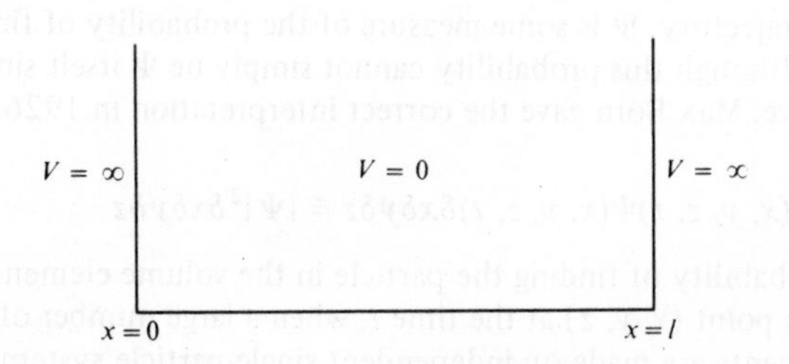


Figure 2.11 Linear box.

where $A$ is a constant and the eigenvalues are (allowed energy values)

$$E_n = \frac{n^2 h^2}{8ml^2} \qquad n = 1, 2, \ldots . \tag{2.60}$$

The application of boundary conditions to the Schrödinger equation naturally results in discrete energy values and hence in quantum numbers.

*The time-dependent equation*

The time-independent equation (2.59) is sufficient for discussing states of constant energy but a more general equation is required for describing other states. It is clearly necessary to eliminate the energy $E$. For solutions of the form (2.58)

$$E\Psi = +i\hbar \frac{\partial \Psi}{\partial t} \tag{2.61}$$

and combining this result with the time independent equation (2.59) the time-dependent equation

$$-\frac{\hbar^2}{2m} \nabla^2 \Psi + V\Psi = i\hbar \frac{\partial \Psi}{\partial t}. \tag{2.62}$$

is obtained.

The derivation of the Schrödinger equations (2.59) and (2.62) given above is not a proof but is given simply to indicate how they were first revealed. A formal derivation based upon the postulates of quantum mechanics is given in the next chapter. These equations are non-relativistic and correctly describe the motion of a particle only at speeds small compared with the velocity of light. Paul Dirac (1928) established the relativistic equation.

*The wave function*

The complex wave function $\Psi(x, y, z, t)$ is a solution of Schrödinger's equation and in quantum mechanics takes the place of the classical particle trajectory. $\Psi$ is some measure of the probability of finding the particle although this probability cannot simply be $\Psi$ itself since $\Psi$ can be negative. Max Born gave the correct interpretation in 1926. He said that

$$\Psi^*(x, y, z, t)\Psi(x, y, z, t)\delta x \delta y \delta z \equiv |\Psi|^2 \delta x \delta y \delta z \tag{2.63}$$

is the probability of finding the particle in the volume element $\delta x \delta y \delta z$ about the point $(x, y, z)$ at the time $t$, when a large number of position measurements are made on independent single-particle systems each of which is described by the wave function $\Psi(x, y, z, t)$.

As the total probability of finding the particle somewhere in space must be unity the wave function must be normalized to unity, i.e.

$$\int \Psi^*(x, y, z, t)\Psi(x, y, z, t)\, d\tau = 1. \tag{2.64}$$

The normalized solutions to the linear box problem are obtained from (2.59) to be

$$\psi_n(x) = \sqrt{\frac{2}{l}} \sin \frac{n\pi x}{l} \qquad n = 1, 2, \ldots.$$

The corresponding probability density

$$p(x) = \psi_n^*(x)\psi_n(x) = \frac{2}{l} \sin^2 \frac{n\pi x}{l}$$

and in the large quantum number limit $n \to \infty$ the average value of $p(x)$ over any interval $\delta x$ is

$$p_A(x) = \frac{1}{l}. \tag{2.65}$$

From Bohr's correspondence principle this should agree with the probability density of a classical particle in the linear box. To establish this, consider a classical particle moving along the $x$-axis with constant speed $v$ between reflecting walls at $x = 0$ and $x = l$. The probability of finding the particle in the element $\delta x$ is proportional to the time the particle spends in this element and so

$$P(x)\delta x = k \frac{\delta x}{v}$$

where $k$ is a constant of proportionality. As

$$\int_0^l P(x)\, dx = 1$$

then

$$k = v/l$$

and so

$$P(x) = 1/l$$

in agreement with (2.65).

*The equivalence of wave and matrix mechanics*

The matrix mechanics of Heisenberg and Dirac was developed from the classical Poisson brackets of Hamiltonian mechanics whereas the wave mechanics of Schrödinger was developed from classical wave dynamics.

The first steps in showing the equivalence of the new matrix and wave mechanics were taken by Schrödinger himself. In 1926 he discovered that there is a very simple connection between Hamiltonian dynamics and wave mechanics.

Consider the classical Hamiltonian $H$ for a 'one-dimensional' particle of mass $m$ moving in a conservative field. If $E$ is the total energy then

$$H = \frac{p_x^2}{2m} + V(x) = E$$

where $P_x$ is the momentum and $V(x)$ is the potential energy. Schrödinger suggested replacing the independent variable by itself and also replacing the momentum $P_x$ by the differential operator $-i\hbar\, \mathrm{d}/\mathrm{d}x$. A quantum mechanical Hamiltonian $\mathscr{H}$ is then defined by

$$\mathscr{H} \equiv -\frac{\hbar^2}{2m}\frac{\mathrm{d}^2}{\mathrm{d}x^2} + V(x)$$

Note that $\mathscr{H}$ is an operator. The Schrödinger time-independent equation (2.59) can then be written

$$\mathscr{H}\psi = E\psi.$$

Schrödinger also realized that the strange result

$$p_x x - x p_x \equiv -i\hbar \tag{2.52}$$

inherent in Heisenberg's theory then followed as

$$p_x x \equiv -i\hbar \frac{\mathrm{d}}{\mathrm{d}x} x \equiv -i\hbar - i\hbar x \frac{\mathrm{d}}{\mathrm{d}x} \equiv -i\hbar + x p_x.$$

Later in the same year Dirac showed the complete equivalence of the matrix mechanics and the wave mechanics. They are both different aspects of the new quantum mechanics (see Chapter 6).

*The uncertainty principle*

In 1927 Heisenberg expounded his famous 'Uncertainty Principle'. In general terms this states that it is impossible to know precisely the exact position of a particle and its momentum simultaneously. When the position is measured the momentum is disturbed and, unlike classical mechanics, the disturbance cannot be allowed for. It can be shown that if $\Delta x$ and $\Delta p_x$ represent the uncertainties in the $x$-coordinate and $x$-

component of linear momentum then

$$\Delta x \Delta p_x \sim \hbar$$

This will be formally derived in Chapter 5 and is a consequence of Heisenberg's relation (2.52). A simple wave-packet 'explanation' will suffice here. Consider the de Broglie-wave packet (2.48) consisting of the superposition of two plane waves. The distance between the maxima of the modulated wave is

$$\Delta x = 2\pi/\Delta k$$

or

$$\Delta x \Delta k = 2\pi.$$

In general, if a group of waves of spread $\Delta x$ and mean wave number $k$ is analysed into its constituent plane waves, then the wave numbers of the constituents will be found to be clustered about the mean value $k$ in a range $\Delta k$ such that

$$\Delta x \Delta k \sim 1.$$

After de Broglie

$$\Delta p_x = \hbar \Delta k$$

and so

$$\Delta x \Delta p_x \sim \hbar.$$

$\Delta x$ is the uncertainty of position of the particle described by the group of waves and $\Delta p_x$ is the corresponding uncertainty in momentum. To specify the position exactly so that $\Delta x = 0$ an infinite number of plane waves would be required and so $\Delta k \to \infty$. On the other hand, to specify the momentum completely only one wave must be used and the wave packet will be of infinite extent so that $\Delta x \to \infty$.

*Further developments*

The new theory developed very rapidly after 1926. In 1927 Dirac gave a quantum mechanical description of the electromagnetic field. Spin was introduced into quantum mechanics by Pauli and the English physicist C. G. Darwin and in 1928 Dirac gave a relativistic theory of quantum mechanics from which spin arose naturally. The reader is referred to Hoffmann's *The Strange Story of the Quantum* for an interesting non-mathematical account of the origins of Quantum Theory.

PROBLEMS

1 Suppose radiation of wave-length $\lambda$ forms standing waves inside a cubical box of side length $a$ with perfectly reflecting walls so that the walls can be taken as nodes. Explain why

$$n_1 \frac{\lambda}{2} = al$$

$$n_2 \frac{\lambda}{2} = am$$

$$n_3 \frac{\lambda}{2} = an$$

where $n_1, n_2, n_3$ are positive integers and where $l, m, n$ are the direction cosines of the radiation direction relative to a set of orthogonal axes coincident with three adjacent sides of the cube. Using the relation

$$l^2 + m^2 + n^2 = 1$$

show that the frequencies of the allowed modes are given by

$$\nu = (n_1^2 + n_2^2 + n_3^2)^{1/2} \frac{c}{2a}$$

where $c$ is the velocity of the radiation.

Construct a set of orthogonal axes corresponding to the variables $n_1c/2a, n_2c/2a, n_3c/2a$. Each of the points in the cubic lattice in the positive octant corresponds to an allowed frequency mode. Deduce that the number of allowed frequencies in the range between $\nu$ and $\nu + \delta\nu$ is

$$\frac{1}{8} \frac{4\pi\nu^2}{\left(\dfrac{c}{2a}\right)^3} \delta\nu$$

and that the total number of allowed modes per unit volume after consideration of polarization directions, is

$$8\pi \frac{\nu^2}{c^3} \delta\nu$$

(This result can be used to obtain the equation (2.1).)

2 Consider the Planck law (2.10) for the radiation emitted by a black-body radiator. Show that in the high frequency limit $\nu \to \infty$ that

$$E(\nu) = \frac{8\pi h\nu^3}{c^3} e^{-h\nu/kT}.$$

This is Wien's experimental law.

3 In free space the electric field flux out of a closed surface $s$ is given by

$$\oint_s \mathscr{E} \cdot \mathbf{n}\, ds = q/\epsilon_0$$

where $\mathscr{E}$ is the electric field, $\mathbf{n}$ is the unit outward normal from the surface, $q$ is the included charge and $\epsilon_0$ is the permittivity of free space. The electrostatic potential $\phi$ is defined to within a constant by the equation $\mathscr{E} = -\text{grad}\ \phi$. Show that $\phi$ cannot have a maximum or minimum value except at points where there is a positive or negative charge respectively. Hence deduce that a free charge cannot be in stable equilibrium at a point unoccupied by charge. This result implies that no stationary arrangement of charges can be in stable equilibrium under their own influence alone.

4 Use the Bohr theory of the hydrogen atom to show that the radius of the first Bohr orbit is

$$a_0 = \frac{e^2}{2Rhc(4\pi\epsilon_0)}$$

where $R$ is the Rydberg constant and that the energy of this state is

$$E = Rhc \simeq 13{\cdot}6 \text{ eV}.$$

5 Consider a free simple rotator composed of two point masses $M$ distance $2a$ apart. Show that if the rotator is free to rotate in a plane about the centre of mass, the total energy is

$$E = Ma^2\omega^2$$

where $\omega$ is the angular velocity. Use the Bohr-Sommerfeld quantization rule to show that $\omega$ can only take the values

$$\omega = \frac{\hbar}{2Ma^2} J$$

where $J$ is a positive integer or zero.

For a diatomic molecule a reasonable estimate for $a$ is the radius of the first Bohr orbit (see problem 4). Show that the allowed energy levels of the rotating system are

$$E = Rhc\left(\frac{m}{2M}\right)J^2$$

where $R$ is the Rydberg constant and $m$ is the electron mass.

6 Consider a plane 'matter' wave of frequency $\nu$ and wave number $k$.

$$\Psi(x, t) = A \exp\left[-i2\pi\left(\nu t - \frac{k}{2\pi}x\right)\right].$$

Rewrite this equation substituting for $\nu$ and $k$ using the relations

$$E = h\nu \qquad \text{and} \qquad p_x = \hbar k$$

where $E$ is the energy of the particle and $p_x$ is the momentum. Deduce that

$$-i\hbar\frac{\partial\Psi}{\partial x} = p_x\Psi$$

and

$$i\hbar\frac{\partial\Psi}{\partial t} = E\Psi.$$

Hence show that the classical energy expression for a free particle

$$E = \frac{1}{2m}p^2$$

can be expressed as the wave equation,

$$-\frac{\hbar^2}{2m}\frac{\partial^2\psi}{\partial x^2} = i\hbar\frac{\partial\psi}{\partial t}.$$

**References**

Bohr, N. *Phil. Mag.* **26** (1913).
Born, M. *Z. f. Phys.* **37**, 863 (1926); *Nature* **119**, 354 (1927).
Broglie, L. de. *Nature* **112** (1923); Thesis, Paris (1924); *Ann. de Physique*, 3, 22 (1925).
Clark, H. *Solid State Physics—an introduction to its theory.* Macmillan, London (1968).
Compton, A. H. *Physics. Rev.* **21**, 483 (1923); **22**, 409 (1923).
Debye, P. *Ann. Physik*, **39**, 789 (1912).
Dirac, P. A. M. *Proc. Roy. Soc.* **A117**, 610 (1928).
Einstein, A. *Ann. Physik.* **22**, 180 (1907); **34**, 170 (1911).
Heisenberg, W. *Z. f. Phys.* **33**, 879 (1925).
Hoffmann, B. *The strange story of the Quantum.* Penguin, London (1959).
Panofsky, W. K. H. and Phillips, *Classical Electricity and Magnetism.* Addison-Wesley, Cambridge, Mass. (2nd edition, 1962).
Planck, M. *Verh. d. deuf. physik. Gesell.* **2**, 237 (1900); *Ann. Physik.* **4**, 553 (1901).
Rutherford, E. *Phil. Mag.* **21**, 669 (1911).

Schrödinger, E. *Ann. Phys.* **79**, 361, 489, 734 (1926); **80**, 437 (1926); **81**, 109 (1926).
Shamos, M. H. *Great experiments in Physics.* Holt, Rinehart & Winston, New York (1960).
Sommerfeld, A. *Ann. Phys.* **51**, 1 (1916).
Thomson, J. J. *Phil. Mag.* **44**, 293 (1897).
Tolansky, S. *Introduction to Atomic Physics.* Longmans, London (1952).
Wilson, W. *Phil. Mag.* **29**, 795 (1915).

CHAPTER 3

# Quantum Mechanics I

## 3.1 Introduction

In the previous chapter the early attempts to account for the breakdown of classical mechanics were briefly outlined. The new theory was clearly not satisfactory and it was left to Schrödinger, Heisenberg and others to develop the quantum mechanics. Historically quantum mechanics was evolved using intuition and analogy with classical mechanics. Even so, there is a basic difference in philosophy between them, and quantum mechanics can certainly not be derived from classical mechanics.

For a system obeying classical mechanics, (this is an approximation, although it is very often a very accurate one), if all the positions and velocities of the particles are known at some instant then the state of the system at some later time is completely determined by Newton's laws. This is not the case for a system obeying quantum mechanics. In fact the position and momentum of a quantum particle can not both be known exactly at the same instant and even the concept of a particle trajectory loses its meaning. There is a limit to what can be known about the state of a quantum system. In quantum mechanics the physical quantities such as energies, momenta etc., which can in principle be directly measured are called observables but not all the observables of a system can be measured simultaneously. If some observable is measured, this act of measurement may disturb the system and change the value of some other observable. This is true for both classical and quantum systems, but there is a basic difference in outlook. The disturbance in a classical system produced by the act of measurement can in principle be allowed for exactly but not in quantum systems. The most complete description possible of the state of a system in quantum mechanics is less detailed than that in classical mechanics. Because of this, the behaviour of the quantum system at some future time is not uniquely determined.

Classical mechanics is based on Newton's three laws. These cannot be proved directly but are verified by the experimental results obtained from 'classical systems'. Similarly, quantum mechanics is based on postulates. At first the postulates seem strange, but their validity is confirmed by the agreement of the theory with experiment. This postulate statement of quantum mechanics is the most effective introduction to the subject and this is given below.

## 3.2 The Postulates

*Postulate 1*

Any state of a system with $n$ degrees of freedom is described as completely as possible by a wave-function $\Psi(q_1, \ldots, q_n, t)$ which depends upon the co-ordinates $q_i$ and the time. $\Psi$ is in general complex and as it has to be physically meaningful it is single-valued. The wave-function may be multiplied by an arbitrary complex number without any essential change in its physical significance. It will be shown that $\Psi$ and its derivative are continuous except at a certain number of points and in general it is also quadratically integrable, i.e.

$$\int \Psi^*\Psi \, d\tau = k^2 \qquad k \text{ finite and real} \tag{3.1a}$$

where $\Psi^*$ is the complex conjugate of $\Psi$ and the field of integration is over all values of the co-ordinates. As the wave-function can be multiplied by a constant without changing its character, it is often convenient to describe the state by

$$\Psi_1 = \frac{1}{k}\Psi.$$

This new wave-function is 'normalized' in the sense

$$\int \Psi_1^*\Psi_1 \, d\tau = 1 \tag{3.1b}$$

and once normalized remains so for all time. $\Psi_1$ remains normalized if it is multiplied by any complex number of modulus unity.

*Definition 1. Linear operator*

An operator transforms one function into another. A well-known example is the differential operator $D \equiv d/dx$, e.g.

$$Dx^3 = 3x^2.$$

An operator $\alpha$ is said to be linear if

$$\alpha(\Psi_1 + \Psi_2) = \alpha\Psi_1 + \alpha\Psi_2 \qquad \text{and} \qquad \alpha(a\Psi) = a\alpha\Psi \tag{3.2}$$

where $\Psi_1, \Psi_2$ are arbitrary functions and $a$ is an arbitrary constant. The operator $D$ is linear.

A simple example of a non-linear operator is the square operator which will be denoted by $S$, i.e.

$$S\Psi = \Psi^2$$

then

$$S(\Psi_1 + \Psi_2) = (\Psi_1 + \Psi_2)^2$$
$$\neq S\Psi_1 + S\Psi_2.$$

*Definition 2. Hermitian operator*

Suppose $\Psi_1$ and $\Psi_2$ are arbitrary quadratically integrable functions. The linear operator $\alpha$ is said to be Hermitian if

$$\int \Psi_1^*(\alpha\Psi_2)\,\mathrm{d}\tau = \int \Psi_2(\alpha\Psi_1)^*\,\mathrm{d}\tau \tag{3.3}$$

the integration being taken over all values of the co-ordinates. Clearly a linear combination of Hermitian operators $\alpha, \beta$ is itself a Hermitian operator and any power of a Hermitian operator $\alpha$ is Hermitian, e.g.

$$\int \Psi_1^*(\alpha + \beta)\Psi_2\,\mathrm{d}\tau = \int \Psi_1^*\alpha\Psi_2\,\mathrm{d}\tau + \int \Psi_1^*\beta\Psi_2\,\mathrm{d}\tau$$
$$= \int \Psi_2(\alpha\Psi_1)^*\,\mathrm{d}\tau + \int \Psi_2(\beta\Psi_1)^*\,\mathrm{d}\tau = \int \Psi_2[(\alpha + \beta)\Psi_1]^*\,\mathrm{d}\tau. \tag{3.4}$$

*Postulate 2*

With every physical observable there is associated a linear operator. This linear operator is Hermitian.

Of course it is essential to determine the precise form of the operator $\alpha$ which represents the observable $A$. It is at this point that analogy with classical mechanics is important. Suppose $\alpha$ and $\beta$ are two quantum operators representing variables $A$ and $B$. The commutator of these two operators is defined to be

$$[\alpha, \beta] \equiv (\alpha\beta - \beta\alpha). \tag{3.5}$$

[Note; the r.h.s. of (3.5) is not in general zero. Suppose $\alpha = x$ and $\beta = \mathrm{d}/\mathrm{d}x$ then

$$\left[x, \frac{\mathrm{d}}{\mathrm{d}x}\right] \equiv \left[x\frac{\mathrm{d}}{\mathrm{d}x} - \frac{\mathrm{d}}{\mathrm{d}x}x\right] \equiv \left[x\frac{\mathrm{d}}{\mathrm{d}x} - 1 - x\frac{\mathrm{d}}{\mathrm{d}x}\right] \equiv -1.$$

Commutators have many of the properties of Poisson brackets, e.g.

$$[\alpha, \beta] \equiv -[\beta, \alpha], \qquad [\alpha, \alpha] \equiv 0.$$

The quantum operators $\alpha, \beta$ are such that their commutator is proportional to the classical Poisson bracket of $A$ and $B$. The constant of proportionality is a universal constant (with the dimensions of 'action', i.e. Joule-sec.) and introduces the Planck constant $h$ into quantum mechanics.

$$[\alpha, \beta] \equiv i\hbar\{A, B\} \qquad \hbar = \frac{h}{2\pi} \qquad i = \sqrt{-1}. \tag{3.6}$$

(If there are any variables remaining on expansion of the Poisson bracket they are to be replaced by operators.)

The form of the operator is not uniquely determined by (3.6) but depends on the 'representation' used. In the commonly used Schrödinger representation any position co-ordinate $q_i$ or the time $t$ is represented by $q_i$ or $t$ respectively. The momentum $p_i$ conjugate to $q_i$ is represented by the operator

$$\frac{\hbar}{i}\frac{\partial}{\partial q_i}.$$

It is easy to verify that the commutator of the operators representing $q_i$ and $p_j$ is

$$\left[q_i, \frac{\hbar}{i}\frac{\partial}{\partial q_j}\right]\Psi = i\hbar\delta_{ij}\Psi \tag{3.7}$$

and by comparison with (1.30) it is seen that (3.6) is indeed satisfied. It is obvious that the operator $q_i$ associated with the co-ordinate $q_i$ is Hermitian. Similarly it is not difficult to show that the operator representing the momentum $p_i$ is Hermitian. On integrating by parts

$$\int \Psi_1^*\left(\frac{\hbar}{i}\frac{\partial}{\partial q_i}\Psi_2\right)\mathrm{d}q_i = \Psi_1^*\frac{\hbar}{i}\Psi_2 + \int \Psi_2\left(-\frac{\hbar}{i}\right)\frac{\partial}{\partial q_i}\Psi_1^*\,\mathrm{d}q_i$$

and if the wave function vanishes at infinity the first term on the r.h.s. is zero and the result is proved. Any classical observable (such as energy), which is a function of the co-ordinates, momenta and time is represented by an operator obtained by substituting

$$q_i \to q_i \qquad p_i \to \frac{\hbar}{i}\frac{\partial}{\partial q_i} \qquad t \to t. \tag{3.8}$$

For example, any function of $p_i$, $f(p_i)$ is represented by

$$f\left(\frac{\hbar}{i}\frac{\partial}{\partial q_i}\right)$$

and similarly for $f(q_i)$. Any functions of conjugate variables must be carefully ordered before converting into an operator. This is necessary because the operators representing the variables $(q_i, p_i)$ do not commute. In classical mechanics it is meaningless to distinguish between $p_i q_i$ and $q_i p_i$ but it is essential to do so in quantum mechanics. The exact order of the factors in a term can often be decided by remembering the condition that the operator be Hermitian. Ambiguity can sometimes be removed by taking the mean of two possible orders of the factors. The classical product $q_i p_i$ (or $p_i q_i$) can be represented by the Hermitian operator

$$\frac{1}{2}\left[q_i \frac{\hbar}{i}\frac{\partial}{\partial q_i} + \frac{\hbar}{i}\frac{\partial}{\partial q_i} q_i\right].$$

It may be that the exact order can only be decided by trial and comparison with experiment. The formulation in terms of rectangular position and momentum co-ordinates is least likely to lead to ambiguity. If necessary the operator can then be transformed to some other generalized co-ordinate system.

For example, consider the Hamiltonian for a particle of mass $m$ moving in a conservative field of force. The classical Hamiltonian is

$$H = \frac{1}{2m}(p_x^2 + p_y^2 + p_z^2) + V(x, y, z)$$

and the corresponding quantum mechanical operator is

$$\mathcal{H} = -\frac{\hbar^2}{2m}\left(\frac{\partial^2}{\partial x^2} \frac{\partial^2}{\partial y^2} + \frac{\partial^2}{\partial z^2}\right) + V(x, y, z).$$

There are other representations apart from the Schrödinger representation and in these the operators will take a different form. Suppose instead of (3.8) the representative operators are taken as

$$q_i \to -\frac{\hbar}{i}\frac{\partial}{\partial q_i} \qquad p_i \to p_i \qquad t \to t. \tag{3.9}$$

The relation (3.6) is still satisfied in this 'momentum representation', and the quantum mechanical Hamiltonian corresponding to the example mentioned above is now

$$\mathcal{H} = \frac{1}{2m}(p_x^2 + p_y^2 + p_z^2) + V\left(i\hbar\frac{\partial}{\partial x}, i\hbar\frac{\partial}{\partial y}, i\hbar\frac{\partial}{\partial z}\right).$$

The form of the potential function may make this representation difficult to use and unless otherwise stated the Schrödinger representation will be used throughout this book.

*Postulate 3*

In classical mechanics, Hamilton's equations determine the time variation of a state. Similarly, the Hamiltonian operator for the system determines the time variation of a quantum mechanical state in that the state wave function $\Psi$ must satisfy Schrödinger's time-dependent equation

$$\mathscr{H}\Psi = i\hbar \frac{\partial \Psi}{\partial t}. \tag{3.10}$$

If the state function is known at some initial time, the equation (3.10) determines $\Psi$ at any other time.

Equation (3.10) is a second order partial differential equation defined for all points where the potential energy is finite. As the second derivatives exist at all points (where $V$ is finite) then the first derivatives must be continuous and so $\Psi$ must be differentiable and continuous. These restrictions on the state functions were mentioned in Postulate one.

*Definition 3. Eigenvalues and eigenfunctions*

If $\Phi$ is a function and $\alpha$ is an operator such that

$$\alpha\Phi = a\Phi \tag{3.11}$$

where $a$ is a constant then $\Phi$ is an eigenfunction of $\alpha$ and $a$ is the corresponding eigenvalue. The totality of all eigenvalues for an operator is called the spectrum of the operator and this may be discrete, continuous or partly discrete and partly continuous. The number of eigenvalues of a quantum operator is in general infinite.

For example, the second-order equation

$$\frac{d^2X}{dx^2} + aX = 0$$

has solutions $X = A \sin\sqrt{a}x + B\cos\sqrt{a}x$ where $A$, $B$ are arbitrary. So, when no boundary conditions are imposed the spectrum of the operator $d^2/dx^2$ is continuous. On the other hand if the boundary conditions,

$$X(0) = 0 \qquad X(L) = 0,$$

are imposed, the spectrum is discrete and the eigenvalues are

$$a_n = \left(\frac{n\pi}{L}\right)^2 \qquad n = 1, 2, \ldots$$

with corresponding eigenfunctions

$$X_n(x) = A \sin \frac{n\pi}{L} x.$$

*Definition 4. Orthogonality*

Functions forming a set are mutually orthogonal if, for all pairs,

$$\int \Phi_n^* \Phi_m \, d\tau = 0 \qquad n \neq m, \tag{3.12a}$$

the integration being over all the values of the co-ordinates. If in addition the functions are normalized to unity (3.1b) then

$$\int \Phi_n^* \Phi_m \, d\tau = \delta_{nm}. \tag{3.12b}$$

The functions are then said to form an orthonormal set.

A well-known example of orthogonal functions is the set

$$\Phi_n(x) = \frac{1}{\sqrt{\pi}} \sin n\pi x$$

the members of which are orthogonal over the range $0 \leqslant x \leqslant 2\pi$, i.e.

$$\int_0^{2\pi} \Phi_n \Phi_m \, dx = \delta_{nm}.$$

*Theorem 1. Reality of eigenvalues*

The eigenvalues of a Hermitian operator are real.

*Proof*

Let the Hermitian operator be $\alpha$ and $\Phi_n$ an eigenfunction belonging to the eigenvalue $a_n$, i.e.

$$\alpha\Phi_n = a_n\Phi_n \quad \text{(i)}$$

$$\therefore \quad \int \Phi_n^* \alpha \Phi_n \, d\tau = a_n \int \Phi_n^* \Phi_n \, d\tau.$$

Take the complex conjugate of (i)

$$(\alpha\Phi_n)^* = a_n^* \Phi_n^*$$

$$\therefore \quad \int (\alpha\Phi_n)^* \Phi_n \, d\tau = a_n^* \int \Phi_n^* \Phi_n \, d\tau.$$

As the operator is Hermitian

$$\int \Phi_n^* \alpha \Phi_n \, d\tau = \int (\alpha\Phi_n)^* \Phi_n \, d\tau$$

$$\therefore \quad a_n = a_n^*$$

and the eigenvalue $a_n$ is real.

*Theorem 2. Orthogonality of eigenfunctions*

Any two eigenfunctions of a Hermitian operator, belonging to different eigenvalues are orthogonal.

*Proof*

Let the Hermitian operator be $\alpha$ and the eigenfunctions be $\Phi_n$, $\Phi_m$ with corresponding eigenvalues $a_n, a_m$, i.e.

$$\alpha\Phi_n = a_n\Phi_n \quad \text{(i)} \qquad \alpha\Phi_m = a_m\Phi_m \quad \text{(ii)}$$

From (i)

$$\int \Phi_m^* \alpha \Phi_n \, d\tau = a_n \int \Phi_m^* \Phi_n \, d\tau.$$

Since $\alpha$ is Hermitian

$$\int \Phi_m^* \alpha \Phi_n \, d\tau = \int \Phi_n (\alpha \Phi_m)^* \, d\tau$$

$$= \int \Phi_n a_m^* \Phi_m^* \, d\tau \qquad \text{(from (ii))}$$

$$= a_m \int \Phi_m^* \Phi_n \, d\tau$$

as $a_m$ is real.

$$\therefore \quad (a_n - a_m) \int \Phi_m^* \Phi_n d\tau = 0$$

and as $a_n \neq a_m$ then

$$\int \Phi_m^* \Phi_n \, d\tau = 0 \qquad n \neq m.$$

*Postulate 4*

Suppose the wave function $\Psi(q_1, \ldots, q_n, t)$ describing a state is an eigenfunction of the operator $\alpha$ representing the observable A. i.e.

$$\alpha \Psi = a \Psi.$$

Then in this state an exact measurement of the observable will yield precisely the constant real value $a$. (As $\alpha$ is Hermitian $a$ is real.)

*The time-independent Schrödinger equation*

When the potential energy of the system is not a function of time, the time-independent Schrödinger equation (3.10) has variable separable solutions of the form

$$\Psi(q_1, \ldots, q_n, t) = \psi(q_1, \ldots, q_n) T(t). \tag{3.13}$$

(Note, the Greek $\psi$ is used to designate the spatial part of the state function $\Psi$.) Substitute (3.13) into (3.10) and divide through by $\Psi$

$$\frac{1}{\psi} \mathscr{H} \psi = \frac{i\hbar}{T} \frac{dT}{dt} = E. \tag{3.14}$$

$E$ is a separation constant independent of $q_i$ and $t$, i.e.

$$\Psi_j = \psi_j \, e^{-\frac{i}{\hbar} E_j t} \tag{3.15}$$

with

$$\mathscr{H} \psi_j = E_j \psi_j \tag{3.16}$$

is a solution of (3.10).

Equation (3.16) is the time-independent Schrödinger equation. $\psi_j$ (also $\Psi_j$) is an eigenfunction of $\mathscr{H}$; $E_j$ is the corresponding eigenvalue and is the energy of the state described by $\Psi_j$. Such states in which the energy has a definite value are called stationary states of the system.

Consider a free particle of mass $m$ moving along the $x$-axis with no force acting. The classical Hamiltonian is $p_x^2/2m$ and the Schrödinger equation governing the motion is

$$-\frac{\hbar^2}{2m}\frac{\partial^2\Psi}{\partial x^2} = i\hbar\frac{\partial\Psi}{\partial t}. \tag{3.17}$$

The spatial parts $\psi$ of the variable separable solutions satisfy

$$-\frac{\hbar^2}{2m}\frac{\mathrm{d}^2\psi}{\partial x^2} = E\psi$$

and so

$$\psi(x) = A\,e^{ikx} + B\,e^{-ikx}$$

with

$$E = \frac{\hbar^2 k^2}{2m}.$$

The boundary conditions would determine the arbitrary constants and also the allowed energy values. The complete variable separable solution is

$$\Psi(x, t) = Ae^{i\left(kx - \frac{E}{\hbar}t\right)} + Be^{-i\left(kx + \frac{E}{\hbar}t\right)}. \tag{3.18}$$

Observe that the two separate solutions are also eigenfunctions of the momentum operator $(\hbar/i)(\mathrm{d}/\mathrm{d}x)$ with eigenvalues $\pm\hbar k$ and represent uniform motion in the positive and negative directions respectively.

*Definition 5. Linear independence*

A set of functions $\{\phi_i\}$ is said to be linearly independent if the relation

$$a_1\phi_1 + a_2\phi_2 + \ldots + a_n\phi_n = 0 \tag{3.19}$$

implies that the constants $a_i = 0$ for all $i$. This means that none of these functions can be expressed in terms of the others.

*Theorem 3*

Any set of mutually orthogonal functions is linearly independent.

*Proof*

Let the orthogonal set of functions be $\{\phi_i\}$ and suppose that (3.19) is satisfied. Multiply equation (3.19) by $\phi_j^*$ and integrate over all values of the co-ordinates.

$$\sum_{i \neq j} a_i \int \phi_j^* \phi_i \, d\tau + a_j \int \phi_j^* \phi_j \, d\tau = 0.$$

As the functions are orthogonal then the first term vanishes and so

$$a_j \int \phi_j^* \phi_j \, d\tau = 0.$$

Clearly this integral is non-zero and so $a_j = 0$. This is true for all $j$ and so the orthogonal functions are linearly independent.

Conversely, if a set of non-orthogonal functions are linearly independent and are also quadratically integrable, it is always possible to form a mutually orthogonal set from them by a suitable linear transformation. This is the Schmidt method. Suppose $\{\theta_i\}$ is the non-orthogonal set, then an orthogonal set $\{\phi_i\}$ can be defined as follows.

Let $\phi_1 = \theta_1$, and then choose $\alpha$ with $\phi_2 = \theta_2 + \alpha\phi_1$ so that $\phi_2$ is orthogonal to $\phi_1$, i.e.

$$\alpha = -\int \phi_1^* \theta_2 \, d\tau \Big/ \int \phi_1^* \phi_1 \, d\tau.$$

Next choose $\beta_1, \beta_2$ with $\phi_3 = \theta_3 + \beta_1\phi_1 + \beta_2\phi_2$ so that $\phi_3$ is orthogonal to $\phi_1$ and $\phi_2$ etc.

*Definition 6. Function space*

The function space defined by the linearly independent set of functions $\{\theta_i\}$ includes all functions of the form

$$b_1\theta_1 + b_2\theta_2 + \ldots + b_n\theta_n \tag{3.20}$$

where $b_i$'s are any complex constants. The number of linearly independent functions used to define this space, gives the dimensionality. Clearly the orthogonal set of functions $\{\phi_i\}$ obtained from $\{\theta_i\}$ by the Schmidt process define the same space. Either of the basis sets $\{\phi_i\}$ or $\{\theta_i\}$ are said to span the space in that any function in the space can be expressed as a linear combination of the members of either set. This definition is in accordance with the more usual ideas of Cartesian space. The unit vectors $\hat{e}_1, \hat{e}_2, \hat{e}_3$ are orthogonal vectors and are also linearly independent. Together they define a three-dimensional space which includes all vectors of the form

$$\mathbf{r} = c_1\hat{e}_1 + c_2\hat{e}_2 + c_3\hat{e}_3.$$

The space defined by the functions $\{\chi_i\}$ is said to be a proper sub-space of the function space $\{\theta_i\}$ if every function in the sub-space belongs to the space $\{\theta_i\}$ and if the converse is not true.

The plane defined by the unit vectors $\hat{e}_1, \hat{e}_2$ is a proper sub-space of the three-dimensional space given by $\hat{e}_1, \hat{e}_2, \hat{e}_3$. All the vectors in the sub-space are of the type

$$\mathbf{r} = c_1\hat{\mathbf{e}}_1 + c_2\hat{\mathbf{e}}_2,$$

and obviously belong to the full Cartesian space. Alternatively, any vector including $\hat{\mathbf{e}}_3$ while belonging to the three-dimensional space does not belong to the sub-space.

*Definition 7. Degeneracy*

There may be several eigenfunctions of an operator belonging to the same eigenvalue. If there are $n$ linearly independent eigenfunctions corresponding to a given eigenvalue, the eigenvalue is said to be $n$-fold degenerate. These eigenfunctions can always be chosen to be orthogonal by the Schmidt process.

*Postulate 5. Expectation value of an operator*

Suppose there are a large number of identical systems each described by the same normalized state function, $\Psi$. If $\Psi$ is an eigenfunction of the operator $\alpha$ representing the observable $A$ then an exact measurement of $A$ will yield the corresponding eigenvalue. In the more general case the state function is not an eigenfunction of $\alpha$. It is now postulated that an exact measurement of $A$ must yield one of the eigenvalues. In general, different values will be obtained for each system and the average value (sometimes called the expectation value), of the results is

$$\langle A \rangle = \int \Psi^* \alpha \Psi \, d\tau. \tag{3.21}$$

*Probability density*

From (3.21), the expectation value of the co-ordinate $q_i$ is

$$\langle q_i \rangle = \int \Psi^* q_i \Psi \, d\tau$$

if $\Psi$ is normalized. This implies that $\Psi^*\Psi \, d\tau$ represents the probability that the co-ordinates lie between $q_i$ and $q_i + dq_i$. That is, $\Psi^*\Psi$ is a probability density.

For a single particle system, the wave function depends upon the co-ordinates of the particle $x$, $y$, $z$ and $\Psi^*\Psi$ represents the probability density of the particle. It is no longer possible to state with certainty the position of the particle.

*Postulate 6. Completeness of eigenfunction sets*

Postulate 1 says that every state of a system is described as completely as possible by a wave function. At any instant, all such state functions belong to some function space which is characteristic of the system. (This space is an abstract Hilbert space. The reader is referred to *Quantum mechanics,* by G. L. Trigg.) As the Hamiltonian operator has an infinite number of linearly independent eigenfunctions each of which can represent a state of the system, then it is clear that the function space is of infinite dimension.

Suppose $\Psi_j$ is a state function of the system and is also an eigenfunction for the operator representing some observable of the system. It is now postulated that the infinite set of all such orthogonal eigenfunctions span the system function space. That is, the set of eigenfunctions is complete in that any state function can be expressed as a linear combination of them. For the moment it will be assumed that the operator spectrum is discrete and

$$\Psi = \sum_j c_j \Psi_j. \tag{3.22}$$

If the eigenfunctions are quadratically integrable and are normalized the expansion co-efficients

$$c_j = \int \Psi_j^* \Psi \, d\tau. \tag{3.23}$$

Equation (3.22) shows that any state can be expressed as a superposition of other states.

The reader will no doubt have met the concept of a complete set of functions in the theory of Fourier series. The functions $\frac{1}{2}$, $\sin nx$, $\cos nx$, $n = 1, 2, \ldots$ form a complete set in that any piecewise continuous function $f(x)$ with period $2\pi$ can be expressed as a linear combination of them, i.e.

$$f(x) = \frac{a_0}{2} + \sum_1^\infty (a_n \cos nx + b_n \sin nx)$$

with

$$a_n = \frac{1}{\pi} \int_{-\pi}^{\pi} f(x) \cos nx \, dx \quad \text{and} \quad b_n = \frac{1}{\pi} \int_{-\pi}^{\pi} f(x) \sin nx \, dx.$$

*Theorem 4*

From the previous postulate, any state function $\Psi$ can be expressed as a linear combination of the orthonormal eigenfunctions $\{\Psi_i\}$ of the operator $\alpha$ representing the observable $A$ (3.22). The corresponding eigenvalues are $a_i$. An exact measurement of $A$ must yield one of these eigenvalues and the probability of observing $a_i$ is

$$P_i = c_i^* c_i = |c_i|^2. \tag{3.24}$$

*Proof*

Expand $\Psi$ as a linear combination of the orthonormal eigenfunctions $\{\Psi_j\}$. From (3.21) the expectation value of $A$ is

$$\langle A \rangle = \int \sum_i c_i^* \Psi_i^* \alpha \sum_j c_j \Psi_j \, d\tau = \sum_i \sum_j c_i^* c_j a_j \int \Psi_i^* \Psi_j \, d\tau = \sum_i |c_i|^2 a_i.$$

So $|c_i|^2$ can be taken as the probability of observing the eigenvalue $a_i$. Note that if $\Psi$ is normalized

$$\int \Psi^* \Psi \, d\tau = \sum_i |c_i|^2 = 1. \tag{3.25}$$

If the value $a_i$ is obtained on measuring $A$ then assuming the eigenvalue is non-degenerate, the act of measurement must have disturbed the system changing the state function from $\Psi$ to the eigenfunction $\Psi_i$. In general this function will change with time but if the observable $A$ is immediately measured again the value $a_i$ will once more be obtained.

If the eigenvalue is degenerate, (3.24) must be summed over the degenerate eigenfunctions.

*Continuous spectrum*

The equations (3.22–3.25) assume the operator has discrete eigenvalues. There are operators however which have eigenvalues running continuously over all or part of the spectrum. Suppose $\alpha$ is an operator for which the eigenvalues are continuous. (An example is the position operator.) The generalization of (3.22) for an arbitrary wave-function $\Psi$ is

$$\Psi = \int c(a) \Psi_a \, da \tag{3.26}$$

where the integration is over all permissible values for the eigenvalue $a$.

The problem of finding the expansion coefficients is complicated by the fact that the eigenfunctions $\Psi_a$ are not quadratically integrable. If the wave-function $\Psi$ is quadratically integrable then from (3.26)

$$\int \Psi^* \Psi \, d\tau = \int \left[ \int c^*(a) \Psi_a^* \, da \right] \Psi \, d\tau = \int c^*(a) \left[ \int \Psi_a^* \Psi \, d\tau \right] da \tag{3.27}$$

assuming the order of integration can be reversed. It would seem reasonable that $|c(a)|^2 \, da$ should represent the probability that the observable has a value between $a$ and $a + da$ and as the sum of all probabilities is unity (cf. (3.25))

$$\int |c(a)|^2 \, da = 1. \tag{3.28}$$

If $\Psi$ is normalized to unity

$$\int c^*(a) \left[ \int \Psi_a^* \Psi \, d\tau \right] da = \int c^*(a) c(a) \, da$$

$$\therefore \quad c(a) = \int \Psi_a^* \Psi \, d\tau. \tag{3.29}$$

Compare this result with (3.23) for a discrete spectrum.

The reader is asked to compare (3.26) and (3.29) with a function $f(x)$ satisfying $\int_{-\infty}^{\infty} |f(x)|^2 \, dx$ finite, and its Fourier transform $g(y)$.

$$f(x) = \frac{1}{\sqrt{2\pi}} \int_{-\infty}^{\infty} g(y)\, e^{ixy}\, dy$$

with

$$g(y) = \frac{1}{\sqrt{2\pi}} \int_{-\infty}^{\infty} e^{-ixy} f(x)\, dx.$$

It is now shown that the eigenfunctions of an operator with a continuous spectrum are not quadratically integrable. From (3.26) and (3.29)

$$c(a) = \int \Psi_a^* \left[ \int c(a') \Psi_{a'}\, da' \right] d\tau. \qquad (3.30)$$

Reversing the order of integration

$$c(a) = \int c(a') \left[ \int \Psi_a^* \Psi_{a'}\, d\tau \right] da'. \qquad (3.31)$$

(Strictly this step is not permissible since the inner integral in (3.31) is divergent.) For this result to hold true for arbitrary wave-functions then the value of $c(a)$ cannot depend upon the value of $c(a')$, $a' \neq a$, i.e.

$$\int \Psi_a^* \Psi_{a'}\, d\tau = 0 \qquad a' \neq a \qquad (3.32)$$

In addition, to satisfy (3.30) and (3.31)

$$\int \left[ \int \Psi_a^* \Psi_{a'}\, d\tau \right] da' = 1. \qquad (3.33)$$

These results imply

$$\int \Psi_a^* \Psi_a\, d\tau = \infty. \qquad (3.34)$$

The 'improper' function with these properties is the Dirac delta function $\delta(x)$ 'defined' by

$$\delta(x) = 0 \qquad x \neq 0$$

$$\delta(0) = \infty$$

$$\int_{-\infty}^{\infty} \delta(x)\, dx = 1. \qquad (3.35)$$

Clearly the delta function is not analytic in the usual sense and a more formal definition may be given as the limit of a sequence of analytical functions satisfying (3.35). There is not a unique sequence with this property but a useful one is

$$\delta(x) = \lim_{\epsilon \to \infty} \frac{\sin^2 \epsilon x}{\pi x^2 \epsilon} \qquad (3.36)$$

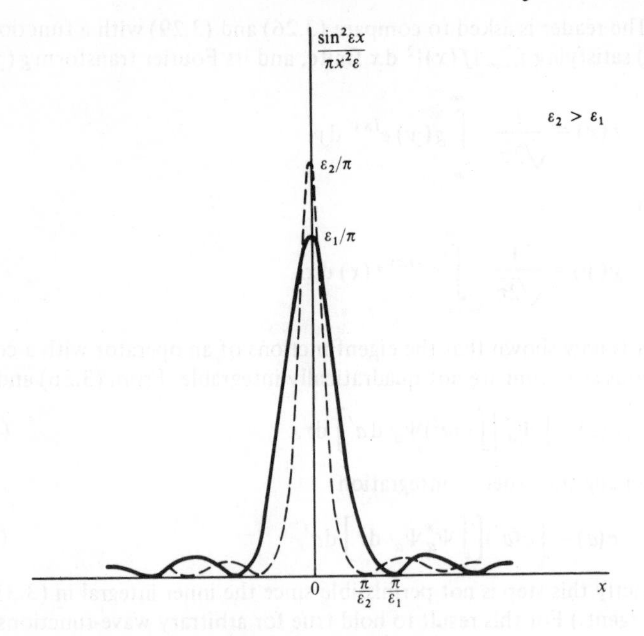


Figure 3.1 **Delta function sequence.**

The reader is asked to confirm that (3.35) is satisfied. Note that

$$\int_{-\infty}^{\infty} \frac{\sin^2 \beta}{\beta^2} \, \mathrm{d}\beta = \pi.$$

Strictly the delta function is only meaningful under an integral sign and the limiting process taken after the integral is evaluated.

From (3.32), (3.33), (3.34), (3.35),

$$\int \Psi_a^* \Psi_{a'} \, \mathrm{d}\tau = \delta\,(a' - a). \tag{3.37}$$

Equation (3.37) is the generalization of (3.12b). The eigenfunctions for different eigenvalues are still orthogonal but they are not quadratically integrable and cannot be normalized to unity.

The corresponding result in Fourier transform theory is

$$\frac{1}{2\pi} \int_{-\infty}^{\infty} e^{-ixy} e^{ixy'} \, \mathrm{d}x = \delta\,(y - y').$$

PROBLEMS

1 Classically, the angular momentum of a particle with respect to the origin is $\mathbf{L} = \mathbf{r} \wedge \mathbf{p}$ where $\mathbf{p}$ is the linear momentum. Show that in the

Schrödinger representation the $x$-component of the angular momentum operator is

$$\mathscr{L}_x \equiv \frac{\hbar}{i}\left(y\frac{\partial}{\partial z} - z\frac{\partial}{\partial y}\right).$$

Confirm that the commutator

$$[\mathscr{L}_x, \mathscr{L}_y] \equiv i\hbar\mathscr{L}_z.$$

2 The classical Hamiltonian for an electron (charge $-e$) in an electromagnetic field is

$$H = \frac{1}{2m}(\mathbf{p} + e\mathbf{A})^2 - e\phi. \qquad (1.45)$$

When expanded this becomes

$$H = \frac{p^2}{2m} + \frac{e}{m}\mathbf{p}\,.\,\mathbf{A} + \frac{e^2}{2m}A^2 - e\phi. \quad \text{(a)}$$

Alternatively it could be written

$$H = \frac{p^2}{2m} + \frac{e}{2m}(\mathbf{p}\,.\,\mathbf{A} + \mathbf{A}\,.\,\mathbf{p}) + \frac{e^2}{2m}A^2 - e\phi. \quad \text{(b)}$$

Write down the corresponding operators using the substitutions (3.8) and show that (b) produces a Hermitian operator whereas (a) does not. The operator obtained from (b) is the correct quantum-mechanical Hamiltonian.

3 The classical Hamiltonian for a linear harmonic oscillator is

$$H = \frac{1}{2m}p^2 + \frac{1}{2}m\omega^2x^2.$$

Derive the Hamiltonian operator in both the Schrödinger and momentum representations and write down the corresponding time-independent Schrödinger equations. Explain why the eigenfunctions of these two equations are identical in form.

4 Schrödinger's time-independent equation for a particle moving along the $x$-axis in a conservative field is

$$-\frac{\hbar^2}{2m}\frac{\mathrm{d}^2\psi}{\mathrm{d}x^2} + V(x)\psi = E\psi$$

where $V(x)$ is the potential energy and $E$ is the energy. Suppose there is a potential well defined by

$$V = -W \qquad -t \leqslant x \leqslant t$$

$$V = 0 \qquad |x| > t.$$

Integrate the Schrödinger's equation across this well. Show that if $V \to -\infty$ as $t \to 0$ in such a manner that $2Vt = -k$ (finite) then

$$\left(\frac{d\psi}{dx}\right)_r - \left(\frac{d\psi}{dx}\right)_l = -\frac{2m}{\hbar^2} k\psi$$

$$\left(\frac{d\psi}{dx}\right)_r \quad \text{and} \quad \left(\frac{d\psi}{dx}\right)_l$$

are the right hand and left hand derivatives at the origin. This result illustrates that the derivative of the wave-function is not continuous at a singularity in the potential.

5 Consider the functions

$$\psi_1(x) = e^{-x}, \qquad \psi_2(x) = e^{-2x}, \qquad \psi_3(x) = e^{-3x}, \qquad \text{etc.}$$

defined in the interval $0 < x < \infty$. Show that these functions are quadratically integrable in this interval and normalize them. Using the Schmidt method obtain an orthonormal set of functions.

6 Show that the function space $(x^3, y^3, x^3 + y^3)$ has the dimension two and not three.

Similarly show that the functions space $(xy^2, yz, z^2x, x(y^2 - z^2))$ has dimension three and not four.

7 A periodic function with period $2\pi$ is defined by

$$f(x) = x \qquad -\pi < x < \pi$$

Show that the associated Fourier series is

$$\sum_{n=1}^{\infty} (-1)^{n+1} \frac{2}{n} \sin nx.$$

8 Suppose $\{\phi_i(x)\}$ is a finite set of $n$ real orthonormal functions in the interval $(a, b)$, i.e.

$$\int_a^b \phi_i \phi_j \, dx = \delta_{ij}.$$

Let $f(x)$ be some real function and define 'Fourier constants' $c_i$ by

$$c_i = \int_a^b f(x)\phi_i(x) \, dx \qquad i = 1, \ldots, n.$$

Consider the integral

$$J \equiv \int_a^b [f(x) - L_n(x)]^2 \, dx$$

where $L_n(x)$ is some linear combination of the orthonormal functions and is written

$$L_n(x) = \sum_{i=1}^{n} \gamma_i \phi_i(x).$$

Show that the values of the constants $\gamma_i$ that make $J$ least are

$$\gamma_i = c_i.$$

It may be assumed that the integral

$$\int_a^b f^2(x) \, dx$$

exists.

Explain why

$$c_1^2 + c_2^2 + \ldots + c_n^2 \leqslant \int_a^b f^2(x) \, dx.$$

This result shows that the sum of the squares of the Fourier coefficients converges.

9 The general form for a second order, linear homogeneous equation is

$$f(x) \frac{d^2 X}{dx^2} + g(x) \frac{dX}{dx} + [n(x) + \lambda k(x)] X = 0$$

where $\lambda$ is a constant. Show that by multiplying through by the integrating factor

$$\frac{1}{f} e^{\int^x \frac{g}{f} \, dx}$$

this equation can be put into the self-adjoint form

$$\frac{d}{dx}\left[ r(x) \frac{dX}{dx} \right] + [q(x) + \lambda p(x)] X = 0. \qquad \text{(a)}$$

10 Suppose $X_1$ and $X_2$ are solutions of the self-adjoint equation (equation (a) in question 9) corresponding to different 'eigenvalues'

$\lambda_1$ and $\lambda_2$ and that both solutions satisfy the same boundary conditions

$$a_1 X + a_2 \frac{\mathrm{d}X}{\mathrm{d}x} = 0 \qquad \text{at } x = a$$

$$b_1 X + b_2 \frac{\mathrm{d}X}{\mathrm{d}x} = 0 \qquad \text{at } x = b.$$

By multiplying the equation satisfied by $X_1$ through by $X_2$ and that satisfied by $X_2$ through by $X_1$ integrating from $x = a$ to $x = b$ and subtracting the two, show that provided $\lambda_1 \neq \lambda_2$

$$\int_a^b p(x) X_1 X_2 \, \mathrm{d}x = 0.$$

That is, the two solutions are orthogonal with respect to the weight function $p(x)$ over the range $(a, b)$.

11 Legendre's differential equation is

$$(1 - x^2)\frac{\mathrm{d}^2 y}{\mathrm{d}x^2} - 2x\frac{\mathrm{d}y}{\mathrm{d}x} + l(l+1)y = 0.$$

Put this equation into self-adjoint form (question 9), and verify that $r(x) = 1 - x^2$. Observe that $r(1) = r(-1) = 0$. When $l = n$ (integer) a solution of the equation finite at $x = \pm 1$ is the Legendre polynomial $P_n(x)$. Following a method similar to that used in question 10 show that

$$\int_{-1}^{+1} P_n(x) P_m(x) \, \mathrm{d}x = 0 \qquad n \neq m.$$

Note that no boundary conditions need be imposed upon the solutions in this case.

12 The Dirac delta function has the property

$$\int_{-\infty}^{\infty} \delta(x) \, \mathrm{d}x = 1.$$

Suppose $\delta'(x)$ is defined as the 'derivative' of $\delta(x)$. Show that this derivative has the property

$$\int_{-\infty}^{\infty} \delta'(x) F(x) \, \mathrm{d}x = -F'(0).$$

13 Using equation (3.29) substitue for $c(a)$ in (3.26). Hence derive the 'closure relation'

$$\int \Psi_a^*(q') \Psi_a(q) \, \mathrm{d}a = \delta(q_1' - q_1)\delta(q_2' - q_2) \ldots \delta(q_n' - q_n).$$

14 In the Schrödinger representation, the operator representing the co-ordinate $q$ is itself. The corresponding eigenvalue equation is

$$q\Psi_{q'} = q'\Psi_{q'}.$$

$q'$ denotes the actual value of $q$. (Of course $q$ has a continuous spectrum.)
Explain why the eigenfunctions are of the form

$$\Psi_{q'} = \delta(q - q').$$

Using (3.26) expand an arbitrary wave function $\Psi(q)$ in terms of the co-ordinate eigenfunctions. Show that the expansion coefficient

$$c(q') = \Psi(q').$$

Deduce the probability that the value of the co-ordinate lies between $q'$ and $q' + \mathrm{d}q'$ is

$$|\Psi(q')|^2 \, \mathrm{d}q'$$

($\Psi^*\Psi$ has already been shown to be the probability density.)

15 The quantum mechanical expectation value for the kinetic energy of a particle of mass $m$ is

$$\langle T \rangle = -\frac{\hbar^2}{2m} \int \psi^* \nabla^2 \psi \, \mathrm{d}\tau$$

where $\psi$ is the state function and the integral is over all space. Substitute for the Laplacian operator using the identity

$$\text{div (A grad B)} = A\nabla^2 B + \text{grad } A \,.\, \text{grad } B$$

where $A, B$ are scalar-functions.
If the wave function falls off faster than $r^{-1}$ confirm that $\langle T \rangle$ is real and positive. Deduce that the allowed energies of a particle satisfy

$$E \geqslant \langle V \rangle$$

where $V$ is the potential energy.
(Wave functions often fall off as $e^{-\alpha r}$. Hint: use the divergence theorem

$$\left. \int \text{div } \mathbf{a} \, \mathrm{d}\tau = \oint_S \mathbf{a} . \, \mathrm{d}\mathbf{S} \right).$$

**References and Further Reading**

Landau, L. D. and Lifshitz E. M. *Quantum mechanics.* (Non-relativistic theory.) Pergamon, London (1959).
Mandl, M. A. *Quantum mechanics.* Butterworths, London (1957).
Trigg, G. L. *Quantum mechanics.* D. Van Nostrand, New York (1964).

CHAPTER 4

# Schrödinger's Equation

## 4.1 The Classical Limit

The fundamental equation of classical mechanics is the Hamilton–Jacobi equation,

$$H\left(q_1, \ldots, q_n, \frac{\partial S}{\partial q_1}, \ldots, \frac{\partial S}{\partial q_n}, t\right) = -\frac{\partial S}{\partial t}. \tag{1.62}$$

For a conservative system, the Hamiltonian is time-independent and the equation reduces to

$$H\left(q_1, \ldots, q_n, \frac{\partial S}{\partial q_1}, \ldots, \frac{\partial S}{\partial q_n}\right) = E \tag{1.63}$$

where $E$ is the constant energy of the system.

There is an obvious superficial relationship between these equations and the basic equations of quantum mechanics. The time-independent Schrödinger equation which every state function must satisfy is

$$\mathscr{H}\left(q_1, \ldots, q_n, \frac{\hbar}{i}\frac{\partial}{\partial q_1}, \ldots, \frac{\hbar}{i}\frac{\partial}{\partial q_n}, t\right)\Psi = -\frac{\hbar}{i}\frac{\partial \Psi}{\partial t}. \tag{3.10}$$

When the Hamiltonian does not depend on the time there are stationary states for which the energy is defined and they are obtained by solving the time-independent Schrödinger equation

$$\mathscr{H}\left(q_1, \ldots, q_n, \frac{\hbar}{i}\frac{\partial}{\partial q_1}, \ldots, \frac{\hbar}{i}\frac{\partial}{\partial q_n}\right)\psi_j = E_j\psi_j. \tag{3.16}$$

Quantum mechanics includes classical mechanics as a limiting case and the Hamilton–Jacobi equation is valid when $\hbar$ is small relative to other quantities with the same dimensions (i.e. as $\hbar \to 0$).

A similar situation occurs in electrodynamics where wave optics includes geometrical optics as the limiting case when the wave-length is small compared to the dimensions of the physical problem (i.e. as $\lambda \to 0$). In a dielectric medium with zero charge density and zero current density, the components of the electric field **E** and the magnetic induction **B** satisfy the wave equation

$$\nabla^2 f - \frac{1}{u^2}\frac{\partial^2 f}{\partial t^2} = 0 \tag{4.1}$$

where $u(x, y, z)$ is the velocity of propagation of the wave. When the velocity is a constant there are plane wave solutions of the type

$$f = Be^{i(\mathbf{k}\,.\,\mathbf{r} - \omega t + \alpha)}. \tag{4.2}$$

The direction of propagation is normal to the wave surface that is the locus of points which have the same phase at a given time. In the more general case there are solutions of the form

$$f = B(x, y, z, t)e^{i\Phi(x, y, z, t)} \tag{4.3}$$

where the phase (eikonal) $\Phi$ does not have the simple form of (4.2). In the limiting case when $\Phi$ changes by large amounts in small distances (corresponds to $\lambda \to 0$) substitution of (4.3) in (4.1) shows that the phase must satisfy the eikonal equation

$$\left(\frac{\partial\Phi}{\partial x}\right)^2 + \left(\frac{\partial\Phi}{\partial y}\right)^2 + \left(\frac{\partial\Phi}{\partial z}\right)^2 = \frac{1}{[u^2]}\left(\frac{\partial\Phi}{\partial t}\right)^2. \tag{4.4}$$

This is the fundamental equation of geometrical optics. For waves of a single frequency $\omega$

$$\Phi(x, y, z, t) = \phi(x, y, z) - \omega t \tag{4.5}$$

and the eikonal equation is

$$\left(\frac{\partial\phi}{\partial x}\right)^2 + \left(\frac{\partial\phi}{\partial y}\right)^2 + \left(\frac{\partial\phi}{\partial z}\right)^2 = \frac{\omega^2}{u^2}. \tag{4.6}$$

This equation is analogous to the Hamilton–Jacobi equation for a particle of mass $m$ in a conservative field,

$$\left(\frac{\partial S}{\partial x}\right)^2 + \left(\frac{\partial S}{\partial y}\right)^2 + \left(\frac{\partial S}{\partial z}\right)^2 = 2m(E - V). \tag{4.7}$$

In geometrical optics the paths of rays are specified by Fermat's principle. One statement of this is that the ray-path is such that the difference between its phases at the fixed end points takes a stationary value. In classical mechanics the motion of a particle is governed by Hamilton's principle. All this suggests a correspondence between the phase of a wave and Hamilton's characteristic function $S$.

The wave function for a single quantum particle satisfies Schrödinger's equation

$$-\frac{\hbar^2}{2m}\nabla^2\Psi + V\Psi = -\frac{\hbar}{i}\frac{\partial \Psi}{\partial t}. \tag{4.8}$$

In analogy to (4.3) try a solution of the form

$$\Psi(x, y, z, t) = B(x, y, z, t)e^{i\Phi(x, y, z, t)} \tag{4.9}$$

and write

$$\Phi = \frac{S(x, y, z, t)}{\hbar}. \tag{4.10}$$

The constant $\hbar$ is introduced as the phase is dimensionless whereas $S$ and $\hbar$ have dimensions of action e.g. Joule secs. Schrödinger's equation becomes

$$-\frac{\hbar^2}{2m}\nabla^2 B - \frac{i\hbar}{m}\text{grad } B \,.\, \text{grad } S - \frac{i\hbar}{2m}B\nabla^2 S + \frac{B}{2m}(\text{grad } S)^2 + VB$$
$$= -\frac{\partial S}{\partial t}B + i\hbar\frac{\partial B}{\partial t}. \tag{4.11}$$

In the limit of small $\hbar$, this equation becomes

$$\left(\frac{\partial S}{\partial x}\right)^2 + \left(\frac{\partial S}{\partial y}\right)^2 + \left(\frac{\partial S}{\partial z}\right)^2 + V2m = -2m\frac{\partial S}{\partial t}. \tag{4.12}$$

In the same approximation, the time-independent Schrödinger equation becomes

$$\left(\frac{\partial S}{\partial x}\right)^2 + \left(\frac{\partial S}{\partial y}\right)^2 + \left(\frac{\partial S}{\partial z}\right)^2 = 2m(E - V) \tag{4.13}$$

which is the Hamilton–Jacobi equation (4.7). In the limit that terms in $\hbar$ can be ignored the wave function changes according to classical rules. From equation (3.6), if $\hbar$ can be replaced by zero then all commutators are zero. This means that quantum-mechanical operators may be replaced by classical variables.

The brief treatment above has indicated how quantum mechanics includes classical mechanics. Historically the approach was quite the reverse. Schrödinger argued that if geometrical optics required extension to include wave optical effects such as diffraction, then perhaps in analogy, classical mechanics could also be extended. This lead to his theory of wave mechanics.

## 4.2 One-Dimensional Problems

In these idealized problems a single particle is confined to move in one dimension, along the $x$-axis, say. They are worth consideration since they bring out the essential elements of a quantum mechanical solution using only elementary mathematical techniques. In fact only a few real quantum mechanical problems have a rigorous analytic solution.

*Potential barrier of infinite width*

A simple example of this type of problem is a potential step function. The potential energy of the particle is defined by

$$\begin{aligned} V(x) &= 0 \qquad x < 0 \\ &= V_0 \qquad x > 0. \end{aligned} \tag{4.14}$$

It is desired to find the eigenfunction solutions of the time-independent Schrödinger equation

$$-\frac{\hbar^2}{2m}\frac{\mathrm{d}^2\psi}{\mathrm{d}x^2} + V(x)\psi = E\psi. \tag{4.15}$$

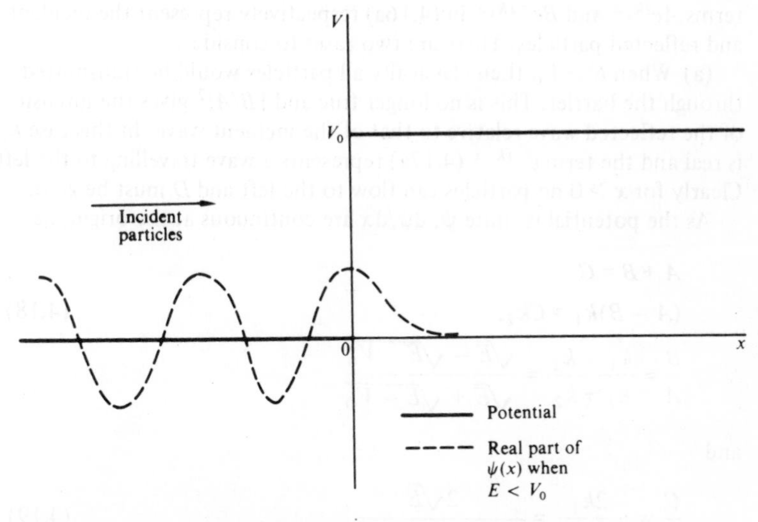


Figure 4.1 Step function potential.

To the left of the origin the potential energy is zero and there are no forces acting. The particles are free and the eigenfunctions are

$$\psi(x) = Ae^{ik_1x} + Be^{-ik_1x} \qquad x < 0 \tag{4.16a}$$

with

$$E = \hbar^2 k_1^2/2m \tag{4.16b}$$

where $A$, $B$ are constants.

To the right of the origin

$$-\frac{\hbar^2}{2m}\frac{d^2\psi}{dx^2} + (V_0 - E)\psi = 0 \qquad x > 0.$$

The eigenfunctions are

$$\psi(x) = Ce^{ik_2x} + De^{-ik_2x} \qquad x > 0 \tag{4.17a}$$

with

$$(E - V_0) = \hbar^2 k_2^2/2m. \tag{4.17b}$$

$C$, $D$ are also constants.

Solutions of this problem exist only for $E > 0$. Then the energy spectrum is continuous and the eigenfunctions are not quadratically integrable although $\psi$ can still be normalized over a finite volume. If a flux of non-interacting particles flows to the right from minus infinity $\psi^*\psi$ can be chosen to represent the particle density. Some particles may be reflected by the potential barrier and some transmitted. The terms $Ae^{ik_1x}$ and $Be^{-ik_1x}$ in (4.16a) respectively represent the incident and reflected particles. There are two cases to consider.

(a) When $E > V_0$ then classically all particles would be transmitted through the barrier. This is no longer true and $|B/A|^2$ gives the intensity of the reflected wave relative to that of the incident wave. In this case $k_2$ is real and the term $e^{-ik_2x}$ (4.17a) represents a wave travelling to the left. Clearly for $x > 0$ no particles can flow to the left and $D$ must be zero.

As the potential is finite $\psi$, $d\psi/dx$ are continuous at the origin, i.e.

$$A + B = C$$
$$(A - B)k_1 = Ck_2. \tag{4.18}$$

$$\therefore \quad \frac{B}{A} = \frac{k_1 - k_2}{k_1 + k_2} = \frac{\sqrt{E} - \sqrt{E - V_0}}{\sqrt{E} + \sqrt{E - V_0}}$$

and

$$\frac{C}{A} = \frac{2k_1}{k_1 + k_2} = \frac{2\sqrt{E}}{\sqrt{E} + \sqrt{E - V_0}}. \tag{4.19}$$

(The absolute values of $A$, $B$, $C$ may be fixed by normalizing $\psi$.) In the limit $E \to \infty$, $B = 0$ and $C = A$. Quantum mechanics is then in agreement with classical mechanics and all particles are transmitted.

(b) When $E < V_0$ then classically no particles would be transmitted. This is also true in quantum mechanics. In this case $k_2^2$ is negative. Only the positive imaginary root is allowed as the negative imaginary value for $k_2$ represents a divergent solution that increases exponentially with $x$. That is $D$ is again zero. Equations (4.18) and (4.19) still apply (with $k_2 = i\,|k_2|$) and $B/A$ is a complex number of modulus unity. Then the reflected wave has an intensity $|B/A|^2 = 1$ corresponding to a reflection of all particles. Even so there is a non-zero probability that a particle will penetrate the barrier to a depth $x$. From (4.17a) this probability is

$$|\psi(x)|^2 = |A|^2 \frac{4E}{V_0} \exp\left[-2\sqrt{2m(V_0 - E)}\,x/\hbar\right] \qquad x > 0.$$

Classically this could imply that the particle has negative kinetic energy for $x > 0$. This surprising unphysical result is explained in quantum mechanics by considering the uncertainty principle (Chapter 5).

A useful limiting case occurs as the potential step is increased without bound. From (4.18) and (4.19)

$$\lim_{V_0 \to \infty} B = -A \qquad \text{and} \qquad \lim_{V_0 \to \infty} C = 0.$$

In this case the wave function becomes

$$\psi(x) = 2iA \sin k_1 x \qquad x < 0, \qquad \psi(x) = 0 \qquad x > 0. \tag{4.20}$$

The wave function vanishes at a boundary where the potential is infinite. This result is valid in three dimensions.

*Tunnel effect*

Another problem where the energy spectrum is continuous occurs in the motion of particles past a potential hill. As an example consider a rectangular potential barrier

$$\begin{aligned} V(x) &= 0 && x < 0 \\ &= V_0 && 0 < x < a \\ &= 0 && x > a. \end{aligned} \tag{4.21}$$

Again solutions exist only for $E > 0$. (Note: the minimum potential energy is zero.)

The most interesting solutions occur when the energy of the particles, incident from the left, is less than the height of the potential barrier. Classically none of these particles would be transmitted but this is not

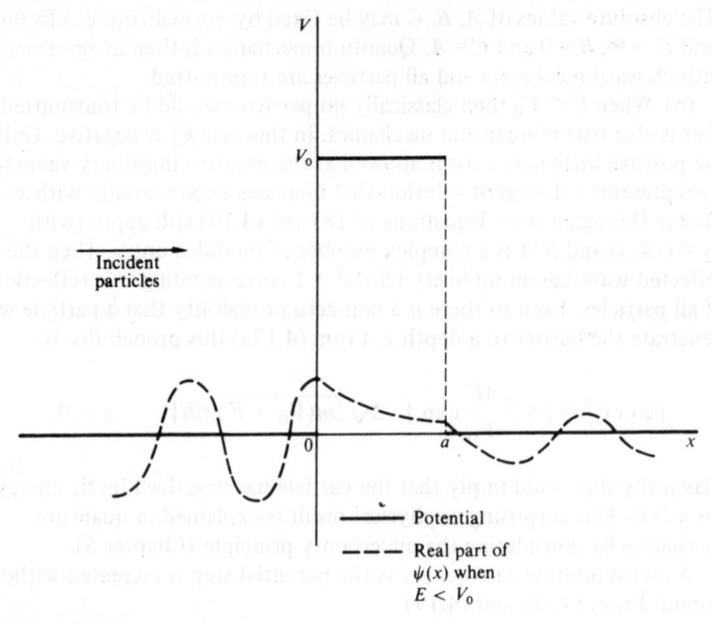


Figure 4.2 Rectangular potential barrier.

the case in quantum mechanics. The Schrödinger equations to be solved are

$$-\frac{\hbar^2}{2m}\frac{\mathrm{d}^2\psi}{\mathrm{d}x^2} = E\psi \qquad \begin{array}{l} x<0 \\ x>a \end{array}$$

$$-\frac{\hbar^2}{2m}\frac{\mathrm{d}^2\psi}{\mathrm{d}x^2} + V_0\psi = E\psi \qquad 0<x<a. \tag{4.22}$$

When $E < V_0$, the solutions are

$$\begin{aligned} \psi(x) &= Ae^{ikx} + Be^{-ikx} && x<0 \\ \psi(x) &= Ce^{\alpha x} + De^{-\alpha x} && 0<x<a \\ \psi(x) &= Fe^{ikx} && x>a \end{aligned}$$

with

$$k = \frac{1}{\hbar}(2mE)^{1/2} \qquad \text{and} \qquad \alpha = \frac{1}{\hbar}[2m(V_0 - E)]^{1/2}. \tag{4.23}$$

$\alpha$ is real. Note that there is no reflected wave for $x > a$ and so there is no term involving $e^{-ikx}$.

The boundary conditions are that the wave function and its derivative be continuous at $x = 0$ and $x = a$. These determine the values of the constants relative to $A$. (As in the previous problem the absolute values may be fixed by normalizing $\psi$.) The continuity of $\psi$ and its derivative at $x = 0$ give

$$A + B = C + D$$

$$ik(A - B) = \alpha(C - D).$$

Similarly at $x = a$

$$Ce^{\alpha a} + De^{-\alpha a} = Fe^{ika}$$

$$\alpha(Ce^{\alpha a} - De^{-\alpha a}) = ikFe^{ika}.$$

These equations can be written

$$\frac{B}{A} - \frac{C}{A} - \frac{D}{A} = -1$$

$$-ik\frac{B}{A} - \alpha\frac{C}{A} + \alpha\frac{D}{A} = -ik$$

$$e^{\alpha a}\frac{C}{A} + e^{-\alpha a}\frac{D}{A} - e^{ika}\frac{F}{A} = 0$$

$$\alpha e^{\alpha a}\frac{C}{A} - \alpha e^{-\alpha a}\frac{D}{A} - ike^{ika}\frac{F}{A} = 0.$$

Solving for $F/A$

$$\frac{F}{A} = \frac{4i\alpha k}{e^{ika}\,[4i\alpha k \cosh \alpha a - (\alpha^2 - k^2)2 \sinh \alpha a]}. \tag{4.24}$$

The number of particles transmitted through the barrier is proportional to $|F/A|^2$, i.e.

$$\left|\frac{F}{A}\right|^2 = \frac{4}{\left[4 \cosh^2 \alpha a + \left(\frac{\alpha}{k} - \frac{k}{\alpha}\right)^2 \sinh^2 \alpha a\right]}.$$

If $\alpha a$ is large then both $\cosh \alpha a$ and $\sinh \alpha a$ behave as $e^{\alpha a}/2$ and

$$\left|\frac{F}{A}\right|^2 = \frac{16\alpha^2 k^2}{(\alpha^2 + k^2)^2}\, e^{-2\alpha a}. \tag{4.25}$$

The number of particles transmitted falls off exponentially as

$$\exp\left\{-2\left[\frac{2m}{\hbar}(V_0 - E)\right]^{1/2} a\right\}.$$

Note that in the case that $h$ can be neglected (i.e. $h \to 0$) the classical limit of no transmission is obtained.

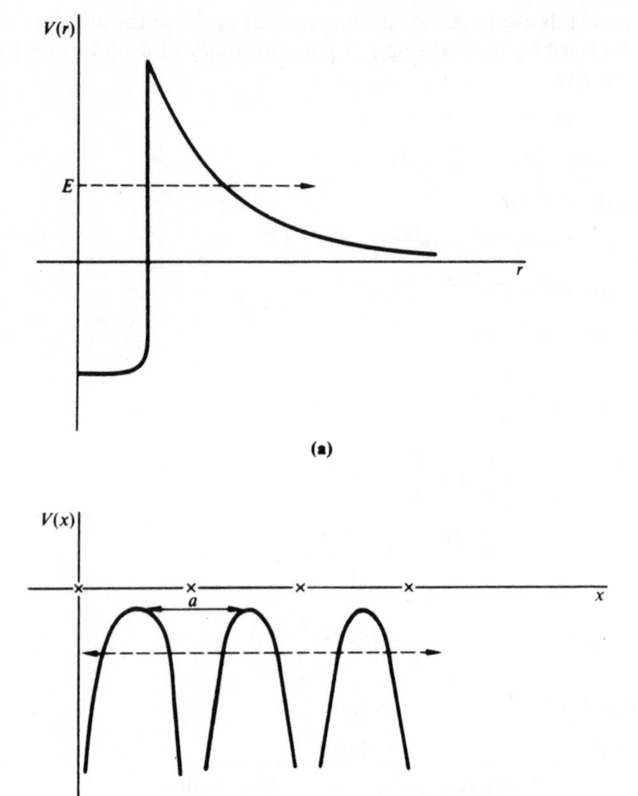


Figure 4.3 (a) The potential function for an $\alpha$-particle in a uranium nucleus. (b) The potential function in a crystal.

Tunnelling of particles occurs in radioactive decay. In the case of uranium an $\alpha$-particle experiences the repulsive nuclear electrostatic field up to about $10^{-14}$ m from the nucleus. Inside this distance the forces are attractive and the $\alpha$-particle is in a potential well. It is observed that $\alpha$-particles are emitted with energies less than that necessary to penetrate the barrier of the nuclear field. For example in the case of $^{238}_{92}U$ the height of the potential barrier is about 30 MeV whereas the kinetic energy of the emitted $\alpha$-particles is only 4·2 MeV.

The smaller the mass of a particle the greater the probability of transmission. This explains why in metals electrons can move from atom to atom through the potential barriers. If the barrier height is taken $\sim 1$ eV and $a \sim 5$ Å the barrier is not very transparent and the electron has a probability of about $10^{-2}$ of penetrating at each 'collision' with the barrier. However if it is considered that the electron is oscillating inside the potential well with wave-length equal to twice the Bohr radius ($a_0 = 5{\cdot}32 \times 10^{-11}$ m) then from de Broglie's hypothesis (2.42) the electron velocity will be $h/2m\,a_0$ and will collide $h/2m\,a_0^2 \sim 10^{17}$ times per second with the barrier. Such an electron will escape through the barrier after $\sim 10^{-15}$ seconds. Clearly it cannot be considered that such electrons are bound to particular atoms but rather that they travel throughout the crystal.

In the case when $E > V_0$ it can be shown in a similar manner to the above analysis, unlike the classical case, that not all particles are transmitted. Some are reflected.

### *Harmonic oscillator*

Consider a particle of mass $m$ oscillating in one dimension. Classically if the force acting is $-kx$ then the potential energy of the system is $kx^2/2$. The Hamiltonian operator is

$$\mathscr{H} = -\frac{\hbar^2}{2m}\frac{d^2}{dx^2} + \frac{1}{2}kx^2 \tag{4.26}$$

and the time-independent Schrödinger equation is

$$-\frac{\hbar^2}{2m}\frac{d^2\psi}{dx^2} + \frac{1}{2}kx^2\psi = E\psi. \tag{4.27}$$

It is usual to introduce the non-dimensional quantities

$$\gamma = \frac{2\sqrt{mE}}{\hbar\sqrt{k}} \qquad y = \left(\frac{mk}{\hbar^2}\right)^{1/4} x \tag{4.28}$$

and then

$$\frac{d^2\psi}{dy^2} + (\gamma - y^2)\,\psi = 0. \tag{4.29}$$

The potential energy tends to infinity as $|x| \to \infty$ and its least value is zero. Solutions exist for $E > 0$ and the wave function must vanish as $|x| \to \infty$. It is well known that (4.29) possesses solutions which tend to zero as $|x| \to \infty$ if and only if $\gamma$ takes one of the values $(1 + 2n)$ where $n$ is a positive integer. The corresponding solutions are

$$\psi(y) = A\,e^{-y^2/2}\,H_n(y) \tag{4.30}$$

where $H_n(y)$ is the Hermite polynomial of degree $n$. (See *Special functions of mathematical physics and chemistry*, by Sneddon.) The energy spectrum is discrete and so the eigenfunctions are quadratically integrable. $A$ is a normalization constant. The first four polynomials are

$$H_0(y) = 1$$

$$H_1(y) = 2y$$

$$H_2(y) = 4y^2 - 2$$

$$H_3(y) = 8y^3 - 12y. \tag{4.31}$$

The wave function $\psi_n(y)$ is odd or even as $n$ is and has $n$ nodes.

Corresponding to the allowed values of $\gamma$ the discrete eigenvalues are

$$E_n = \left(n + \frac{1}{2}\right) \hbar \sqrt{\frac{k}{m}} \qquad n = 0, 1, 2, \ldots. \tag{4.32}$$

The analogous classical oscillator has an angular frequency $\omega = (k/m)^{1/2}$ and in terms of this frequency

$$E_n = (n + \tfrac{1}{2})\hbar\omega \qquad n = 0, 1, 2, \ldots. \tag{4.33}$$

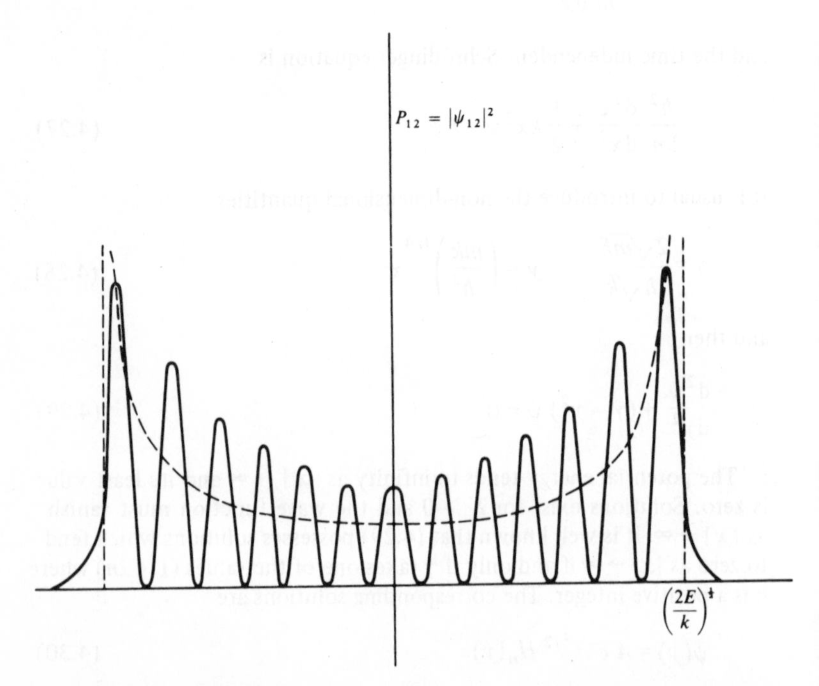


Figure 4.4 Probability distribution for $n = 12$. The classical value is represented by the dashed curve.

The term $n\hbar\omega$ gives the Planck series of energy levels; the energy of the ground state is not zero but takes the value $\hbar\omega/2$. This is called the 'zero point' energy and is a manifestation of the uncertainty principle to be discussed in the next chapter. Classically a particle oscillator would be in the lowest energy state when at rest at the centre of the oscillation. But this gives a precise value to $x$ and in quantum mechanics it is not then possible to give precise values to momentum and energy. For large quantum numbers the wave function probability density approximates to the classical probability curve for an oscillator with the same total energy.

In this classical case the probability is least at the midpoint of the oscillation where the particle velocity is greatest. This is an illustration of the correspondence principle discussed in Chapter 2.

All the oscillator eigenfunctions, except the ground state, have one or more nodes. The presence of nodes in a wave function leads to a classical paradox. As the probability density is zero at a node then apparently the particle must pass from one side to the other at infinite speed as it oscillates. The quantum mechanical answer to the 'node paradox' is that the momentum and position of a particle cannot be known simultaneously and so this trajectory problem does not arise. (For a further discussion see Kemble, p. 86.)

The simplest application of the linear oscillator is in the discussion of the vibrational energy levels of diatomic molecules.

## 4.3 Three-Dimensional Problems

*Particle in a box*

Consider a cube of side $a$ with the origin of the co-ordinate system at a corner of the box. Inside the box the potential energy of an electron is zero and outside the potential is infinite. This is essentially the model for a metal suggested by Sommerfeld in 1928.

The time-independent Schrödinger equation is

$$\frac{\partial^2\psi}{\partial x^2} + \frac{\partial^2\psi}{\partial y^2} + \frac{\partial^2\psi}{\partial z^2} + k^2\psi = 0 \tag{4.34}$$

inside the box, where the electron kinetic energy is

$$E = \hbar^2 k^2/2m. \tag{4.35}$$

As $E > 0$ then $k$ must be real.

The partial differential equation (4.34) is solved by the method of separation of variables. By writing

$$\psi(x, y, z) = X(x)Y(y)Z(z) \tag{4.36}$$

equation (4.34) becomes

$$\frac{1}{X}\frac{d^2X}{dx^2}+\frac{1}{Y}\frac{d^2Y}{dy^2}+\frac{1}{Z}\frac{d^2Z}{dz^2}+k^2=0. \tag{4.37}$$

The first term does not include $y$ or $z$, the second does not involve $x$ or $z$ and the third does not involve $x$ or $y$. This means that

$$\frac{1}{X}\frac{d^2X}{dx^2}+k_x^2=0$$

$$\frac{1}{Y}\frac{d^2Y}{dy^2}+k_y^2=0$$

$$\frac{1}{Z}\frac{d^2Z}{dz^2}+k_z^2=0 \tag{4.38}$$

where $k_x, k_y, k_z$ are constants such that

$$k^2=k_x^2+k_y^2+k_z^2. \tag{4.39}$$

Each of the equations (4.38) has a general solution of the form

$$X(x)=A\sin k_x x+B\cos k_x x \qquad k_x^2>0 \qquad \text{(a)}$$

$$X(x)=Ax+B \qquad k_x^2=0 \qquad \text{(b)}$$

$$X(x)=A\sinh|k_x|x+B\cosh|k_x|x \qquad k_x^2<0 \qquad \text{(c)} \tag{4.40}$$

where $A$, $B$ are arbitrary constants and of course are not the same in the three cases (a), (b) and (c).

The boundary conditions require that the wave function vanishes at the box walls where the potential is infinite. The only solution which can satisfy these conditions is (4.40a), and the wave function (4.36) is then composed of terms of the type

$$\begin{matrix}\sin\\ \cos\end{matrix}\, k_x x \; \begin{matrix}\sin\\ \cos\end{matrix}\, k_y y \; \begin{matrix}\sin\\ \cos\end{matrix}\, k_z z. \tag{4.41}$$

This wave function must vanish at all points on the cube surface and the only product satisfying this condition is

$$A\sin k_x x\sin k_y y\sin k_z z$$

with

$$k_x=\frac{l\pi}{a},\qquad k_y=\frac{m\pi}{a},\qquad k_z=\frac{n\pi}{a} \tag{4.42}$$

where $l, m, n$ are integers. The electron energy takes the discrete values

$$E = \frac{h^2}{8ma^2}(l^2 + m^2 + n^2). \tag{4.43}$$

It is very important to note that it is the *boundary* conditions that quantize the energy. In general the energy spectrum is partly or wholly discrete if a region of space is surrounded by a potential wall. In the present problem the electron is localized within a cube by an infinite potential barrier whereas in the previous problem (oscillator) $V(x) \to \infty$ as $|x| \to \infty$. The energy spectrum is wholly discrete in both these cases of a finite and infinite domain.

However if the potential is such that $V(x) < 0$ inside some closed region and $V(x) = 0$ outside this region there is a discrete set of bound states for $E < 0$ and a continuous spectrum of states for $E > 0$. The wave functions for these latter states are not quadratically integrable and represent 'infinite' motion of the system.

More generally, if the potential tends to a finite value at infinity there will be a continuous energy spectrum for values greater than this.

This problem is the basis of the free-electron model for solids. Although this application is apparently crude it is surprisingly useful.

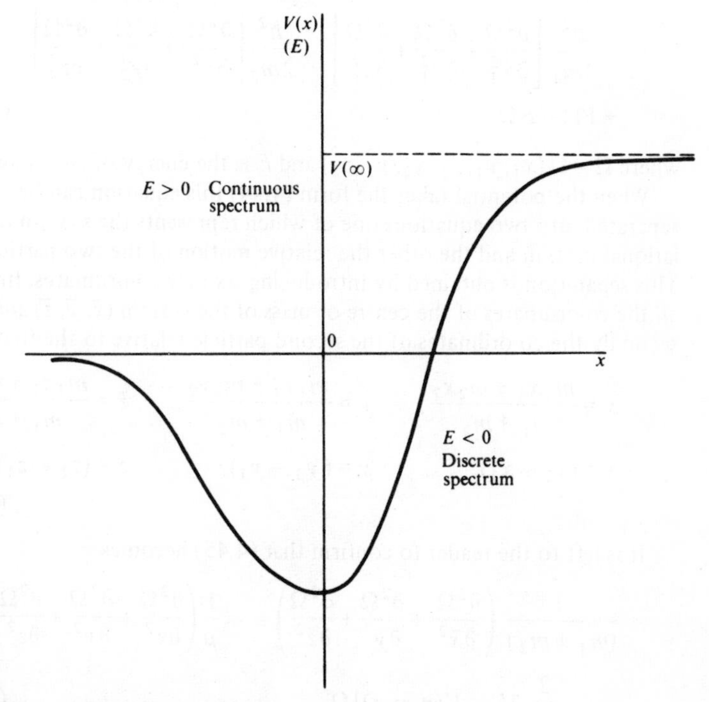


Figure 4.5 A finite potential well.

*Rigid rotator*

It is convenient at this stage to discuss the system of two interacting point particles the results of which will not only be needed in the present problem but also in the following treatment of the hydrogen atom. Suppose the particles have masses $m_1, m_2$ and are at the positions $(x_1, y_1, z_1)$ and $(x_2, y_2, z_2)$ respectively and suppose that the potential energy is a function of the relative displacement of the particles, i.e.

$$V = V(x_2 - x_1, y_2 - y_1, z_2 - z_1). \tag{4.44}$$

The classical Hamiltonian for this two-body problem is

$$H = \frac{1}{2m_1}P_1^2 + \frac{1}{2m_2}P_2^2 + V$$

where $P_1, P_2$ are the momenta of the particles. The corresponding operator is obtained by substituting

$$P_i^2 \equiv -\hbar^2\left(\frac{\partial^2}{\partial x_i^2} + \frac{\partial^2}{\partial y_i^2} + \frac{\partial^2}{\partial z_i^2}\right) \qquad i = 1, 2.$$

The time-independent Schrödinger equation is

$$-\frac{\hbar^2}{2m_1}\left[\frac{\partial^2\Omega}{\partial x_1^2} + \frac{\partial^2\Omega}{\partial y_1^2} + \frac{\partial^2\Omega}{\partial z_1^2}\right] - \frac{\hbar^2}{2m_2}\left[\frac{\partial^2\Omega}{\partial x_2^2} + \frac{\partial^2\Omega}{\partial y_2^2} + \frac{\partial^2\Omega}{\partial z_2^2}\right]$$
$$+ V\Omega = E\Omega \tag{4.45}$$

where $\Omega = \Omega(x_1, y_1, z_1, x_2, y_2, z_2)$ and $E$ is the energy of the system.

When the potential takes the form (4.44) this equation can be separated into two equations one of which represents the system translational motion and the other the relative motion of the two particles. This separation is obtained by introducing six new co-ordinates, first of all the co-ordinates of the centre of mass of the system $(\bar{x}, \bar{y}, \bar{z})$ and secondly the co-ordinates of the second particle relative to the first.

$$\bar{x} = \frac{m_1x_1 + m_2x_2}{m_1 + m_2}, \qquad \bar{y} = \frac{m_1y_1 + m_2y_2}{m_1 + m_2}, \qquad \bar{z} = \frac{m_1z_1 + m_2z_2}{m_1 + m_2}$$
$$x = (x_2 - x_1), \qquad y = (y_2 - y_1), \qquad z = (z_2 - z_1). \tag{4.46}$$

It is left to the reader to confirm that (4.45) becomes

$$\frac{1}{(m_1 + m_2)}\left(\frac{\partial^2\Omega}{\partial\bar{x}^2} + \frac{\partial^2\Omega}{\partial\bar{y}^2} + \frac{\partial^2\Omega}{\partial\bar{z}^2}\right) = -\frac{1}{\mu}\left(\frac{\partial^2\Omega}{\partial x^2} + \frac{\partial^2\Omega}{\partial y^2} + \frac{\partial^2\Omega}{\partial z^2}\right)$$
$$-\frac{2}{\hbar^2}[E - V(x, y, z)]\Omega \tag{4.47}$$

where the 'reduced mass' of the system is

$$\mu = \frac{m_1 m_2}{m_1 + m_2} \tag{4.48}$$

It is now possible to separate this equation by writing

$$\Omega(\bar{x}, \bar{y}, \bar{z}, x, y, z) = f(\bar{x}, \bar{y}, \bar{z})\psi(x, y, z) \tag{4.49}$$

When this expression is substituted into (4.47) and the equation divided by the product $f\psi$ it is found that the left hand side is independent of $(x, y, z)$ and the right hand side is independent of $(\bar{x}, \bar{y}, \bar{z})$. Each of these parts must be equal to a separation constant which is conveniently taken as $-2E_t/\hbar^2$. The resulting two equations are

$$\frac{\partial^2 f}{\partial \bar{x}^2} + \frac{\partial^2 f}{\partial \bar{y}^2} + \frac{\partial^2 f}{\partial \bar{z}^2} + \frac{2}{\hbar^2}(m_1 + m_2)E_t f = 0 \tag{4.50}$$

$$\frac{\partial^2 \psi}{\partial x^2} + \frac{\partial^2 \psi}{\partial y^2} + \frac{\partial^2 \psi}{\partial z^2} + \frac{2\mu}{\hbar^2}[(E - E_t) - V]\psi = 0 \tag{4.51a}$$

Equation (4.50) is simply Schrödinger's equation for a particle of mass $(m_1 + m_2)$ moving in field-free space. The parameter $E_t$ represents the translation energy of the system. Equation (4.51a) is the Schrödinger equation for a particle with mass equal to the reduced mass of the system moving in the potential $V(x, y, z)$. The parameter $E_s = (E - E_t)$ represents the vibrational and rotational energy of the system. It is often useful to express (4.51a) in spherical polar co-ordinates.

$$x = r \sin\theta \cos\phi, \qquad y = r \sin\theta \sin\phi, \qquad z = r\cos\theta.$$

Then

$$\frac{1}{r^2}\frac{\partial}{\partial r}\left(r^2 \frac{\partial \psi}{\partial r}\right) + \frac{1}{r^2 \sin^2\theta}\frac{\partial^2 \psi}{\partial \phi^2} + \frac{1}{r^2 \sin\theta}\frac{\partial}{\partial \theta}\left(\sin\theta \frac{\partial \psi}{\partial \theta}\right)$$
$$+ \frac{2\mu}{\hbar^2}[E_s - V]\psi = 0. \tag{4.51b}$$

Consider now the case of the rigid rotator where the masses are kept at a fixed distance $a$ apart and where $V$ may be taken as zero. The Schrödinger equation (4.51b) then represents a mass moving over the surface of a sphere of radius $a$, i.e.

$$\frac{1}{\sin^2\theta}\frac{\partial^2 \psi}{\partial \phi^2} + \frac{1}{\sin\theta}\frac{\partial}{\partial \theta}\left(\sin\theta \frac{\partial \psi}{\partial \theta}\right) + \frac{2I}{\hbar^2}E_s\psi = 0 \tag{4.52}$$

where $I = \mu a^2$. The wave function depends only upon $\theta$ and $\phi$. This equation may be separated by writing

$$\psi(\theta, \phi) = \Theta(\theta)\Phi(\phi) \tag{4.53}$$

and dividing by the product $\Theta\Phi$.

$$\frac{1}{\sin^2\theta\Phi}\frac{d^2\Phi}{d\phi^2}+\frac{1}{\sin\theta\Theta}\frac{d}{d\theta}\left(\sin\theta\frac{d\Theta}{d\theta}\right)+\frac{2I}{h^2}E_s=0.$$

This can be written as the two equations

$$\frac{d^2\Phi}{d\phi^2}+m^2\Phi=0 \tag{4.54}$$

$$\frac{d^2\Theta}{d\theta^2}+\frac{d\Theta}{d\theta}\cot\theta+\frac{2I}{\hbar^2}E_s\Theta-\frac{m^2}{\sin^2\theta}\Theta=0 \tag{4.55}$$

where $m^2$ is the separation constant. The solution of (4.54) is

$$\Phi=A\sin m\phi+B\cos m\phi. \tag{4.56}$$

Since a wave function must be single-valued and continuous $\Phi(\phi+2\pi)=\Phi(\phi)$ and so $m$ must be restricted to integer values. If the substitution $\alpha=\cos\theta$ is used in (4.55) then the $\Theta$-equation becomes

$$(1-\alpha^2)\frac{d^2\Theta}{d\alpha^2}-2\alpha\frac{d\Theta}{d\alpha}+\left\{\lambda-\frac{m^2}{1-\alpha^2}\right\}\Theta=0 \qquad -1\leqslant\alpha\leqslant 1 \tag{4.57}$$

where

$$\lambda=2IE_s/\hbar^2.$$

This second-order equation has two solutions but in general these are both infinite at $\alpha=\pm1$ (i.e. $\theta=0,\pi$). Such a solution is not physically acceptable. But if $\lambda$ takes one of the values

$$\lambda=K(K+1) \qquad K=0,1,2,\ldots \tag{4.58}$$

with

$$K\geqslant|m|$$

then one of the solutions is finite in the range $-1\leqslant\alpha\leqslant1$ while the other is not. The equation (4.57) is Legendre's associated equation (see Sneddon's book) and the acceptable solution is the associated Legendre function $P_k^m(\alpha)$, i.e.

$$\Theta(\theta)=P_K^m(\cos\theta). \tag{4.59}$$

So the physical restrictions on the wave functions require the energy to take the discrete values

$$E_s=\frac{K(K+1)\hbar^2}{2I} \qquad K=0,1,2,\ldots. \tag{4.60}$$

This simple model is useful in the theory of band structure of molecules.

*The infra red spectrum of diatomic molecules*

The more rapid electron motion in a molecule can be separated from the nuclear motion of a molecule (Born–Oppenheimer approximation) and the molecular energy (apart from translational energy) can be considered to be composed of three parts

(a) the electron energy,
(b) the vibrational energy of the component atoms,
(c) the rotational energy of the molecule as a whole.

Each electron energy level has a fine structure of vibrational energy levels and these in turn have a rotational fine structure. This section is concerned with the fine structure due to the atomic vibrational and molecular rotational levels.

Classically, an accelerating electric charge produces an electromagnetic field. Molecular vibrations and rotations involve oscillations of charges and produce emission (absorption) of radiation. The frequency of this radiation is equal to the frequency of the corresponding molecular motion. For diatomic molecules the vibrational frequencies are in the infrared region $\sim 10^{13}$–$10^{14}$ Hz and the rotational frequencies are lower. Infrared and microwave spectroscopy of gases are powerful methods for helping to determine molecular structure.

A diatomic molecule can be regarded as a dumb-bell rotator-oscillator. In the previous discussion of the rigid rotator the potential function $V(r, \theta, \phi)$ in (4.51) was taken to be zero. A simple model for diatomic molecules is to assume the simple harmonic potential

$$V(r) = \tfrac{1}{2}k(r - r_0)^2$$

where $r_0$ is the equilibrium separation of the nuclei and $k$ is a force constant. A detailed analysis of the rotator-oscillator (e.g. Pauling and Wilson, p. 263) shows that to second-order terms the energy levels of the system are given by

$$E_{n,K} = (n + \tfrac{1}{2})\hbar\omega + K(K+1)\frac{\hbar^2}{2I} - K^2(K+1)^2\frac{\hbar^4}{2\omega^2 I^3}$$

$$n = 0, 1, 2, \ldots$$
$$K = 0, 1, 2, \ldots$$

where $\omega$ is the frequency of oscillation of the molecule and $I$ is the moment of inertia. The first term is due to the molecular vibrations (4.33) and the second term is due to the rotation of the molecule (4.60). The third term is a correction term arising from the centrifugal stretching of the non-rigid molecule during the rotation. The corresponding eigenfunctions are

$$\psi_{n,K}(r, \theta, \phi) = R_n(r)P_K^m(\cos\theta)\,e^{im\phi}$$

and $R_n(r)$ has the general form of (4.30).

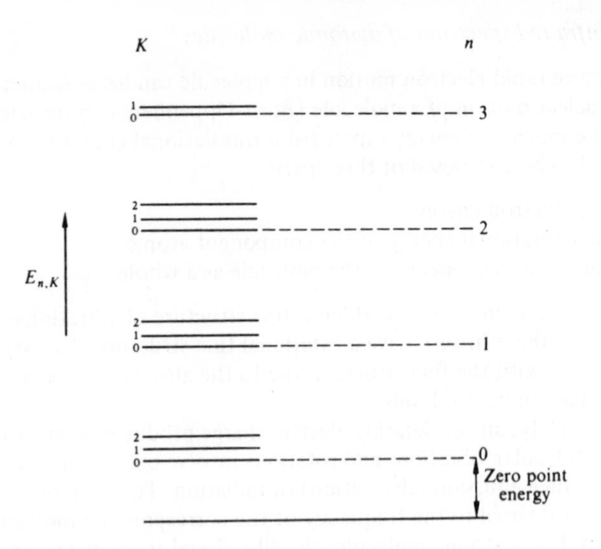


Figure 4.6 The energy levels $E_{n,\,K}$ for the non-rigid rotator-oscillator.

The infrared spectrum observed depends upon the allowed energy transitions and these are determined by the selection rules. In Chapter 8 it is shown using the dipole approximation that a transition of a charge from a state $i$ to a state $f$ is possible if the integral (8.55)

$$\int \chi_f^* \mathbf{r} \chi_i \, \mathrm{d}\tau$$

is non-zero where $\chi_i$ and $\chi_f$ are the initial and final state functions and $\mathbf{r}$ is the position of the charge.

The properties of Hermite functions (Chapter 6, problem 14) restrict the allowed transitions for a harmonic oscillator to those between neighbouring energy levels so that the selection rule is

$$\delta n = \pm 1.$$

Consequently the emission or absorption spectrum of a harmonic oscillator should consist of one sharp line with a frequency equal to that of the oscillator.

Analysis of the rotator-oscillator eigenfunctions shows that the selection rules are

$$\delta n = 0, \pm 1, \pm 2, \ldots \qquad \delta K = \pm 1.$$

The transitions with $\delta K = +1$ form the $R$-branch and those with $\delta K = -1$ form the $P$-branch. The infra-red absorption spectrum (no electron transition) of heteronuclear molecules such as HCl conform to these rules. Each vibrational spectrum line has a fine structure (Fig. 4.7).

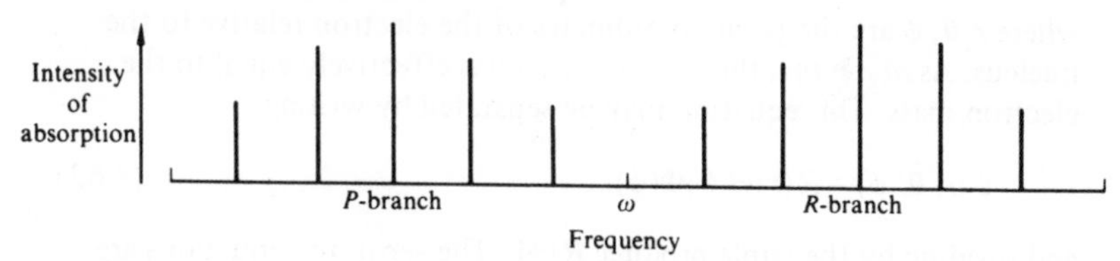


Figure 4.7

Note that the rule $\delta K = \pm 1$ means that there is no central line to the structure.

There is no infra-red absorption spectrum for homonuclear diatomic molecules such as $H_2$ or $O_2$. This is of course to be expected as in the vibrations of a homonuclear molecule there is no net movement of charge.

It can be shown that when the electron motion is taken into account the above selection rules are correct on the assumption that the electrons have zero angular momentum about the nuclear-axis. ($\Sigma$ state.) This is the case for almost all diatomic molecules (NO is an exception) in the ground state. However, if the electrons do have a non-zero angular momentum about the nuclear axis then the transitions with $\delta K = 0$ are allowed and form the *Q*-branch.

Transitions involving a change of electron energy are very complex and in this case homonuclear molecules can show a vibration-rotation spectra. For further information the reader is referred to *Introduction to Quantum Mechanics*, by Pauling and Wilson, and *The Wave Mechanics of Atoms, Molecules and Ions*, by Schutte.

### *Central field Coulomb potential*

In this final section of Chapter 4, an outline treatment is given for the hydrogen atom problem. The first part of the previous section is of importance. Consider the motion of a single electron of mass $m_1$ and charge $-e$ in the coulomb field of a nucleus of mass $m_2$ and charge $Ze$. The potential energy of the system depends only upon the separation of the particles and is

$$V(r) = -\frac{Ze^2}{4\pi\epsilon_0 r}$$

where $\epsilon_0$ is the permitting of free space.

The analysis from (4.44) to (4.51) applies and the energy (excluding the translation energy) $E_s$ of the system satisfies

$$\frac{1}{r^2}\frac{\partial}{\partial r}\left(r^2\frac{\partial\psi}{\partial r}\right) + \frac{1}{r^2\sin^2\theta}\frac{\partial^2\psi}{\partial\phi^2} + \frac{1}{r^2\sin\theta}\frac{\partial}{\partial\theta}\left(\sin\theta\frac{\partial\psi}{\partial\theta}\right)$$

$$+\frac{2\mu}{\hbar^2}\left(E_s + \frac{Ze^2}{4\pi\epsilon_0 r}\right)\psi = 0 \qquad (4.61)$$

where $r, \theta, \phi$ are the polar co-ordinates of the electron relative to the nucleus. As $m_2 \gg m_1$, the reduced mass $\mu$ is effectively equal to the electron mass. This equation may be separated by writing

$$\psi(r, \theta, \phi) = R(r)\Theta(\theta)\Phi(\phi) \tag{4.62}$$

and dividing by the triple produce $R\Theta\Phi$. The separated equations are

$$\frac{d^2\Phi}{d\phi^2} + m_l^2\Phi = 0 \tag{4.63}$$

$$\frac{d^2\Theta}{d\theta^2} + \cot\theta\,\frac{d\Theta}{d\theta} + \left\{l(l+1) - \frac{m_l^2}{\sin^2\theta}\right\}\Theta = 0 \tag{4.64}$$

$$\frac{1}{r^2}\frac{d}{dr}\left(r^2\frac{dR}{dr}\right) + \left\{\frac{2\mu}{\hbar^2}\left(E_s + \frac{Ze^2}{4\pi\epsilon_0 r}\right) - \frac{l(l+1)}{r^2}\right\}R = 0 \tag{4.65}$$

where $m_l^2$ and $l(l+1)$ are the separation constants (see rigid rotator).

As $\Phi$ must be single-valued and continuous the constant $m_l$ must be an integer. Equation (4.64) is the transformed Legendre's associated equation and as the wave functions must be bounded in the range $0 \leqslant \theta \leqslant \pi$ the required solution is

$$\Theta = P_l^{m_l}(\cos\theta) \tag{4.66}$$

with $l$ a positive integer or zero and $l \geqslant |m_l|$.

The radial equation (4.65) may be solved by writing

$$\alpha^2 = -\frac{2\mu}{\hbar^2}E_s, \qquad \beta = \frac{\mu Ze^2}{4\pi\epsilon_0\hbar^2\alpha} \tag{4.67}$$

and changing the independent variable to $\rho = 2\alpha r$. Then

$$\frac{d^2R}{d\rho^2} + \frac{2}{\rho}\frac{dR}{d\rho} + \left\{-\frac{1}{4} - \frac{l(l+1)}{\rho^2} + \frac{\beta}{\rho}\right\}R = 0. \tag{4.68}$$

Observe that for bound states $E_s < 0$ and $\alpha$ is real. The physical requirements of single-valuedness and continuity have been imposed on the wave-function but no boundary conditions have been applied up to this point. These boundary conditions are

$$\text{(i) } \psi \to 0 \text{ as } r \to \infty \qquad \text{(ii) } \psi \text{ is finite as } r \to 0. \tag{4.69}$$

Equation (4.68) is discussed in books on special functions (e.g. Sneddon), and solutions that satisfy $R \to 0$ as $\rho \to \infty$ and as $\rho \to 0$ exist only if $\beta$ is an integer $n$, such that $n \geqslant l + 1$. The solutions are the Laguerre functions

$$R_{nl}(\rho) = e^{-\rho/2}\rho^l L_{n+1}^{2l+1}(\rho) \qquad n \geqslant l + 1 \tag{4.70}$$

where $L_{n+1}^{2l+1}$ is an associated Laguerre polynomial. The restriction on $\beta$ results in a discrete energy spectrum

$$E_s = -Z^2 e^4 \mu / 2\hbar^2 n^2 (4\pi\epsilon_0)^2 \qquad n = 1, 2, \ldots. \tag{4.71}$$

The quantum number $n$ determines the energy and the degenerate wave-functions associated with this energy are

$$\psi_{nlm_l}(r, \theta, \phi) = A_{nlm_l} R_{nl}(2\alpha r) P_l^{m_l}(\cos\theta) e^{im_l\phi} \tag{4.72}$$

$$l = 0, 1, \ldots, \qquad (n-1)$$

$$m_l = -l, -(l-1), \ldots, 0, \ldots, +l$$

where $A_{nlm_l}$ is a normalization constant chosen so that

$$\int_{\text{all space}} \psi^* \psi \, d\tau = 1.$$

The number of degenerate wave-functions belonging to the $n$th energy level is

$$\sum_{l=0}^{n-1} (2l+1) = n^2.$$

The quantum numbers $l$ and $m_l$ are related to the angular momentum of the system and the degeneracy with respect to $m_l$ is associated with the rotational symmetry of the atom and occurs in any central field problem (see Chapter 9). However, the degeneracy with respect to $l$ is a property peculiar to the coulomb potential and is not readily explained in terms of the atomic symmetry (see Kemble, *Quantum mechanics*, p. 312).

The ground state (lowest state) of the electron occurs when $n = 1$ and the single normalized wave function is

$$\psi_{100} \equiv (1s) = \frac{1}{\sqrt{\pi}} \left(\frac{Z}{a_0}\right)^{3/2} e^{-\rho/2} \qquad a_0 = \frac{\hbar^2 4\pi\epsilon_0}{\mu e^2}, \; \rho = \frac{2Zr}{na_0} \tag{4.73}$$

The next level corresponds to $n = 2$ and has four degenerate wave functions associated with it.

$$\psi_{200} \equiv (2s) = \frac{1}{4\sqrt{2\pi}} \left(\frac{Z}{a_0}\right)^{3/2} (2-\rho) e^{-\rho/2}$$

$$\psi_{210} \equiv (2p_z) = \frac{1}{4\sqrt{2\pi}} \left(\frac{Z}{a_0}\right)^{3/2} \rho e^{-\rho/2} \cos\theta$$

$$\psi_{21\pm1} = \frac{1}{8\sqrt{\pi}} \left(\frac{Z}{a_0}\right)^{3/2} \rho e^{-\rho/2} \sin\theta \, e^{\pm i\phi}. \tag{4.74}$$

From the last two it is possible to construct the functions

$$\begin{matrix}(2p_x)\\ \\ (2p_y)\end{matrix} = \frac{1}{4\sqrt{2\pi}}\left(\frac{Z}{a_0}\right)^{3/2} \rho e^{-\rho/2} \sin\theta \begin{matrix}\cos\phi\\ \\ \sin\phi\end{matrix} \qquad (4.75)$$

The formula (4.71) gives the basic structure of the atomic line spectrum but it does not give the fine structure. This is essentially a relativistic phenomenon and is connected with the concept of electron spin.

Positive values of the total energy correspond to unbound states in which the electron kinetic energy is greater than the potential energy. The energy spectrum is then continuous and the wave functions are not normalizable.

## PROBLEMS

1 Consider the one-dimensional Schrödinger equation (4.15). If $\psi_1$, $\psi_2$ are two different eigenfunctions corresponding to the same value of the energy show that

$$\frac{d^2\psi_1}{dx^2}\psi_2 - \frac{d^2\psi_2}{dx^2}\psi_1 = 0.$$

Integrate the equation to obtain

$$\frac{d\psi_1}{dx}\psi_2 - \psi_1\frac{d\psi_2}{dx} = \text{constant}.$$

If the spectrum is discrete both $\psi_1$ and $\psi_2$ vanish at infinity. By integrating a second time show that $\psi_1$ and $\psi_2$ are linearly dependent. This result shows that in a one-dimensional problem the energy levels of a discrete spectrum are not degenerate.

2 Suppose the potential in a one-dimensional Schrödinger equation is symmetric, i.e.

$$V(-x) = V(x).$$

By carrying out the change of variable $x \to -x$ show that in the non-degenerate case the eigenfunctions are either odd or even, i.e.

$$\psi(-x) = \mp\psi(x).$$

3 Carry out the analysis of the tunnel effect in the case $E > V_0$ (4.22). Show that contrary to classical predictions some particles are reflected.

4 With the notation of (4.26) show that the frequency of a classical oscillator is $\omega = (k/m)^{1/2}$. From energy considerations show that the

maximum displacement of the particle is $x_0 = (2E/k)^{1/2}$ where $E$ is the total energy of the oscillation.

Define $P(x)\,\mathrm{d}x$ to be the probability that the particle is in the element $\mathrm{d}x$. This probability must be proportional to time $\mathrm{d}t$ spent in $\mathrm{d}x$. Using the harmonic solution $x = x_0 \sin \omega t$ together with the relation $\mathrm{d}t = \mathrm{d}x/\dot{x}$ show that

$$P(x)\,\mathrm{d}x = \frac{\mathrm{d}x}{\pi\left[\dfrac{2E}{k} - x^2\right]^{1/2}}.$$

(See Fig. 4.4.)

5 The potential in a three-dimensional problem is defined by

$$V = -V_0 \quad r \leqslant a, \qquad V = 0 \quad r > a \qquad V_0 > 0.$$

($r$ is a spherical polar co-ordinate.)

Show that the spherically symmetric bound solutions ($E < 0$) which satisfy the boundary conditions

(i) $\psi$ is finite at $r = 0$
(ii) $\psi$ and $\mathrm{d}\psi/\mathrm{d}r$ are continuous at $r = a$.
(iii) $\psi \to 0$ as $r \to \infty$

are of the form

$$\psi = \frac{A}{r} e^{-\alpha r} \qquad r > a \qquad \alpha = \sqrt{\frac{-2mE}{\hbar^2}}$$

$$\psi = \frac{B}{r} \sin \beta r \qquad r \leqslant a \qquad \beta = \sqrt{\frac{2m}{\hbar^2}(V_0 + E)}.$$

($m$ is the electron mass.)

Show that a necessary condition for a non-trivial solution is

$$\alpha a = -\beta a \cot(\beta a).$$

$\alpha$ and $\beta$ must also satisfy.

$$(\alpha a)^2 + (\beta a)^2 = \frac{2m}{\hbar^2} V_0 a^2.$$

Illustrate these conditions graphically.

Explain why bound solutions exist only if

$$\sqrt{\frac{2mV_0}{\hbar^2}}\, a > \frac{\pi}{2}.$$

(The spherically symmetric Laplacian is

$$\frac{1}{r^2}\frac{\partial}{\partial r}\left(r^2 \frac{\partial}{\partial r}\right).)$$

6 The potential in a one-dimensional problem is defined by

$$V = -V_0 \quad |x| \leqslant a, \qquad V = 0 \quad |x| > a, \qquad V_0 > 0.$$

The correct boundary conditions on the eigenfunctions are

(i) $\psi \to 0$ as $|x| \to \infty$ (Bound states).
(ii) $\psi$ and $d\psi/dx$ are continuous at $|x| = a$.

Solve the time-independent Schrödinger equation.

Show that if

$$\sqrt{\frac{2m|E|}{\hbar^2}}\, a \gg 1$$

then the simplified boundary conditions $\psi = 0$ at $|x| = a$ may be used. Verify that the eigenfunctions are either odd or even.

7 The potential inside a cylinder of radius $a$ and length $l$ is $-V_0 (V_0 > 0)$ and is zero outside.

The co-ordinate axes may be chosen so that the $z$-axis coincides with that of the cylinder with the origin at the centre of the lower plane end. Schrödinger's equation in cylindrical polars $(\rho, \phi, z)$ is

$$-\frac{\hbar^2}{2m}\left[\frac{1}{\rho}\frac{\partial}{\partial \rho}\left(\rho \frac{\partial}{\partial \rho}\right) + \frac{1}{\rho^2}\frac{\partial^2}{\partial \phi^2} + \frac{\partial^2}{\partial z^2}\right]\psi - [V_0 + E]\psi = 0$$

$$\rho < a \qquad 0 < z < l \qquad 0 < \phi < 2\pi.$$

For $|E|$ sufficiently large the boundary conditions reduce to $\psi = 0$ on the cylinder boundary. Use the method of separation of variables to show that the solutions finite and single-valued within the cylinder are of the form

$$\psi = A \sin \frac{n\pi z}{l} J_r(\alpha\rho) \begin{matrix} \cos r\phi \\ \sin r\phi \end{matrix}$$

where $n$ is a positive integer, $r$ is a positive integer or zero and $J_r(\alpha a) = 0$. Find the corresponding eigenvalues. (The Bessel function $J_r(x)$ is a solution of

$$\frac{dy^2}{dx^2} + \frac{1}{x}\frac{dy}{dx} + \left(1 - \frac{r^2}{x^2}\right)y = 0.)$$

8 Suppose two interacting particles have masses $m_1$ and $m_2$ and are at the positions $(x_1, y_1, z_1)$ and $(x_2, y_2, z_2)$ respectively, and suppose that the potential energy is a function of the distance between the particles.

The classical Lagrangian is

$$L = \frac{m_1}{2}(\dot{x}_1^2 + \dot{y}_1^2 + \dot{z}_1^2) + \frac{m_2}{2}(\dot{x}_2^2 + \dot{y}_2^2 + \dot{z}_2^2) - V.$$

Carry out the change of variables (4.46). Find the momenta conjugate to the six new co-ordinates and derive the classical Hamiltonian

$$H = \frac{1}{2(m_1 + m_2)}(P_x^2 + P_y^2 + P_z^2) + \frac{1}{2\mu}\left(P_r^2 + \frac{P_\theta^2}{r^2} + \frac{P_\phi^2}{r^2 \sin^2\theta}\right) + V(r)$$

where $\mu$ is the reduced mass (4.48).

9 The classical Hamiltonian for a rigid rotator is (see Chapter 1).

$$H(\theta, \phi, P_\theta, P_\phi) = \frac{1}{2I}\left(P_\theta^2 + \frac{1}{\sin^2\theta} P_\phi^2\right)$$

where $I$ is the moment of inertia. An alternative way of writing this is

$$H = \frac{1}{2I \sin^2\theta}[(\sin\theta P_\theta)^2 + P_\phi^2].$$

Carry out the quantum substitutions

$$p_j \to \frac{\hbar}{i}\frac{\partial}{\partial q_j}.$$

Which one of the above Hamiltonians gives the correct Hamiltonian operator used in (4.52)? Verify that the correct operator is Hermitian. Comment on the results of this question.

10 In the solution of the simple harmonic oscillator the wave function can be written in the form (4.30) where $H(y)$ satisfies Hermite's differential equation

$$H'' - 2yH' + (\gamma - 1)H = 0.$$

Use the method of solution in series to obtain the first three terms of the two linearly independent solutions.

The physically acceptable solution must be composed of a finite number of terms. Find this solution and explain why

$$\gamma = 1 + 2n, \qquad n = 0, 1, 2, \ldots$$

if $y^n$ is to be the highest power in the solution.

**References**

Kemble, E. C. *The fundamental principles of quantum mechanics with elementary applications.* Dover Publications, New York (1958).

Pauling, L. and Wilson, E. B. *Introduction to Quantum Mechanics.* McGraw-Hill, New York (1935).

Schutte, C. J. H. *The wave mechanics of atoms, molecules and ions.* Edward Arnold, London (1968).

Sneddon, I. N. *Special functions of mathematical physics and chemistry.* Oliver & Boyd, Edinburgh (2nd edition, 1961).

CHAPTER 5

# Quantum Mechanics II

## 5.1 Complete Description of a State

It has been mentioned that in a quantum mechanical system not all observables are simultaneously measurable. The $x$-co-ordinate of a particle and the $x$-component of its linear momentum cannot both be known exactly at the same instant. However there do exist sets of observables that can be measured simultaneously. For example it is possible to know simultaneously the energy and the total angular momentum of a particle in a central field.

The state of a system is known as completely as possible when the values of all the observables that can be simultaneously measured are known.

*Compatibility*

If two observables $A$ and $B$ are simultaneously measurable in a particular state then if either is measured, an unique result is obtained. This implies that the state function is an eigenfunction of the representative operators $\alpha$ and $\beta$, (Postulates four and five).

The observables $A$ and $B$ are said to be compatible if there exists a complete set of state functions, each of which is an eigenfunction of the corresponding operators.

The following two theorems illustrate the connection between compatible observables and commuting operators and lead naturally to a discussion of degeneracy.

*Theorem 1*

If the operators $\alpha$ and $\beta$ have a complete set of simultaneous eigenfunctions then the operators commute.

*Proof*

Let $\psi_i$ be one of the complete set of eigenfunctions of the operators $\alpha$ and $\beta$ with corresponding eigenvalues $a_i$ and $b_i$

$$\alpha\psi_i = a_i\psi_i \qquad \beta\psi_i = b_i\psi_i. \tag{5.1}$$

Since the set of simultaneous eigenfunctions $\{\psi_i\}$ is complete an arbitrary wave-function can be expanded

$$\psi = \sum_j c_j \psi_j. \tag{5.2}$$

Consider the effect of the commutator $\alpha\beta - \beta\alpha$ on $\psi$.

$$(\alpha\beta - \beta\alpha)\psi = (\alpha\beta - \beta\alpha)\sum_j c_j\psi_j. \tag{5.3}$$

From (5.1) this can be written

$$\begin{aligned}(\alpha\beta - \beta\alpha)\psi &= \sum_j c_j(b_j a_j - a_j b_j)\psi_j \\ &= 0.\end{aligned} \tag{5.4}$$

Since $\psi$ is arbitrary then

$$(\alpha\beta - \beta\alpha) \equiv 0. \tag{5.5}$$

The commutator representing the co-ordinate $q_i$ and its conjugate momentum $p_i$ is

$$\left[q_i, \frac{\hbar}{i}\frac{\partial}{\partial q_i}\right] \equiv i\hbar \tag{5.6}$$

and clearly $q_i$ and $p_i$ are not compatible.

*Theorem 2*

If the operators $\alpha$ and $\beta$ commute, there exists a complete set of functions which are simultaneously the eigenfunctions for both $\alpha$ and $\beta$. This implies that the observables $A$ and $B$ are compatible and this theorem is the converse of theorem one.

*Proof*

Consider first the case when the operator $\alpha$ has non-degenerate eigenvalues. It will be shown that every eigenfunction of $\alpha$ is also an eigenfunction of $\beta$ if $\alpha$ and $\beta$ commute. Let $\psi_i$ be one of the complete set of non-degenerate eigenfunctions for the operator $\alpha$,

$$\alpha\psi_i = a_i\psi_i. \tag{5.7}$$

Operate on this equation with $\beta$ and as $\alpha$ and $\beta$ commute

$$\beta\alpha\psi_i = a_i(\beta\psi_i),$$

$$\alpha(\beta\psi_i) = a_i(\beta\psi_i). \tag{5.8}$$

The function $\beta\psi_i$ is an eigenfunction of $\alpha$ with eigenvalue $a_i$. As the eigenstates of $\alpha$ are by assumption non-degenerate then clearly

$$\beta\psi_i = (\text{constant})\,\psi_i. \tag{5.9}$$

This implies that $\psi_i$ is an eigenfunction of $\beta$ and the result is proved.

Suppose now that $\alpha$ has degenerate eigenvalues. Let $\psi_i$ be one of the $n$ linearly independent eigenfunctions spanning the $n$-dimensional sub-space associated with the degenerate eigenvalue $a$.

$$\alpha\psi_i = a\psi_i \qquad i = 1, \ldots, n. \tag{5.10}$$

Let this sub-space be $V(a)$. Any eigenfunction of $\alpha$ with eigenvalue $a$ must lie in this sub-space and vice-versa. Operate on (5.10) with $\beta$ and as $\alpha$ and $\beta$ commute

$$\alpha(\beta\psi_i) = a(\beta\psi_i) \qquad i = 1, \ldots, n \tag{5.11}$$

$(\beta\psi_i)$ is an eigenfunction of $\alpha$ with eigenvalue $a$ and must belong to $V(a)$.

$$\beta\psi_i = \sum_{j=1}^{n} a_{ij}\,\psi_j. \tag{5.12}$$

That is, $V(a)$ is 'invariant' under $\beta$ and this is true for each eigenvalue sub-space of $\alpha$.

Since the eigenfunctions of $\alpha$ form a complete set then any eigen-function of $\beta$, with eigenvalue $b$ say, can be expanded

$$\sum_{i=1}^{n} k_i\psi_i + \phi \tag{5.13}$$

where the sum represents a function in $V(a)$ and $\phi$ is a linear combination of the eigenfunctions of $\alpha$ outside $V(a)$. Since each eigenvalue sub-space of $\alpha$ is invariant under $\beta$ then the linearly independent functions

$$\chi = \sum_{i=1}^{n} k_i\psi_i \tag{5.14}$$

and $\phi$ are each eigenfunctions of $\beta$.

$$\beta\phi = b\phi \tag{5.15}$$

$$\beta\chi = b\chi. \tag{5.16}$$

From (5.14) $\chi$ is simultaneously an eigenfunction of $\alpha$ with

$$\alpha\chi = a\chi. \tag{5.17}$$

The function $\phi$ can be further decomposed into its linearly independent components in the other eigenvalue sub-spaces of $\alpha$, although some of these components could be zero. So each of the eigenfunctions of $\beta$ can be constructed from a single eigenvalue sub-space of $\alpha$. Since $\beta$ is an operator representing an observable its eigenfunctions form a complete set and the theorem is proved.

In particular there must be $n$ simultaneous eigenfunctions of $\alpha$ and $\beta$ that span $V(a)$. They all correspond to the eigenvalue $a$ of $\alpha$ but in general will belong to different eigenvalues of $\beta$.

As an illustration, consider the operators representing the three cartesian components of linear momentum of a particle.

$$\frac{\hbar}{i}\frac{\partial}{\partial x}, \qquad \frac{\hbar}{i}\frac{\partial}{\partial y}, \qquad \frac{\hbar}{i}\frac{\partial}{\partial z}. \tag{5.18}$$

These operators commute with one another and by theorem two there must be a complete set of functions which are simultaneously eigenfunctions of all three. The eigenfunctions are

$$\psi(x, y, z) = \exp\left\{\frac{i}{\hbar}(p_x x + p_y y + p_z z)\right\} \quad -\infty < \begin{matrix} p_x \\ p_y \\ p_z \end{matrix} < \infty \tag{5.19}$$

where $p_x, p_y, p_z$ are the eigenvalues.

*Degeneracy*

For a free particle the wave-function (5.19) satisfies Schrödinger's time-independent equation and is a possible state function. The values of the momentum components $p_x, p_y, p_z$ determine the wave-function, apart from the usual arbitrary wave factor, and the three compatible momentum observables form what is called a complete set.

The commutation of operators is closely related to the occurrence of degeneracies. In proving theorem two it was shown that from the set of eigenfunctions $\psi_1, \ldots, \psi_n$ of $\alpha$ which span the $n$-dimensional sub-space of the degenerate eigenvalue $a$, it is possible to construct $n$ linearly independent functions

$$\chi^j = \sum_{i=1}^{n} k_i^j \psi_i \qquad j = 1, \ldots, n \tag{5.20) [(5.14)}$$

which are simultaneously eigenfunctions of $\alpha$ and $\beta$ if the commutator $[\alpha, \beta]$ is zero.

$$\alpha\chi^j = a\chi^j \tag{5.21)[(5.17)}$$

$$\beta\chi^j = b^j\chi^j. \tag{5.22)[(5.16)}$$

The superscript $j$ is introduced to distinguish the $n$-eigenvalues $b^j$.

As the operators $\alpha, \beta$ commute, the associated observables $A$ and $B$ are compatible and may be measured simultaneously. If on such a measurement the values $a, b^j$ are obtained then immediately after, the state function is essentially $\chi^j$. If the eigenvalue $b^j$ is non-degenerate the wave-function is determined completely. On the other hand, if $b^j$ is, say, $m$-fold degenerate $(m \leqslant n)\chi^j$ must be a function in the associated $m$-dimensional sub-space. In this case, experience shows that another observable $C$ may be found whose operator $\gamma$ commutes with both $\alpha$ and $\beta$ and so may also be measured. After this measurement is carried out the wave-function is specified more completely. In this way the dimensionality of the sub-space, in which the wave-function is known to be, is reduced. More compatible observables may be found and measured until the wave-function is specified completely.

Eventually a 'complete set' of commuting observables is found so that when the eigenvalues of each are specified, the wave-function is known completely, apart from a phase factor. No other observable can be measured unless it is a function of those already in the set.

For a given system there may be several distinct complete sets of observables. All the members of one such set do not commute with all the other members of any other set although they may have some members in common.

There is a very interesting alternative but equivalent way of discussing degeneracies. This involves a study of the symmetry properties of the system and is dealt with in Chapter 9.

### *Angular momentum quantum numbers*

It was shown in section 4.3 that the motion of an electron in the field of a nucleus is essentially described by the Schrödinger equation

$$-\frac{\hbar^2}{2\mu}\nabla^2\psi + V(r)\psi = E\psi, \qquad (5.23)[(4.61)]$$

where the nucleus is at the origin, $\mu$ is the reduced electron mass and $E$ is the internal energy of the system. The potential of interaction is spherically symmetric and is a function of $r$ only. It will be shown below that in this central field problem three observables form a complete set.

Classically, the angular momentum about the origin of a particle with linear momentum $\mathbf{p}$ at the position $\mathbf{r}$ is

$$\mathbf{L} = \mathbf{r} \wedge \mathbf{p}. \qquad (5.24)$$

The $z$-component of this orbital angular momentum is

$$L_z = xp_y - yp_x. \qquad (5.25)$$

The quantum mechanical operator $\mathscr{L}_z$ representing this observable is obtained using the substitutions (3.8).

$$\mathscr{L}_z \equiv \frac{\hbar}{i}\left(x\frac{\partial}{\partial y} - y\frac{\partial}{\partial x}\right). \tag{5.26}$$

(As there are no non-commuting factors in (5.26) there is no ambiguity.) Similarly, operators may be obtained for the other components.

$$\mathscr{L}_x \equiv \frac{\hbar}{i}\left(y\frac{\partial}{\partial z} - z\frac{\partial}{\partial y}\right)$$

$$\mathscr{L}_y \equiv \frac{\hbar}{i}\left(z\frac{\partial}{\partial x} - x\frac{\partial}{\partial z}\right). \tag{5.27}$$

It is easily shown that the operators (5.26), (5.27), do not commute with one another and so the observables $L_x, L_y, L_z$ are not compatible (Theorem 1). For example

$$\mathscr{L}_x\mathscr{L}_y - \mathscr{L}_y\mathscr{L}_x \equiv \left(\frac{\hbar}{i}\right)^2\left[\left(y\frac{\partial}{\partial z} - z\frac{\partial}{\partial y}\right)\left(z\frac{\partial}{\partial x} - x\frac{\partial}{\partial z}\right)\right.$$
$$\left. - \left(z\frac{\partial}{\partial x} - x\frac{\partial}{\partial z}\right)\left(y\frac{\partial}{\partial z} - z\frac{\partial}{\partial y}\right)\right]$$
$$\equiv -\hbar^2\left(y\frac{\partial}{\partial x} - x\frac{\partial}{\partial y}\right)$$

i.e.

$$\mathscr{L}_x\mathscr{L}_y - \mathscr{L}_y\mathscr{L}_x \equiv i\hbar\mathscr{L}_z. \tag{5.28}$$

Similarly,

$$\mathscr{L}_y\mathscr{L}_z - \mathscr{L}_z\mathscr{L}_y \equiv i\hbar\mathscr{L}_x$$
$$\mathscr{L}_z\mathscr{L}_x - \mathscr{L}_x\mathscr{L}_z \equiv i\hbar\mathscr{L}_y. \tag{5.29}$$

However, all three angular momentum operators do commute with the Hamiltonian of (5.23). In particular, if all derivatives concerned are continuous,

$$\mathscr{L}_z\nabla^2 - \nabla^2\mathscr{L}_z \equiv 0 \tag{5.30}$$

and

$$\mathscr{L}_zV(r) \equiv \frac{\hbar}{i}\left(x\frac{\partial}{\partial y} - y\frac{\partial}{\partial x}\right)V(r)$$

$$\equiv \frac{\hbar}{i}\left[x\frac{\mathrm{d}V}{\mathrm{d}r}\frac{y}{r} - y\frac{\mathrm{d}V}{\mathrm{d}r}\frac{x}{r}\right] + V(r)\frac{\hbar}{i}\left(x\frac{\partial}{\partial y} - y\frac{\partial}{\partial x}\right)$$

i.e.

$$\mathscr{L}_z V(r) - V(r)\mathscr{L}_z \equiv 0. \tag{5.31}$$

As the Hamiltonian operator is

$$\mathscr{H} \equiv -\frac{\hbar^2}{2\mu}\nabla^2 + V(r)$$

then

$$\mathscr{L}_z \mathscr{H} - \mathscr{H}\mathscr{L}_z \equiv 0. \tag{5.32}$$

From theorem two the sets of observables $(H, L_x)$, $(\mathrm{H}, L_y)$, $(H, L_z)$ are compatible.

The magnitude of the total angular momentum is

$$L^2 = L_x^2 + L_y^2 + L_z^2 \tag{5.33}$$

and the operator representing this observable is

$$\mathscr{L}^2 \equiv \mathscr{L}_x^2 + \mathscr{L}_y^2 + \mathscr{L}_z^2. \tag{5.34}$$

As $\mathscr{L}_x$, $\mathscr{L}_y$, $\mathscr{L}_z$ all commute with $\mathscr{H}$ then so does $\mathscr{L}^2$, e.g.

$$\mathscr{L}_z^2 \mathscr{H} \equiv \mathscr{L}_z(\mathscr{H}\mathscr{L}_z) \equiv \mathscr{H}\mathscr{L}_z^2$$

$$\therefore \quad \mathscr{L}^2\mathscr{H} - \mathscr{H}\mathscr{L}^2 \equiv 0. \tag{5.35}$$

Similarly $\mathscr{L}_x$, $\mathscr{L}_y$, $\mathscr{L}_z$ all commute with $\mathscr{L}^2$, e.g.

$$\begin{aligned}\mathscr{L}_z\mathscr{L}^2 &\equiv \mathscr{L}_z\mathscr{L}_x^2 + \mathscr{L}_z\mathscr{L}_y^2 + \mathscr{L}_z\mathscr{L}_z^2 \\ &\equiv (i\hbar\mathscr{L}_y + \mathscr{L}_x\mathscr{L}_z)\mathscr{L}_x + (-i\hbar\mathscr{L}_x + \mathscr{L}_y\mathscr{L}_z)\mathscr{L}_y + \mathscr{L}_z^3 \\ &\equiv i\hbar(\mathscr{L}_y\mathscr{L}_x - \mathscr{L}_x\mathscr{L}_y) + \mathscr{L}_x(\mathscr{L}_x\mathscr{L}_z + i\hbar\mathscr{L}_y) \\ &\quad + \mathscr{L}_y(\mathscr{L}_y\mathscr{L}_z - i\hbar\mathscr{L}_x) + \mathscr{L}_z^3 \\ &\equiv \mathscr{L}_x^2\mathscr{L}_z + \mathscr{L}_y^2\mathscr{L}_z + \mathscr{L}_z^3\end{aligned}$$

i.e.

$$\mathscr{L}_z\mathscr{L}^2 - \mathscr{L}^2\mathscr{L}_z \equiv 0. \tag{5.36}$$

Clearly, then the sets of observables

$$\begin{aligned}&H, L^2, L_x \qquad \text{(i)} \\ &H, L^2, L_y \qquad \text{(ii)} \\ &H, L^2, L_z \qquad \text{(iii)}\end{aligned} \tag{5.37}$$

are compatible and in fact each is complete.

As in section 4.3 it is useful to work in spherical polar co-ordinates defined by

$$x = r\sin\theta\cos\phi \qquad y = r\sin\theta\sin\phi \qquad z = r\cos\theta. \tag{5.38}$$

The operators (5.26), (5.27) and (5.34) become

$$\mathscr{L}_x \equiv -\frac{\hbar}{i}\left(\sin\phi\frac{\partial}{\partial\theta} + \cot\theta\cos\phi\frac{\partial}{\partial\phi}\right)$$

$$\mathscr{L}_y \equiv -\frac{\hbar}{i}\left(-\cos\phi\frac{\partial}{\partial\theta} + \cot\theta\sin\phi\frac{\partial}{\partial\phi}\right)$$

$$\mathscr{L}_z \equiv \frac{\hbar}{i}\frac{\partial}{\partial\phi} \tag{5.39}$$

$$\mathscr{L}^2 \equiv -\hbar^2\left\{\frac{1}{\sin\theta}\frac{\partial}{\partial\theta}\left(\sin\theta\frac{\partial}{\partial\theta}\right) + \frac{1}{\sin^2\theta}\frac{\partial^2}{\partial\phi^2}\right\}. \tag{5.40}$$

The eigenfunctions of the Hamiltonian for central field problems are essentially

$$\psi_{nlm_l}(r, \theta, \phi) = R_{n_l}(r)P_l^{m_l}(\cos\theta)e^{im_l\phi}. \qquad (5.41)\ [(4.72)]$$

(In the special case of a Coulomb potential the radial part may be identified as a Laguerre function (4.70), and the energy depends only upon $n$).

Clearly these energy eigenfunctions are also eigenfunctions of the $z$-momentum operator $\mathscr{L}_z$ with eigenvalues $\hbar m_l$. $m_l$ is called the magnetic quantum number. Similarly by comparing (5.40) with (4.64) it is seen that $\psi_{nlm_l}$ is an eigenfunction of the total angular momentum operator $\mathscr{L}^2$ with eigenvalue $\hbar^2 l(l+1)$. $l$ is called the angular momentum number. Measurement of the observables $L^2$ and $L_z$ determines the angular part of the wave-function and measurement of the energy determines the radial part.

Observe that $\psi_{nlm_l}$ is not an eigenfunction of the operators $\mathscr{L}_x$, $\mathscr{L}_y$ except in the case that the quantum number $l = 0$. Then the wave-function depends upon $r$ only and $\mathscr{L}_x$, $\mathscr{L}_y$, $\mathscr{L}_z$ have the common eigenvalue zero. (Note that the commutation rules (5.28), (5.29) are satisfied.)

## 5.2 Constants of the Motion

It is well known that there are important conservation laws in classical mechanics. The rate of change of a classical variable $F(q_1, \ldots, q_n, p_1, \ldots p_n, t)$ is given by

$$\frac{\mathrm{d}F}{\mathrm{d}t} = \{F, H\} + \frac{\partial F}{\partial t} \tag{1.34}$$

where $\{F, H\}$ is the Poisson bracket of $F$ and the Hamiltonian. When $H$ does not depend explicitly on the time $\mathrm{d}H/\mathrm{d}t = 0$ and the Hamiltonian is a constant of the motion. Any time independent variable of the system whose Poisson bracket with the Hamiltonian vanishes, is conserved.

Another example is the conservation of angular momentum of a particle moving in a central force field. The Hamiltonian is

$$H = \frac{1}{2m}(p_x^2 + p_y^2 + p_z^2) + V(r) \tag{5.42}$$

where the potential energy is spherical symmetric when the origin is suitably chosen. The $z$-component of angular momentum is

$$L_z = xp_y - yp_x$$

and it is a simple matter to show that the Poisson bracket

$$\{L_z, H\} = 0, \tag{5.43}$$

and so $L_z$ is a constant of the motion. A similar argument applies for $L_x$ and $L_y$ and consequently the vector angular momentum is conserved.

There are analogous results in quantum mechanics. In a given state described by a normalized wave-function $\Psi(q_1, \ldots q_n, t)$ a general observable does not have a definite value. The expectation value of the observable, $A$ is given by

$$\langle A \rangle = \int \Psi^* \alpha \Psi \, \mathrm{d}\tau \tag{3.21}$$

where $\alpha$ is the representative operator. This expectation value may not be constant in time and the rate at which it changes is given by

$$\frac{\mathrm{d}}{\mathrm{d}t}\langle A \rangle = \int \left[ \Psi^* \frac{\partial \alpha}{\partial t} \Psi + \frac{\partial \Psi^*}{\partial t} \alpha \Psi + \Psi^* \alpha \frac{\partial \Psi}{\partial t} \right] \mathrm{d}\tau. \tag{5.44}$$

The state function must satisfy Schrödinger's time-dependent equation.

$$\frac{\partial \Psi}{\partial t} = -\frac{i}{\hbar}\mathscr{H}\Psi \qquad \frac{\partial \Psi^*}{\partial t} = +\frac{i}{\hbar}(\mathscr{H}\Psi)^* \tag{5.45)[(3.10)]$$

and as the Hamiltonian operator is Hermitian

$$\int (\mathscr{H}\Psi)^*(\alpha\Psi)\,\mathrm{d}\tau = \int \Psi^* \mathscr{H}\alpha\Psi\,\mathrm{d}\tau. \tag{5.46}$$

Substitution of (5.45) and (5.46) into (5.44) gives

$$\frac{\mathrm{d}}{\mathrm{d}t}\langle A \rangle = \int \Psi^* \left\{ \frac{\partial \alpha}{\partial t} + \frac{i}{\hbar}[\mathscr{H}\alpha - \alpha\mathscr{H}] \right\} \Psi \, \mathrm{d}\tau. \tag{5.47}$$

This equation is the quantum analogue of (1.34) and may be used to define an operator representing the time dervative of $A$,

$$\frac{\partial \alpha}{\partial t} + \frac{i}{\hbar}(\mathscr{H}\alpha - \alpha\mathscr{H}). \tag{5.48}$$

This illustrates a point already mentioned (3.6) that in the transition from classical mechanics to quantum mechanics Poïsson brackets are replaced by commutators.

If the operator $\alpha$ does not depend explicitly on the time and if it also commutes with the Hamiltonian of the system

$$\frac{d}{dt}\langle A\rangle = 0$$

and the expectation value of the observable $A$ is a constant of the motion, $A$ is said to conserved. In particular, if the state function is also an eigenfunction of $\alpha$, then $A$ has the definite value equal to the eigenvalue in that state. The state function is an eigenfunction of any given representative operator immediately after a measurement of the appropriate observable. However the state function will remain an eigenfunction only if the observable is conserved and the associated quantum numbers are called 'good quantum numbers'.

For most systems the Hamiltonian does not depend explicitly on the time and its expectation value is a constant.

If the state function is an eigenfunction of the Hamiltonian operator at some instant it will remain so. Such a state is called a stationary state and the wave-function has the form

$$\Psi(q_1, \ldots, q_n, t) = \psi(q_1, \ldots, q_n)e^{-iEt/\hbar} \tag{5.49}$$

It is always possible to measure conserved observables simultaneously with the Hamiltonian. In the central field problem the observables ($H$, $L^2$, $L_z$) form a complete set with values which are constants of the motion. The quantum numbers $n$, $l$, $m_l$ are good quantum numbers.

## 5.3 Uncertainty principle

Consider two observables $A$ and $B$ with representative operators $\alpha$ and $\beta$. If $\alpha$ and $\beta$ commute both observables may be measured simultaneously. If the operators $\alpha$, $\beta$ do not commute they do not have a complete set of simultaneous eigenfunctions, (Theorem one). Define $\gamma$ by

$$(\alpha\beta - \beta\alpha) \equiv i\gamma. \tag{5.50}$$

There are two possible cases.

(i) If $\gamma$ is a Hermitian operator it may be possible to measure $A$ and $B$ simultaneously in certain states. If $\psi_i$ is a simultaneous eigenfunction of $\alpha$ and $\beta$ then it must also be an eigenfunction of $\gamma$ with zero eigenvalue. Even though there may be an infinite number of such eigenfunctions they cannot form a complete set.

(ii) If $\gamma$ is a real constant then

$$\gamma\psi_i = 0$$

has no non-trivial solutions and $\alpha, \beta$ do not have any eigenfunctions in common. A well-known case is that of the position co-ordinate and conjugate momentum

$$(q_i\hat{p}_i - \hat{p}_i q_i) \equiv i\hbar.$$

The position and momentum of a particle can never be known simultaneously. Any measurement of the position must disturb the momentum and vice-versa.

Even so, it is possible to obtain a relation between the uncertainties in the measurements.

*Schwarz inequality*

If $\psi_1$ and $\psi_2$ are two quadratically integrable functions then

$$\left[\int \psi_1^*\psi_1 \, d\tau\right]\left[\int \psi_2^*\psi_2 \, d\tau\right] \geqslant \left|\int \psi_2^*\psi_1 \, d\tau\right|^2. \tag{5.51}$$

This is analogous to the result

$$|\mathbf{A}|^2 |\mathbf{B}|^2 \geqslant |\mathbf{A} \cdot \mathbf{B}|^2$$

where **A, B** are vectors in three-dimensional Euclidean space. Hilbert space is a unitary space and the integral $\int \psi_2^*\psi_1 \, d\tau$ represents the 'scalar' product of $\psi_2^*$ and $\psi_1$.

*Proof*

Let $A$ be a real constant defined such that the function $\chi_2 = A\psi_2$ is normalized to unity.

$$A^2 = 1\Big/\int |\psi_2|^2 \, d\tau.$$

Consider the function

$$\Phi = \psi_1 - \chi_2 \int \chi_2^*\psi_1 \, d\tau. \tag{5.52}$$

Since $\int |\Phi|^2 \, d\tau \geqslant 0$ then

$$\int |\psi_1|^2 \, d\tau - 2\left|\int \chi_2^*\psi_1 \, d\tau\right|^2 + \left|\int \chi_2^*\psi_1 \, d\tau\right|^2 \geqslant 0$$

i.e.

$$\int |\psi_1|^2 \, d\tau \geqslant \left|\int \chi_2^*\psi_1 \, d\tau\right|^2.$$

Substitute for $\chi_2$ to obtain

$$\int |\psi_1|^2 \, d\tau \int |\psi_2|^2 \, d\tau \geqslant \left|\int \psi_2^*\psi_1 \, d\tau\right|^2. \tag{5.53}$$

*Uncertainty relation*

When a system is in a state described by the normalized function $\psi$ the uncertainty $(\Delta A)$ in a measurement of the observable $A$ is defined by

$$(\Delta A)^2 = \int \psi^* [\alpha - \langle A \rangle]^2 \psi \, \mathrm{d}\tau \tag{5.54}$$

where the expectation value of $A$ is defined as usual by

$$\langle A \rangle = \int \psi^* \alpha \psi \, \mathrm{d}\tau$$

(5.54) may be expressed in another form by expanding $\psi$ in terms of the eigenfunctions of $\alpha$. Writing

$$\psi = \sum_j c_j \psi_j \qquad \text{with} \qquad \alpha \psi_i = a_i \psi_i$$

$$(\Delta A)^2 = \sum_i |c_i|^2 [a_i - \langle A \rangle]^2$$

$(\Delta A)^2$ is the variance of the possible results of measuring $A$.

In the Schwarz inequality write

$$\psi_1 = [\alpha - \langle A \rangle] \psi \qquad \text{and} \qquad \psi_2 = [\beta - \langle B \rangle] \psi$$

where $\beta$ is the operator representing the observable $B$. The integral

$$\int |\psi_1|^2 \, \mathrm{d}\tau = \int [(\alpha - \langle A \rangle)\psi]^* [(\alpha - \langle A \rangle)\psi] \, \mathrm{d}\tau$$

$$= \int \psi^* (\alpha - \langle A \rangle)^2 \psi \, \mathrm{d}\tau$$

since $(\alpha - \langle A \rangle)$ is Hermitian, i.e.

$$\int |\psi_1|^2 \, \mathrm{d}\tau = (\Delta A)^2.$$

Similarly

$$\int |\psi_2|^2 \, \mathrm{d}\tau = (\Delta B)^2.$$

Since the square of the modulus of a complex quantity is greater or equal to the square of its imaginary part then

$$\left| \int \psi_2^* \psi_1 \, \mathrm{d}\tau \right|^2 \geqslant \tfrac{1}{4} \left| \int \psi_2^* \psi_1 \, \mathrm{d}\tau - \int \psi_1^* \psi_2 \, \mathrm{d}t \right|^2$$

$$\geqslant \tfrac{1}{4} \left| \int \psi^* (\alpha\beta - \beta\alpha) \psi \, \mathrm{d}\tau \right|^2.$$

The Schwarz inequality (5.51) gives

$$(\Delta A)^2 (\Delta B)^2 \geqslant \tfrac{1}{4} \left| \int \psi^* (\alpha\beta - \beta\alpha) \psi \, \mathrm{d}\tau \right|^2. \tag{5.55}$$

If the state function $\psi$ is an eigenfunction of both $\alpha$ and $\beta$ then

$$(\Delta A)^2 (\Delta B)^2 = 0$$

as expected.

In the more general case when $[\alpha, \beta] = i\gamma$

$$(\Delta A)(\Delta B) \geqslant \tfrac{1}{2}\left|\int \psi^* \gamma \psi \, d\tau\right|. \tag{5.56}$$

This is a general statement of the Heisenberg uncertainty relation. In particular, the uncertainty relation for the momentum and position of a particle is

$$(\Delta q_i)(\Delta p_i) \geqslant \frac{\hbar}{2}. \tag{5.57}$$

This confirms a point already mentioned in Chapter 4 that when $\hbar$ can be neglected, classical mechanics can be used.

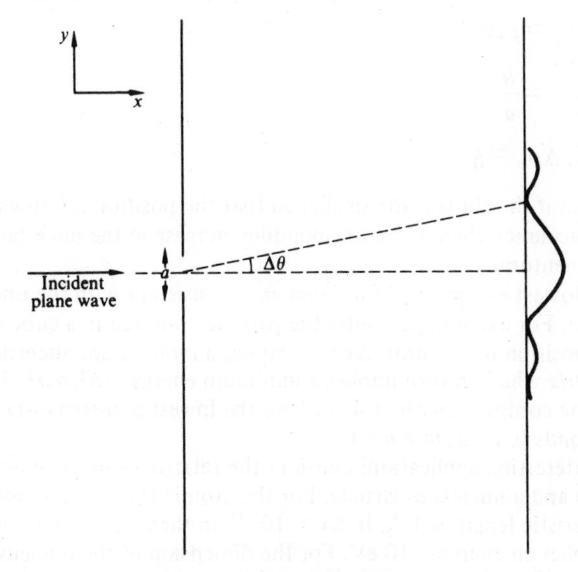


Figure 5.1 Illustration of the uncertainty principle.

There are several simple experiments which illustrate the uncertainty principle. Suppose a particle with a known momentum is incident normal to a screen which contains a single slit. The wave-function of the particle is the plane wave

$$\psi = e^{ikx}$$

and the momentum is $p_x = \hbar k$. (The co-ordinate axes are shown in (5.1).) Before reaching the screen, the position of the particle is completely uncertain but the component of momentum parallel to the screen is known to be zero. After passing through the slit the $y$-co-ordinate is known with an uncertainty essentially equal to the slit width $a$, i.e.

$$\Delta y = a.$$

In discovering this information about the particle position an element of uncertainty is introduced into the conjugate component of momentum $p_y$. After passing through the slit the 'particle wave' undergoes diffraction and there is a finite probability for the particle to be deflected through an angle $\Delta\theta$. It is well known that the first minimum in the diffraction pattern occurs when

$$\Delta\theta \simeq \frac{\lambda}{a}.$$

$\lambda$ is the 'wave-length' of the particle wave and is equal to $h/p$. A measure of the uncertainty in the momentum is clearly

$$\begin{aligned}\Delta p_y &\simeq p\Delta\theta \\ &\simeq \frac{h}{a}\end{aligned}$$

$$\therefore \quad \Delta y \Delta p_y \simeq h. \tag{5.58}$$

Note that if the slit is made smaller so that the position is known with greater accuracy there is a corresponding increase in the uncertainty of the momentum.

The lowest energy level for a system is consistent with the uncertainty principle. For example, consider the particle confined in a cube of side $a$. The position uncertainty $\Delta x \sim a$ implies a momentum uncertainty $\Delta p_x \sim h/a$ which in turn implies a minimum energy $\sim h^2/ma^2$. This should be compared with (4.43) where the lowest non-trivial-state corresponds to $l = 1, m = n = 0$.

As interesting applications consider the relative energies of an atomic electron and a nuclear $\alpha$-particle. For the atomic electron a sensible characteristic length is 1 Å. If $\Delta x \sim 10^{-10}$ m then $\Delta p_x \sim 10^{10}\, h$ and this implies an energy $\sim$10 eV. For the dimension of the nucleus take $\Delta x \sim 10^{-14}$ m then $\Delta p_x \sim 10^{14}\, h$ and this implies an energy $\sim$1 MeV. Atomic electrons do indeed have energies sensibly measured in terms of eV and the kinetic energy of emitted $\alpha$-particles is of the order of MeV.

## 5.4 Wave Packets

A 'wave packet' is a wave-function that attempts to give an almost classical description to a particle so that it is possible to give some meaning even if uncertain, to the momentum and position simultaneously. The one-dimensional wave-function $B \exp\,[i(kx - \omega t)]$ represents a particle with momentum $\hbar k$ and energy $\hbar\omega$. The momentum eigenvalues form a continuous spectrum and a superposition of them is represented by an integral.

$$\Psi(x, t) = \int_{-\infty}^{\infty} B(k)\, e^{i(kx - \omega t)}\, dk \tag{5.59}$$

$|B(k)|^2\, \mathrm{d}k$ represents the probability that a measurement of the momentum will give a value such that

$$\hbar\left(k - \frac{\mathrm{d}k}{2}\right) < p < \hbar\left(k + \frac{\mathrm{d}k}{2}\right).$$

If the wave-function $\Psi(x, t)$ is to represent a particle with a characteristic momentum it is necessary that $B(k)$ be vanishingly small except for a small range of values of $k$.

Suppose $B(k)$ has an appreciable value only near $k = k_0$. The energy and hence the 'frequency' $\omega$ are functions of $k$ and may be expanded in Taylor series about $k_0$

$$\omega(k) = \omega_0 + (k - k_0)\frac{\mathrm{d}\omega}{\mathrm{d}k_0} + \frac{1}{2}(k - k_0)^2 \frac{\mathrm{d}^2\omega}{\mathrm{d}k_0^2} + \ldots \qquad k \sim k_0$$

where

$$\omega_0 = \omega(k_0), \frac{\mathrm{d}\omega}{\mathrm{d}k_0} = \left(\frac{\mathrm{d}\omega}{\mathrm{d}k}\right)_{k_0} \qquad \text{and} \qquad \frac{\mathrm{d}^2\omega}{\mathrm{d}k_0^2} = \left(\frac{\mathrm{d}^2\omega}{\mathrm{d}k^2}\right)_{k_0}. \tag{5.60}$$

The wave-functions becomes

$$\Psi(x, t) \simeq e^{i(k_0 x - \omega_0 t)} \int_{-\infty}^{\infty} B(k) \exp\left[i(k - k_0)\left(x - \frac{\mathrm{d}\omega}{\mathrm{d}k_0} t\right) - \frac{i}{2}(k - k_0)^2 \frac{\mathrm{d}^2\omega}{\mathrm{d}k_0^2} t\right] \mathrm{d}k. \tag{5.61}$$

If $\omega(k)$ is a linear function $\mathrm{d}^2\omega/\mathrm{d}k_0^2 = 0$ and the wave packet does not disperse but travels without change of shape. In this case defining

$$f(x) = \int_{-\infty}^{\infty} B(k)\, e^{i(k - k_0)x}\, \mathrm{d}k \tag{5.62}$$

then

$$\Psi(x, t) \simeq f\left(x - \frac{\mathrm{d}\omega}{\mathrm{d}k_0} t\right) e^{i(k_0 x - \omega_0 t)}. \tag{5.63}$$

In particular

$$\Psi(x, 0) = f(x)\, e^{ik_0 x}$$

Since $B(k)$ is appreciable only in the neighbourhood of $k = k_0$, $f(x)$ must be a relatively slowly varying function of $x$ and is non-zero in the neighbourhood of $x = 0$. Equation (5.63) then represents a plane wave with a modulated amplitude

$$f\left(x - \frac{\mathrm{d}\omega}{\mathrm{d}k} t\right)$$

and the 'wave packet' moves with a group velocity

$$V_g = \frac{d\omega}{dk}. \tag{5.64}$$

For a free particle

$$E = \frac{\hbar^2 k^2}{2m} \qquad \text{and so} \qquad \frac{d\omega}{dk_0} = \frac{\hbar k_0}{m} = \frac{p_0}{m}.$$

The packet velocity is equal to the velocity corresponding to the mean value of the wave number $k_0$.

The wave packet description makes it possible to associate a mean position and momentum with a particle but the finite size of the packet makes it impossible to specify the position exactly and the spread in the $k$-values needed to construct the packet make it impossible to specify the momentum exactly. This is in accordance with the uncertainty principle.

The above treatment is readily extended to three dimensions and the wave packet is then

$$\Psi(\mathbf{r}, t) = \iiint_{-\infty}^{\infty} B(\mathbf{k})\, e^{i(\mathbf{k}\,.\,\mathbf{r} - \omega t)}\, d\mathbf{k} \tag{5.65}$$

with an associated velocity

$$\mathbf{V}_g = \text{grad}_{\mathbf{k}}\, \omega(k). \tag{5.66}$$

*Ehrenfest's theorem*

In 1927 Ehrenfest showed that the motion of a wave packet obeys Newton's laws. He showed that if $\langle x \rangle$ and $\langle p \rangle$ are the average values of the position and momentum for the wave packet then

$$m \frac{d}{dt} \langle x \rangle = \langle p_x \rangle \tag{5.67}$$

and if $V$ is the potential energy of the particle

$$\frac{d}{dt} \langle p_x \rangle = -\left\langle \frac{\partial V}{\partial x} \right\rangle. \tag{5.68}$$

*Proof*

Let $\Psi(\mathbf{r}, t)$ be the state function describing the motion of a particle. Assume it is quadratically integrable and normalized to unity. The expectation value of the $x$-co-ordinate is defined by

$$\langle x \rangle = \int \Psi^* x \Psi \, d\tau \tag{5.69}$$

and the average value of the $x$-component of the velocity is (5.48)

$$\frac{\mathrm{d}}{\mathrm{d}t}\langle x\rangle = \frac{i}{\hbar}\int \Psi^*[\mathscr{H}x - x\mathscr{H}]\Psi\,\mathrm{d}\tau.$$

The Hamiltonian operator

$$\mathscr{H} \equiv -\frac{\hbar^2}{2m}\nabla^2 + V(\mathbf{r})$$

and so

$$\frac{\mathrm{d}}{\mathrm{d}t}\langle x\rangle = -\frac{i\hbar}{2m}\int [\Psi^*(\nabla^2 x\Psi) - \Psi^* x\nabla^2\Psi]\,\mathrm{d}\tau$$

$$= -\frac{i\hbar}{2m}\int \Psi^*\left[x\nabla^2\Psi + 2\frac{\partial\Psi}{\partial x} - x\nabla^2\Psi\right]\mathrm{d}\tau$$

$$= \frac{1}{m}\int \Psi^*\left(\frac{\hbar}{i}\frac{\partial\Psi}{\partial x}\right)\mathrm{d}\tau$$

i.e.

$$\frac{\mathrm{d}}{\mathrm{d}t}\langle x\rangle = \frac{1}{m}\langle p_x\rangle. \tag{5.70}$$

The rate of change of the expectation value of the $x$-component of momentum is

$$\frac{\mathrm{d}}{\mathrm{d}t}\langle p_x\rangle = \int \Psi^*\left[\mathscr{H}\frac{\partial}{\partial x} - \frac{\partial}{\partial x}\mathscr{H}\right]\Psi\,\mathrm{d}\tau$$

$$= \int \Psi^*\left(V\frac{\partial}{\partial x} - \frac{\partial}{\partial x}V\right)\Psi\,\mathrm{d}\tau$$

$$= \int \Psi^*\left(-\frac{\partial V}{\partial x}\right)\Psi\,\mathrm{d}\tau \tag{5.71}$$

i.e.

$$\frac{\mathrm{d}}{\mathrm{d}t}\langle p_x\rangle = -\left\langle\frac{\partial V}{\partial x}\right\rangle. \tag{5.72}$$

For a sharply defined wave packet $\langle x\rangle$ and $\langle p_x\rangle$ can be associated with the classical meaning of position and momentum of the particle and $-\langle\partial V/\partial x\rangle$ can be identified with the classical force component. The packet then behaves like a classical particle.

## 5.5 Probability Current Density

In the previous section it has been shown that the operator representing the velocity is the momentum operator divided by the mass, i.e. $(\hbar/im)$ grad (5.70). This operator can be used to find the expectation value of the velocity for a given state function.

A related quantity is the probability current vector **S** which can be used to find the probability that a particle will cross a given surface in unit time. Suppose $\Psi(\mathbf{r}, t)$ is the state function representing a one-particle system. The probability that the particle is in the finite volume $\Omega$ is

$$p(\Omega) = \int_{\Omega} \Psi^* \Psi \, d\tau. \tag{5.73}$$

The rate of change of this probability is

$$\frac{dp(\Omega)}{dt} = \int_{\Omega} \left( \frac{\partial \Psi^*}{\partial t} \Psi + \Psi^* \frac{\partial \Psi}{\partial t} \right) d\tau \tag{5.74}$$

and as $\Psi$ satisfies Schrödinger's time-dependent equation

$$\frac{dp(\Omega)}{dt} = \frac{i}{\hbar} \int [\Psi(\mathscr{H}\Psi)^* - \Psi^*(\mathscr{H}\Psi)] \, d\tau.$$

The Hamiltonian operator is

$$\mathscr{H} \equiv -\frac{\hbar^2}{2m} \nabla^2 + V$$

and so

$$\Psi(\mathscr{H}\Psi)^* - \Psi^*(\mathscr{H}\Psi) = -\frac{\hbar^2}{2m} [\Psi \nabla^2 \Psi^* - \Psi^* \nabla^2 \Psi].$$

Using the vector identity

$$\operatorname{div} [\Psi \operatorname{grad} \Psi^* - \Psi^* \operatorname{grad} \Psi] = \Psi \nabla^2 \Psi^* - \Psi^* \nabla^2 \Psi$$

the rate of change of $p(\Omega)$ is

$$\frac{dp(\Omega)}{dt} = -\frac{i\hbar}{2m} \int \operatorname{div} [\Psi(\operatorname{grad} \Psi^*) - \Psi^* \operatorname{grad} \Psi] \, d\tau.$$

A vector function of position **S** may be defined by

$$\mathbf{S} = \frac{i\hbar}{2m} [\Psi \operatorname{grad} \Psi^* - \Psi^* \operatorname{grad} \Psi] \tag{5.75}$$

and then

$$\frac{dp(\Omega)}{dt} = -\int_{\Omega} \operatorname{div} \mathbf{S} \, d\tau. \tag{5.76}$$

The volume integral can be transformed to a surface integral by Gauss's theorem, i.e.

$$\frac{\mathrm{d}p(\Omega)}{\mathrm{d}t} = -\oint_A \mathbf{S}.\mathrm{d}\mathbf{A} \tag{5.77}$$

where $A$ is the surface enclosing the volume $\Omega$ and the direction of $\mathrm{d}\mathbf{A}$ is along the outward normal. Clearly from (5.77), the integral of **S** over the surface $A$ is the probability that the particle will cross the surface going outwards in unit time. **S** is the probability current density. From (5.74) and (5.76)

$$\frac{\partial}{\partial t}(\Psi^*\Psi) + \mathrm{div}\,\mathbf{S} = 0. \tag{5.78}$$

This equation is analogous to the equations of continuity of hydrodynamics and electrodynamics. Observe (5.75) that if $\Psi$ is a real wave-function, the probability current density is zero.

Consider the energy eigenfunctions for a plane wave moving in the $x$-direction. The wave-function

$$\psi(x, y, z) = \sqrt{N}\, e^{ikx}$$

represents a density of $N$ particles per unit volume. The probability current density vector has a non-zero component only in the $x$-direction and

$$\mathbf{S}_x = \frac{\hbar k}{m} N.$$

This is the expected result as the particles have momentum $\hbar k$ and this expression gives the number of particles per second crossing unit area perpendicular to the $x$-axis.

A more interesting example is obtained by considering the central field eigenfunctions. From (5.41) these have the form

$$\psi(r, \theta, \phi) = f(r, \theta)\, e^{im_l\phi}$$

where $f(r, \theta)$ is a real function of $r$ and $\theta$ such that $\psi$ is normalized to unity. The vector **S** has a non-zero component only in the $\phi$-direction i.e. in the direction $\partial\mathbf{r}/\partial\phi$ about the $z$-axis. A simple calculation show that

$$\mathbf{S}_\phi = \frac{\hbar m_l}{\mu r \sin\theta} |\psi|^2 \tag{5.79}$$

where $\mu$ is the reduced electron mass. The spherically symmetric $s$-states have zero current density. For the $p$, d states etc. equation (5.79) represents an electric current circulating about the $z$-axis and it can be shown (Slater, p. 163) that the associated magnetic moment is

$$\frac{e\hbar}{2\mu} m_l. \tag{5.80}$$

PROBLEMS

1 A function of the operator $\alpha$ can be defined by

$$f(\alpha) \equiv \sum_{n=0}^{\infty} k_n \alpha^n.$$

Show that the eigenfunctions of $\alpha$ are also eigenfunctions of $f(\alpha)$ and express the eigenvalue of the function operator in terms of the eigenvalues of $\alpha$.

Explain why if $\alpha$ is Hermitian then $f(\alpha)$ is also Hermitian.

If the commutator of the two operators $\alpha, \beta$ is zero show that the commutator $[f(\alpha), \beta]$ is zero also.

2 The angular momentum about the origin of a particle with linear momentum $\mathbf{p}$ is given by

$$\mathbf{L} = \mathbf{r} \wedge \mathbf{p}.$$

By differentiating $\mathbf{L}$ with respect to time show that in the case of a central field when the force is directed towards the origin that $\mathbf{L}$ is a constant of the motion.

3 In the case of a system of two particles the operator representing the $x$-component of the angular momentum about the origin is

$$\mathscr{L}_z \equiv \frac{\hbar}{i} \sum_{n=1}^{2} \left( x_n \frac{\partial}{\partial y_n} - y_n \frac{\partial}{\partial x_n} \right)$$

and similarly for $\mathscr{L}_x$ and $\mathscr{L}_y$ (5.26), (5.27).

If the masses of the particles are $m_1, m_2$ the $x$-co-ordinate of the centre of mass is

$$x = \frac{m_1 x_1 + m_2 x_2}{m_1 + m_2}$$

with similar expressions for $y$ and $z$ (4.46). The co-ordinates of the second particle relative to the first are

$$\alpha = (x_2 - x_1), \qquad \beta = (y_2 - y_1), \qquad \gamma = (z_2 - z_1).$$

Verify that the given expression for $\mathscr{L}_z$ can be transformed to give

$$\mathscr{L}_z \equiv \frac{\hbar}{i}\left[\left(x\frac{\partial}{\partial y} - y\frac{\partial}{\partial x}\right) + \left(\alpha\frac{\partial}{\partial \beta} - \beta\frac{\partial}{\partial \alpha}\right)\right].$$

Explain the significance of the two terms in the angular momentum.

4 Show that if $\alpha$ is an operator and $\Psi_1$ and $\Psi_2$ are two quadratically integrable state functions then

$$\frac{\mathrm{d}}{\mathrm{d}t}\left[\int \Psi_1^* \alpha \Psi_2 \,\mathrm{d}\tau\right] = \int \Psi_1^* \left[\frac{\partial \alpha}{\partial t} + \frac{i}{\hbar}(\mathscr{H}\alpha - \alpha\mathscr{H})\right]\Psi_2 \,\mathrm{d}\tau.$$

(Compare equation (5.47).)

5 Show that if $\alpha$ and $\beta$ are two Hermitian operators and $\psi_1$ and $\psi_2$ are two quadratically integrable wave-functions then

$$\int \psi_1^*(\alpha\beta\psi_2)\,\mathrm{d}\tau = \int (\beta\alpha\psi_1)^*\psi_2\,\mathrm{d}\tau.$$

Hence show that

$$\int \psi_1^*[\alpha\beta - \beta\alpha]\,\psi_2\,\mathrm{d}\tau = \int [(\beta\alpha - \alpha\beta)\psi_1]^*\psi_2\,\mathrm{d}\tau.$$

Deduce that if an operator $\gamma$ is defined by

$$(\alpha\beta - \beta\alpha) \equiv i\gamma$$

then $\gamma$ is Hermitian (5.50).

6 It can be shown that for a linear harmonic oscillator the expectation value of the potential energy is equal to the expectation value of the kinetic energy. In particular, for the lowest energy state

$$\left[\left\langle \frac{1}{2}kx^2 \right\rangle\right] \times \left[\left\langle \frac{p^2}{2m} \right\rangle\right] = \left(\frac{\hbar\omega}{4}\right)^2 \qquad \omega = \left(\frac{k}{m}\right)^{1/2}.$$

As the expectation values of $x$ and $p$ are clearly both zero then the uncertainties in $x$ and $p$ are respectively given by

$$(\Delta x)^2 = \int \psi^* x^2 \psi\,\mathrm{d}\tau, \qquad (\Delta p)^2 = \int \psi^* \left(\frac{\hbar}{i}\frac{\mathrm{d}}{\mathrm{d}x}\right)^2 \psi\,\mathrm{d}\tau.$$

Show that in the lowest energy state

$$(\Delta x)(\Delta p) = \frac{\hbar}{2}$$

in agreement with the certainty principle. Confirm that for the higher states

$$(\Delta x)(\Delta p) > \frac{\hbar}{2}.$$

7 The function $\psi_{nlm_l}$ (5.41) is an eigenfunction of $\mathscr{L}_z$ with eigenvalue $\hbar m$. The commutator of $\mathscr{L}_x$ and $\mathscr{L}_y$ is

$$\mathscr{L}_x \mathscr{L}_y - \mathscr{L}_y \mathscr{L}_x = i\hbar \mathscr{L}_z. \qquad (5.28)$$

Using equation (5.55) show that

$$(\Delta L_x)(\Delta L_y) \geqslant \frac{\hbar^2 |m|}{2}.$$

8 Verify that the result of adding the two travelling waves

$$y_1 = A \cos(k_1 x - \omega_1 t) \qquad y_2 = A \cos(k_2 x - \omega_2 t)$$

is

$$y = y_1 + y_2 = 2A \cos\left(\frac{\delta k}{2} x - \frac{\delta \omega}{2} t\right) \cos(k_0 x - \omega_0 t)$$

where

$$k_0 = \frac{k_1 + k_2}{2}, \qquad \omega_0 = \frac{\omega_1 + \omega_2}{2}, \qquad \delta k = k_1 - k_2,$$

$$\delta\omega = \omega_1 - \omega_2.$$

This represents a wave with average frequency $\omega_0$ with a modulated amplitude. Verify that the modulation moves with a velocity $\delta\omega/\delta k$. (Compare (5.63) and (5.64).)

9 Verify that the wave packet

$$\Psi(x, t) = \frac{A}{a\left[a^2 + \frac{i\hbar t}{m} 2\right]} e^{-x^2/(a^2 + \frac{i\hbar t}{m} 2)}$$

is a solution of Schrödinger's time-dependent equation in one-dimension for a free particle. Show that the expectation values of the position and momentum are both zero and derive the following expression for the variance

$$(\Delta x)^2 = \left[\left(\frac{a}{2}\right)^2 + \left(\frac{\hbar t}{ma}\right)^2\right].$$

This shows that a wave packet (in a dispersive medium where $E(k)$) spreads in time. You may assume the definite integrals,

$$\int_{-\infty}^{\infty} x^2 e^{-x^2}\,\mathrm{d}x = \sqrt{\pi}/2 \qquad \int_{-\infty}^{\infty} e^{-x^2}\,\mathrm{d}x = \sqrt{\pi}.$$

10 Prove the vector identity

$$\mathrm{div}(\psi\ \mathrm{grad}\ \psi^* - \psi^*\ \mathrm{grad}\ \psi) = \psi\ \nabla^2\psi^* - \psi^*\nabla^2\psi$$

where $\psi$ is a scalar function of position.

11 Let $\psi_{nlm_l}$ be an eigenfunction of the central field problem. The current density at a point in space is (from (5.79))

$$j_\phi = \frac{e\hbar m_l}{\mu r \sin\theta}\ |\psi_{nlm_l}|^2.$$

The magnetic moment associated with a current carrying loop is equal to the current in the loop multiplied by the area of the loop. By first of all considering the magnetic moment in an annulus of small cross-section area about the $z$-axis and then summing over all such annuli show that the total magnetic moment is

$$M = \frac{e\hbar}{2\mu}\, m_l.$$

12 The Hamiltonian operator is, in spherical polar co-ordinates,

$$\mathscr{H} \equiv -\frac{\hbar^2}{2\mu}\left[\frac{1}{r^2}\frac{\partial}{\partial r}\left(r^2\frac{\partial}{\partial r}\right) + \frac{1}{r^2\sin\theta}\frac{\partial}{\partial\theta}\left(\sin\theta\frac{\partial}{\partial\theta}\right) + \frac{1}{r^2\sin^2\theta}\frac{\partial^2}{\partial\phi^2}\right] + V(r,\theta,\phi)$$

where $\mu$ is the particle mass.

Explain why, if the potential $V$ is independent of $\theta$ and $\phi$ the energy eigenfunctions can also be chosen to be eigenfunctions of the total angular momentum operator $\mathscr{L}^2$.

Show that the radial part $R(r)$ of the variable separable eigenfunction of an energy state with angular momentum quantum number $l$, satisfies

$$-\frac{\hbar^2}{2\mu}\frac{1}{r^2}\frac{\mathrm{d}}{\mathrm{d}r}\left(r^2\frac{\mathrm{d}R}{\mathrm{d}r}\right) + \frac{\hbar^2}{2\mu}\frac{l(l+1)}{r^2}R + VR = ER$$

where $E$ is the state energy.

Substitute $R(r) = \chi(r)/r$ and obtain

$$-\frac{\hbar^2}{2\mu}\frac{\mathrm{d}^2\chi}{\mathrm{d}r^2} + \left[\frac{l(l+1)}{2\mu r^2}\hbar^2 + V(r)\right]\chi = E\chi.$$

Suppose

$$V(r) = -V_0 \exp[-r/a], \qquad V_0 > 0$$

By substituting $z = \exp[-r/2a]$ show that the radial wave-functions for bound states with $l = 0$ are of the form

$$R(r) = \frac{A}{r} J_\nu (k e^{-r/2a})$$

where $A$ is a constant and

$$J_\nu(k) = 0.$$

[$J_\nu$ is the Bessel function of order $\nu$.]

## References

Slater, J. C. *Quantum Theory of Atomic Structure*, McGraw-Hill (1960).

CHAPTER 6

# Matrix Representations

## 6.1 Introduction

In the previous chapters the theory of quantum mechanics has included differential operators and differential equations. There is an equivalent alternative description in terms of matrices. Matrix mechanics was developed by Heisenberg, Born and Jordan and actually preceded the Schrödinger wave mechanics already described. In this chapter matrix mechanics will be deduced from the Schrödinger approach.

An eigenvalue problem can be interpreted in terms of solving a set of linear equations. Suppose $\alpha$ is a Schrödinger differential operator representing an observable $A$ and let $\psi$ be an eigenfunction with eigenvalue $a$;

$$\alpha\psi = a\psi. \tag{6.1}$$

$\psi$ can be expanded in terms of the complete set of orthonormal eigenfunctions belonging to some other operator, say $\beta$; a discrete spectrum is assumed, i.e.

$$\psi = \sum_j c_j \phi_j \qquad \text{with} \qquad \beta\phi_j = b_j \phi_j. \tag{6.2}$$

With this expansion for $\psi$, equation (6.1) becomes

$$\sum_j c_j \alpha\phi_j = a \sum_j c_j \phi_j. \tag{6.3}$$

If this equation is premultiplied by $\phi_l^*$ and integrated over all the space then

$$\sum_j c_j \alpha_{lj} = a c_l \tag{6.4}$$

where

$$\alpha_{lj} = \int \phi_l^* \alpha \phi_j \, \mathrm{d}\tau.$$

Equation (6.4) is true for all $l$ and represents a set of linear equations.

$$\begin{aligned}
\alpha_{11}c_1 + \alpha_{12}c_2 + \ldots &= ac_1 \\
\alpha_{21}c_1 + \alpha_{22}c_2 + \ldots &= ac_2 \\
\alpha_{31}c_1 + \alpha_{32}c_2 + \ldots &= ac_3 \\
&\vdots
\end{aligned} \tag{6.5}$$

As the operator $\beta$ has an infinite number of eigenvalues then (6.5) includes an infinite number of equations. Sets of equations such as (6.5) can conveniently be represented in terms of matrices and it is useful at this point to make a brief digression to remind the reader of the essentials of matrix algebra.

## 6.2 Matrix Algebra

A matrix is a rectangular array which obeys certain rules of algebra, e.g.

$$\mathbf{A} = \begin{pmatrix} \alpha_{11} & \alpha_{12} \cdots & \alpha_{1n} \\ \alpha_{21} & \alpha_{22} \cdots & \alpha_{2n} \\ \vdots & & \\ \alpha_{m1} & \alpha_{m2} \cdots & \alpha_{mn} \end{pmatrix} \tag{6.6}$$

$\alpha_{ij}$ is called an element of the matrix $\mathbf{A}$, the subscript $i$ denotes the row and the subscript $j$ the column. The matrix (6.6) with $m$ rows and $n$ columns is called an '$m$ by $n$' matrix. If the number of rows is equal to the number of columns then the matrix is square.

A matrix

$$\begin{pmatrix} c_1 \\ c_2 \\ \cdot \\ \cdot \\ \cdot \\ c_n \end{pmatrix}$$

composed of one column is called a column matrix or a column vector. Similarly a matrix with a one row

$$(c_1, c_2, \ldots, c_n)$$

is called a row matrix or row vector.

Two '$m$ by $n$' matrices are equal if and only if corresponding elements are separately equal.

The two fundamental laws of matrix algrebra are addition and multiplication.

*Addition and subtraction*

The sum of two '$m$ by $n$' matrices $\mathbf{\Gamma}$ and $\mathbf{\Lambda}$ with respective elements $\gamma_{ij}$ and $\lambda_{ij}$ is defined by the '$m$ by $n$' matrix $\mathbf{A}$ with elements

$$\alpha_{ij} = \gamma_{ij} + \lambda_{ij} \tag{6.7}$$

and

$$\mathbf{A} = \mathbf{\Gamma} + \mathbf{\Lambda}$$

e.g.

$$\mathbf{\Gamma} = \begin{pmatrix} 0 & 1 \\ 2 & 1 \\ 0 & 3 \end{pmatrix} \quad \mathbf{\Lambda} = \begin{pmatrix} 1 & 3 \\ 0 & 2 \\ 1 & 4 \end{pmatrix}$$

$$\mathbf{A} = \mathbf{\Gamma} + \mathbf{\Lambda} = \begin{pmatrix} 1 & 4 \\ 2 & 3 \\ 1 & 7 \end{pmatrix}$$

Addition is only defined for matrices that have the same number of rows and the same number of columns.

Substraction is the reverse of addition and the matrix

$$\mathbf{\Sigma} = \mathbf{\Gamma} - \mathbf{\Lambda} \tag{6.8}$$

with

$$\sigma_{ij} = \gamma_{ij} - \lambda_{ij}$$

is called the difference of $\mathbf{\Gamma}$ and $\mathbf{\Lambda}$.

From the definition (6.7) it is obvious that the addition of matrices is both commutative and associative, i.e.

$$\mathbf{\Gamma} + \mathbf{\Lambda} = \mathbf{\Lambda} + \mathbf{\Gamma} \tag{6.9}$$

and

$$(\mathbf{A} + \mathbf{\Gamma}) + \mathbf{\Lambda} = \mathbf{A} + (\mathbf{\Gamma} + \mathbf{\Lambda}). \tag{6.10}$$

*Multiplication*

The product of the matrix $\mathbf{\Gamma}$ and the constant d is defined to be the matrix d$\mathbf{\Gamma}$ with elements

$$(\mathrm{d}\mathbf{\Gamma})_{ij} = \mathrm{d}\gamma_{ij}. \tag{6.11}$$

The product $\mathbf{A} = \boldsymbol{\Gamma\Lambda}$ of two matrices is defined only when the number of columns of $\boldsymbol{\Gamma}$ is equal to the number of rows of $\boldsymbol{\Lambda}$. The elements of the product matrix are given by

$$\alpha_{ij} = \sum_l \gamma_{il}\lambda_{lj} \tag{6.12}$$

e.g.

$$\boldsymbol{\Gamma} = \begin{pmatrix} 0 & 1 \\ 2 & 1 \\ 0 & 3 \end{pmatrix} \qquad \boldsymbol{\Lambda} = \begin{pmatrix} 1 & 2 & 3 & 0 \\ 0 & 1 & 0 & 1 \end{pmatrix}$$

$$\mathbf{A} = \boldsymbol{\Gamma\Lambda} = \begin{pmatrix} 0 & 1 & 0 & 1 \\ 2 & 5 & 6 & 1 \\ 0 & 3 & 0 & 3 \end{pmatrix}. \tag{6.13}$$

Observe that the number of rows of $\mathbf{A}$ is equal to the number of rows of $\boldsymbol{\Gamma}$ and that the number of columns of $\mathbf{A}$ is equal to the number of columns of $\boldsymbol{\Lambda}$. Matrix multiplication is not commutative, i.e.

$$\boldsymbol{\Gamma\Lambda} \neq \boldsymbol{\Lambda\Gamma} \qquad \text{in general.} \tag{6.14}$$

For example

$$\boldsymbol{\Gamma} = \begin{pmatrix} 1 & 1 \\ 0 & 1 \end{pmatrix} \qquad \boldsymbol{\Lambda} = \begin{pmatrix} 1 & 2 \\ 1 & 1 \end{pmatrix} \tag{6.15}$$

then

$$\boldsymbol{\Gamma\Lambda} = \begin{pmatrix} 2 & 3 \\ 1 & 1 \end{pmatrix} \qquad \text{and} \qquad \boldsymbol{\Lambda\Gamma} = \begin{pmatrix} 1 & 3 \\ 1 & 2 \end{pmatrix}.$$

However matrix multiplications is associative, i.e.

$$(\mathbf{A}\boldsymbol{\Gamma})\boldsymbol{\Lambda} = \mathbf{A}(\boldsymbol{\Gamma\Lambda}). \tag{6.16}$$

Also multiplication is distributative over addition.

$$\mathbf{A}(\boldsymbol{\Gamma} + \boldsymbol{\Lambda}) = \mathbf{A}\boldsymbol{\Gamma} + \mathbf{A}\boldsymbol{\Lambda}. \tag{6.17}$$

A matrix whose elements are all zero is called the zero or null matrix and is denoted by $\mathbf{0}$.

An important result is that

$$\boldsymbol{\Gamma\Lambda} = \mathbf{0}$$

does not necessarily imply

$$\boldsymbol{\Gamma} = \mathbf{0} \qquad \text{or} \qquad \boldsymbol{\Lambda} = \mathbf{0}$$

e.g.

$$\begin{pmatrix} 1 & 1 \\ 1 & 1 \end{pmatrix} \begin{pmatrix} -1 & 1 \\ 1 & -1 \end{pmatrix} = \begin{pmatrix} 0 & 0 \\ 0 & 0 \end{pmatrix}.$$

It is now clear how sets of equations can be represented by matrices. The equations

$$\begin{array}{l} \alpha_{11}c_1 + \alpha_{12}c_2 + \ldots + \alpha_{1n}c_n = a_1 \\ \ldots\ldots\ldots\ldots\ldots\ldots\ldots\ldots \\ \ldots\ldots\ldots\ldots\ldots\ldots\ldots\ldots \\ \ldots\ldots\ldots\ldots\ldots\ldots\ldots\ldots \\ \alpha_{m1}c_1 + \alpha_{m2}c_2 + \ldots + \alpha_{mn}c_n = a_m \end{array}$$

can be written

$$\mathbf{Ac} = \mathbf{a}$$

where

$$\mathbf{A} = \begin{pmatrix} \alpha_{11} & \alpha_{12} \ldots \alpha_{1n} \\ \vdots & \\ \alpha_{m1} & \alpha_{m2} \ldots \alpha_{mn} \end{pmatrix}$$

and **c** and **a** are the column matrices

$$\mathbf{c} = \begin{pmatrix} c_1 \\ c_2 \\ \cdot \\ \cdot \\ \cdot \\ c_n \end{pmatrix} \quad \mathbf{a} = \begin{pmatrix} a_1 \\ a_2 \\ \cdot \\ \cdot \\ \cdot \\ a_m \end{pmatrix}.$$

*The transpose of a matrix*

Suppose $\mathbf{\Gamma}$ is an '$m$ by $n$' matrix with elements $\gamma_{ij}$. The transpose of this matrix is an '$n$ by $m$' matrix $\tilde{\mathbf{\Gamma}}$ such that

$$(\tilde{\mathbf{\Gamma}})_{ij} = \gamma_{ji} \qquad (6.18)$$

The columns of $\mathbf{\Gamma}$ become the rows of $\tilde{\mathbf{\Gamma}}$ and vice-versa, e.g. if

$$\mathbf{\Gamma} = \begin{pmatrix} 1 & 2 \\ 0 & 1 \\ 2 & 0 \end{pmatrix} \quad \text{then} \quad \tilde{\mathbf{\Gamma}} = \begin{pmatrix} 1 & 0 & 2 \\ 2 & 1 & 0 \end{pmatrix}$$

Clearly the transpose of a transpose is the original matrix, i.e.

$$\widetilde{(\tilde{\mathbf{\Gamma}})} = \mathbf{\Gamma}. \qquad (6.19)$$

The transpose of a product is the product of the transposes in reverse order.

$$
\begin{aligned}
(\widetilde{\mathbf{\Gamma\Lambda}})_{ij} &= (\mathbf{\Gamma\Lambda})_{ji} \\
&= \sum_l (\mathbf{\Gamma})_{jl} (\mathbf{\Lambda})_{li} \\
&= \sum_l (\tilde{\mathbf{\Lambda}})_{il} (\tilde{\mathbf{\Gamma}})_{lj} \\
&= (\tilde{\mathbf{\Lambda}}\tilde{\mathbf{\Gamma}})_{ij}
\end{aligned}
$$

i.e.

$$
(\widetilde{\mathbf{\Gamma\Lambda}}) = \tilde{\mathbf{\Lambda}}\tilde{\mathbf{\Gamma}}. \tag{6.20}
$$

*Square matrices*

A matrix that has the same number of rows and columns is called a square matrix.

If a square matrix is equal to its transpose it is called a symmetric matrix, i.e. if

$$
\mathbf{\Gamma} = \tilde{\mathbf{\Gamma}} \tag{6.21}
$$

e.g.

$$
\mathbf{\Gamma} = \tilde{\mathbf{\Gamma}} = \begin{pmatrix} 1 & 0 & 0 \\ 0 & 2 & 1 \\ 0 & 1 & 3 \end{pmatrix}.
$$

A square matrix whose only non-zero elements lie in the principal diagonal is called a diagonal matrix, e.g.

$$
\begin{pmatrix} 1 & 0 & 0 \\ 0 & 2 & 0 \\ 0 & 0 & 3 \end{pmatrix}.
$$

All diagonal matrices of the same order commute and have diagonal products.

A diagonal matrix whose elements in the principal diagonal are all unity is called a unit matrix and is denoted by **I**. The unit matrix of order three is

$$
\mathbf{I} = \begin{pmatrix} 1 & 0 & 0 \\ 0 & 1 & 0 \\ 0 & 0 & 1 \end{pmatrix}.
$$

*Trace*

The trace (or spur) of a square matrix is the sum of the diagonal elements;

$$\text{Trace } \mathbf{\Gamma} = \sum_i \gamma_{ii}. \tag{6.22}$$

The trace of a product of two matrices does not depend on the order of the factors.

$$\text{Trace } (\mathbf{\Gamma\Lambda}) = \sum_i \left(\sum_j \gamma_{ij}\lambda_{ji}\right) = \sum_j \left(\sum_i \lambda_{ji}\gamma_{ij}\right)$$

i.e.

$$\text{Trace } (\mathbf{\Gamma\Lambda}) = \text{Trace } (\mathbf{\Lambda\Gamma}). \tag{6.23}$$

*The inverse of a matrix*

The inverse of the square matrix $\mathbf{\Gamma}$ is the matrix $\mathbf{\Gamma}^{-1}$ such that

$$\mathbf{\Gamma\Gamma}^{-1} = \mathbf{\Gamma}^{-1}\mathbf{\Gamma} = \mathrm{I}. \tag{6.24}$$

This inverse exists if and only if the matrix $\mathbf{\Gamma}$ is nonsingular; that is if the determinant of the elements of $\mathbf{\Gamma}$ is non-zero. In this case the inverse is obtained from $\mathbf{\Gamma}$ by replacing each element by its cofactor and then transposing the resulting matrix and finally dividing it by (det $\mathbf{\Gamma}$), e.g. if

$$\mathbf{\Gamma} = \begin{pmatrix} 1 & 0 \\ 2 & 2 \end{pmatrix} \quad \text{then} \quad \mathbf{\Gamma}^{-1} = \frac{1}{2}\begin{pmatrix} 2 & 0 \\ -2 & 1 \end{pmatrix} = \begin{pmatrix} 1 & 0 \\ -1 & \frac{1}{2} \end{pmatrix}.$$

If $\mathbf{\Gamma}$ is a diagonal matrix, the inverse is also a diagonal matrix whose non-zero elements are the reciprocals of those of $\mathbf{\Gamma}$.

$$\mathbf{\Gamma} = \begin{pmatrix} \gamma_{11} & 0 \cdot & \cdot & \cdot & 0 \\ 0 & \gamma_{22} & \cdot & \cdot & \cdot \\ \cdot & & \cdot & & \cdot \\ \cdot & & & \cdot & \cdot \\ \cdot & & & \cdot & \cdot \\ 0 & \cdot & \cdot & \cdot & \gamma_{nn} \end{pmatrix} \qquad \mathbf{\Gamma}^{-1} = \begin{pmatrix} 1/\gamma_{11} & 0 & \cdot & \cdot & 0 \\ 0 & 1/\gamma_{22} & \cdot & \cdot & \cdot \\ \cdot & & \cdot & & \cdot \\ \cdot & & & \cdot & \cdot \\ \cdot & & & \cdot & \cdot \\ 0 & \cdot & \cdot & \cdot & 1/\gamma_{nn} \end{pmatrix}. \tag{6.25}$$

The inverse of the inverse of a matrix is the original matrix. To prove this consider

$$\mathbf{\Gamma\Gamma}^{-1} = \mathbf{I}.$$

Put $\mathbf{\Gamma} = \mathbf{\Lambda}^{-1}$ then

$$\mathbf{\Lambda}^{-1}(\mathbf{\Lambda}^{-1})^{-1} = \mathbf{I}.$$

Premultiply this equation by $\mathbf{\Lambda}$ and then

$$(\mathbf{\Lambda}^{-1})^{-1} = \mathbf{\Lambda} \tag{6.26}$$

as required.

The inverse of a product is the product of the inverses in reverse order. If $(\mathbf{\Gamma\Lambda})^{-1}$ is the inverse of $(\mathbf{\Gamma\Lambda})$

$$(\mathbf{\Gamma\Lambda})(\mathbf{\Gamma\Lambda})^{-1} = \mathbf{I}.$$

Premultiply this equation by $\mathbf{\Gamma}^{-1}$.

$$\mathbf{\Lambda}(\mathbf{\Gamma\Lambda})^{-1} = \mathbf{\Gamma}^{-1}.$$

Now multiply by $\mathbf{\Lambda}^{-1}$ and

$$(\mathbf{\Gamma\Lambda})^{-1} = \mathbf{\Lambda}^{-1}\mathbf{\Gamma}^{-1} \tag{6.27}$$

as required.

*Complex matrices*

The matrices that occur in quantum mechanics generally have complex elements. The complex conjugate of the matrix $\mathbf{\Gamma}$ is denoted by $\mathbf{\Gamma}^*$ and has elements such that

$$(\mathbf{\Gamma}^*)_{ij} = (\mathbf{\Gamma})_{ij}^*. \tag{6.28}$$

A square matrix $\mathbf{\Gamma}$ whose transpose is equal to the complex conjugate is called Hermitian or self-adjoint, i.e.

$$\tilde{\mathbf{\Gamma}} = \mathbf{\Gamma}^*. \tag{6.29}$$

Clearly the elements in the principal diagonal of a Hermitian matrix must be real. If all the elements are real, the matrix must be symmetric.

A square matrix $\mathbf{\Gamma}$ whose tranpose is equal to the complex conjugate of the inverse is called a unitary matrix, i.e.

$$\tilde{\mathbf{\Gamma}} = (\mathbf{\Gamma}^{-1})^*. \tag{6.30}$$

A real unitary matrix is called an orthogonal matrix.

*Eigenvalues*

Let $\mathbf{A}$ be a square matrix with $n$ rows and $\mathbf{c}$ a column vector with $n$ elements. In general the product $\mathbf{Ac}$ will produce another column vector. However in some special cases this resultant column vector is simply a multiple of $\mathbf{c}$.

$$\mathbf{Ac} = a\mathbf{c}. \tag{6.31}$$

This is a matrix eigenvalue equation. The null vector $\mathbf{c} = \mathbf{0}$ is a solution for all finite $a$. A value for $a$ for which $\mathbf{c} \neq \mathbf{0}$ is called an eigenvalue of the matrix **A** and **c** is the corresponding eigenvector. Equation (6.31) represents the $n$ equations

$$\begin{aligned} \alpha_{11}c_1 + \alpha_{12}c_2 + \ldots + \alpha_{1n}c_n &= ac_1 \\ &\vdots \\ \alpha_{n1}c_1 + \alpha_{n2}c_2 + \ldots + \alpha_{nn}c_n &= ac_n \end{aligned} \tag{6.32}$$

i.e.

$$\begin{aligned} (\alpha_{11} - a)c_1 + \alpha_{12}c_2 + \ldots + \alpha_{1n}c_n &= 0 \\ &\vdots \\ \alpha_{n1}c_1 + \alpha_{n2}c_2 + \ldots + (\alpha_{nn} - a)c_n &= 0. \end{aligned} \tag{6.33}$$

This system of $n$ homogeneous linear equations has a non-trivial solution if and only if the characteristic determinant is zero.

$$\det(\mathbf{A} - a\mathbf{I}) = \begin{vmatrix} \alpha_{11} - a & \alpha_{12} \ldots & \alpha_{1n} \\ \vdots & & \\ \alpha_{n1} & \alpha_{n2} & \alpha_{nn} - a \end{vmatrix} = 0. \tag{6.34}$$

This determinant can be expanded to give a polynomial of $n$th degree in $a$: the characteristic polynomial of **A**. The $n$ roots of this polynomial are the eigenvalues and may be real or complex. For every eigenvalue, the equations (6.32) have a solution giving the corresponding eigenvectors. However it should be observed that each eigenvector can be multiplied by an arbitrary constant and still remain an eigenvector. Consequently it is always possible to define normalized eigenvectors that satisfy

$$\tilde{\mathbf{c}}^*\mathbf{c} = 1. \tag{6.35}$$

As an illustration consider the two by two matrix

$$\mathbf{A} = \begin{pmatrix} 0 & 1 \\ 1 & 0 \end{pmatrix}.$$

The eigenvectors of **A** are the solutions of the simultaneous equations

$$\begin{aligned} -ac_1 + c_2 &= 0 \\ c_1 - ac_2 &= 0. \end{aligned}$$

These equations have a non-trivial solution only when

$$\begin{vmatrix} -a & 1 \\ 1 & -a \end{vmatrix} = a^2 - 1 = 0$$

giving

$$a = \pm 1.$$

The corresponding normalized eigenvectors are

$$\mathbf{c}_+ = \frac{1}{\sqrt{2}}\begin{pmatrix}1\\1\end{pmatrix} \qquad \mathbf{c}_- = \frac{1}{\sqrt{2}}\begin{pmatrix}1\\-1\end{pmatrix}.$$

*Theorem 1*

The eigenvalues of a Hermitian matrix are real.

*Proof*

Suppose $\mathbf{A}$ is a Hermitian matrix and $a$ is an eigenvalue with non-zero eigenvector $\mathbf{c}$.

$$\mathbf{Ac} = a\mathbf{c} \quad \text{(i)}.$$

Premultiply this equation by $\tilde{\mathbf{c}}^*$

$$\tilde{\mathbf{c}}^*\mathbf{Ac} = a\tilde{\mathbf{c}}^*\mathbf{c} \quad \text{(ii)}.$$

From (i)

$$\mathbf{A}^*\mathbf{c}^* = a^*\mathbf{c}^*.$$

Premultiply by $\tilde{\mathbf{c}}$

$$\tilde{\mathbf{c}}\mathbf{A}^*\mathbf{c}^* = a^*\tilde{\mathbf{c}}\mathbf{c}^*.$$

As $\mathbf{A}$ is Hermitian

$$\tilde{\mathbf{c}}\tilde{\mathbf{A}}\mathbf{c}^* = a^*\tilde{\mathbf{c}}\mathbf{c}^*.$$

Transpose (using (6.20)).

$$\tilde{\mathbf{c}}^*\mathbf{Ac} \qquad a^*\tilde{\mathbf{c}}^*\mathbf{c}$$

and consequently by comparison with (ii).

$$a = a^* \tag{6.36}$$

and the eigenvalue is real.

*Theorem 2*

The eigenvectors of a Hermitian matrix belonging to different eigenvalues are orthogonal.

*Proof*

Let $\mathbf{A}$ be a Hermitian matrix with eigenvectors $\mathbf{c}_n$, $\mathbf{c}_m$ such that

$$\mathbf{Ac}_n = a_n\mathbf{c}_n \quad \text{(i)} \qquad \mathbf{Ac}_m = a_m\mathbf{c}_m \quad \text{(ii)}.$$

Premultiplying (i) by $\tilde{\mathbf{c}}_m^*$

$$\tilde{\mathbf{c}}_m^* \mathbf{A} \mathbf{c}_n = a_n \tilde{\mathbf{c}}_m^* \mathbf{c}_n .$$

Also since $\mathbf{A}$ is Hermitian

$$\tilde{\mathbf{c}}_m^* \mathbf{A} \mathbf{c}_n = \tilde{\mathbf{c}}_m^* \tilde{\mathbf{A}}^* \mathbf{c}_n = (\widetilde{\mathbf{A}^* \mathbf{c}_m^*}) \mathbf{c}_n$$

$$= a_m \tilde{\mathbf{c}}_m^* \mathbf{c}_n$$

as $a_m$ is real.

$$\therefore \quad (a_n - a_m) \tilde{\mathbf{c}}_m^* \mathbf{c}_n = 0$$

and as $a_n \neq a_m$

$$\tilde{\mathbf{c}}_m^* \mathbf{c}_n = 0 \qquad n \neq m. \tag{6.37}$$

That is the eigenvectors $\mathbf{c}_m$, $\mathbf{c}_n$ are orthogonal.

## 6.3 Matrix Mechanics

A quantum mechanical eigenvalue problem can be stated in differential equation form (Schrödinger) or in matrix form. If $\alpha$ is a Schrödinger differential operator representing an observable $A$, the eigenvalues $a$ are given by

$$\alpha \psi_i = a_i \psi_i . \tag{6.1}$$

It has been shown in section 6.1 that this can be represented as a set of simultaneous equations (6.5). Writing

$$\psi = \sum_j c_j \phi_j \qquad \text{with} \qquad \beta \phi_j = b_j \phi_j \tag{6.2}$$

then in matrix notation the equations (6.5) are

$$\mathbf{A} \mathbf{c} = a \mathbf{c} \tag{6.38}$$

where

$$\mathbf{A} = \begin{pmatrix} \alpha_{11} \alpha_{12} \cdots \\ \alpha_{21} \alpha_{22} \\ \cdot \\ \cdot \\ \cdot \end{pmatrix} \qquad \text{with} \qquad \alpha_{lj} = \int \phi_l^* \alpha \phi_j \, d\tau, \tag{6.39}$$

and

$$c = \begin{pmatrix} c_1 \\ c_2 \\ \cdot \\ \cdot \\ \cdot \end{pmatrix}. \tag{6.40}$$

The orthonormal set $\{\phi_j\}$ is called the basis of the representation and the column vector $\mathbf{c}$ is the representative of the wave-function $\psi$.

It is assumed that the operator spectrum is discrete and the matrices (6.38) have a denumerably infinite number of rows and/or columns. All the appropriate results given in section 6.2 for finite matrices will be taken over in the infinite case. The more general case when the spectrum is partly or wholly continuous is outside the scope of this book. The reader is referred to *Mathematical foundations of Quantum Mechanics*, by John von Neumann.

*Hermitian matrices in quantum mechanics*

As the operator $\alpha$ represents an observable it must be Hermitian (postulate 2). Clearly

$$\alpha_{ij} = \alpha_{ji}^*$$

and the matrix $\mathbf{A}$ is Hermitian (6.29). It has already been shown in theorems one and two that such matrices possess real eigenvalues and that their eigenvectors belonging to different eigenvalues are orthogonal.

*The diagonal representation*

The matrix representing an observable can always be put into diagonal form. If the orthonormal basis used to expand the eigenfunctions of (6.1) is taken to be these eigenfunctions themselves, then the diagonal representation is obtained.

$$\alpha_{lj} = \int \psi_l^* \alpha \psi_j \, d\tau = a_j \delta_{lj}.$$

The matrix eigenvalue equation (6.38) becomes

$$\begin{pmatrix} a_1 & 0 & 0 & \cdots \\ 0 & a_2 & 0 & \cdots \\ 0 & 0 & a_3 & \cdots \\ \cdot & & & \\ \cdot & & & \\ \cdot & & & \end{pmatrix} \begin{pmatrix} c_{1n} \\ c_{2n} \\ c_{3n} \\ \cdot \\ \cdot \\ \cdot \end{pmatrix} = a_n \begin{pmatrix} c_{1n} \\ c_{2n} \\ c_{3n} \\ \cdot \\ \cdot \\ \cdot \end{pmatrix}. \tag{6.41}$$

The eigenvectors are

$$\mathbf{c}_1 = \begin{pmatrix} 1 \\ 0 \\ 0 \\ \cdot \\ \cdot \\ \cdot \end{pmatrix}, \quad \mathbf{c}_2 = \begin{pmatrix} 0 \\ 1 \\ 0 \\ \cdot \\ \cdot \\ \cdot \end{pmatrix}, \ldots \mathbf{c}_n = \begin{pmatrix} 0 \\ 0 \\ \vdots \\ 0 \\ 1 \\ 0 \\ \vdots \end{pmatrix} \begin{matrix} \\ \\ \\ n-1 \\ n \\ n+1 \\ \\ \end{matrix}. \tag{6.42}$$

The diagonal matrix representing the Hamiltonian operator for a one-dimensional harmonic oscillator is

$$\mathbf{H} = \hbar\omega \begin{pmatrix} \frac{1}{2} & 0 & 0 & \dots \\ 0 & \frac{3}{2} & 0 & \dots \\ 0 & 0 & \frac{5}{2} & \dots \end{pmatrix}$$

and the eigenfunction with energy $(n + \frac{1}{2})\hbar\omega$ is represented by the column vector

$$\mathbf{c}_n = \begin{pmatrix} 0 \\ 0 \\ \vdots \\ 0 \\ 1 \\ 0 \\ \vdots \end{pmatrix} \begin{matrix} 0 \\ 1 \\ \\ n-1 \\ n \\ n+1 \\ \\ \end{matrix}.$$

*Transformation of basis functions*

An arbitrary wave function $\psi$ can be expanded in terms of a complete orthonormal set of basis functions which span the Hilbert space of the system,

$$\psi = \sum_j c_j\phi_j, \int \phi_i^*\phi_j \, \mathrm{d}\tau = \delta_{ij}. \tag{6.43}$$

The $\phi_j$'s are analogous to unit vectors and the $c_j$'s are the 'components of the vector' **c**.

Of course there are different sets of basis functions apart from $\{\phi_j\}$. Suppose the orthonormal set $\{\chi_j\}$ also spans the Hilbert space.

Clearly the members of the new set can be expanded in terms of the old set, i.e.

$$\chi_i = \sum_p \phi_p u_{pi} \qquad i = 1, 2, \dots \tag{6.44}$$

where $u_{pi}$ are the expansion coefficients and form an infinite matrix **u**. Similarly

$$\phi_i = \sum_q \chi_q v_{qi} \qquad i = 1, 2, \dots \tag{6.45}$$

where $v_{pi}$ form an infinite matrix **v**. Substitution of (6.45) into (6.44) gives

$$\chi_i = \sum_q \chi_q \sum_p v_{qp}u_{pi}$$

$$\therefore \quad \sum_p v_{qp}u_{pi} = \delta_{qi}.$$

In matrix language

$$\mathbf{vu} = \mathbf{I}. \tag{6.46}$$

Similarly, substitution of (6.44) into (6.45) gives

$$\mathbf{uv} = \mathbf{I}. \tag{6.47}$$

From the definition of a matrix inverse (6.24)

$$\mathbf{v} = \mathbf{u}^{-1}. \tag{6.48}$$

As both basis sets are orthonormal, then from (6.44)

$$\int \chi_i^* \chi_j \, \mathrm{d}\tau = \delta_{ij} = \int \sum_p \phi_p^* u_{pi}^* \sum_q \phi_q^* u_{qj} \, \mathrm{d}\tau$$

i.e.

$$\delta_{ij} = \sum_p u_{pi}^* u_{pj}.$$

An element of the transpose matrix $\tilde{\mathbf{u}}$ will be denoted by $\tilde{u}_{pi}$. As

$$u_{pi}^* = \tilde{u}_{ip}^*$$

then

$$\delta_{ij} = \sum_p \tilde{u}_{ip}^* u_{pj}. \tag{6.49}$$

In matrix rotation

$$\mathbf{I} = \tilde{\mathbf{u}}^* \mathbf{u}. \tag{6.50}$$

It has been shown that $\mathbf{u}$ has an inverse and if (6.50) is post-multiplied by $\mathbf{u}^{-1}$

$$\mathbf{u}^{-1} = \tilde{\mathbf{u}}^*. \tag{6.51}$$

This final result shows that the matrix transforming one orthonormal set into another is unitary (see (6.30)).

A change of basic functions is analogous to a rotation of co-ordinate axes in three-dimensional Cartesian space and as is well known such rotations are performed by orthogonal (real unitary) matrices.

*Transformation of a representative*

Obviously the representative of an arbitrary wave function $\psi$ will alter under a change of basis functions. Suppose

$$\psi = \sum_j c_j \phi_j \tag{6.43}$$

in terms of the old set $\{\phi_j\}$ and

$$\psi = \sum_i f_i \chi_i \tag{6.52}$$

in terms of the new set $\{\chi_i\}$. The members of the new set can be expanded in terms of the old set and substituting (6.44) into (6.52)

$$\psi = \sum_i f_i \sum_p \phi_p u_{pi} = \sum_p \left( \sum_i u_{pi} f_i \right) \phi_p. \tag{6.53}$$

Comparison of (6.43) and (6.53) gives

$$c_j = \sum_i u_{ji} f_i \qquad j = 1, 2, \ldots$$

i.e.

$$\mathbf{c} = \mathbf{uf}. \tag{6.54}$$

The old representative **c** is equal to the transformation matrix times the new representative **f**.

*Transformation of an operator matrix*

The infinite square matrix representing an operator $\alpha$ must also alter under a change of basis. $\mathbf{A}^{(1)}$ will be used to represent the matrix with respect to the set $\{\phi_j\}$ and $\mathbf{A}^{(2)}$ for the matrix in the $\{\chi_j\}$ representation. From (6.4)

$$\alpha_{ij}^{(1)} = \int \phi_i^* \alpha \phi_j \, d\tau, \qquad \alpha_{ij}^{(2)} = \int \chi_i^* \alpha \chi_j \, d\tau. \tag{6.55}$$

Using (6.44)

$$\begin{aligned} \alpha_{ij}^{(2)} &= \int \sum_p \phi_p^* u_{pi}^* \alpha \sum_q \phi_q u_{qj} \, d\tau \\ &= \sum_p \sum_q u_{pi}^* u_{qj} \alpha_{pq}^{(1)} = \sum_p \sum_q \tilde{u}_{ip}^* \alpha_{pq}^{(1)} u_{qj} \end{aligned}$$

i.e.

$$\mathbf{A}^{(2)} = \tilde{\mathbf{u}}^* \mathbf{A}^{(1)} \mathbf{u}.$$

As the transformation matrix is unitary

$$\mathbf{A}^{(2)} = \mathbf{u}^{-1} \mathbf{A}^{(1)} \mathbf{u}. \tag{6.56}$$

This is a similarity transformation.

*Theorem 3*

A Hermitian matrix remains Hermitian under a unitary transformation.

*Proof*

Suppose the Hermitian matrix $\mathbf{A}^{(1)}$ is transformed into $\mathbf{A}^{(2)}$ by the unitary transformation $\mathbf{u}$.

$$\mathbf{A}^{(2)} = \mathbf{u}^{-1}\mathbf{A}^{(1)}\mathbf{u}.$$

Transposing

$$\tilde{\mathbf{A}}^{(2)} = \tilde{\mathbf{u}}\tilde{\mathbf{A}}^{(1)}\tilde{\mathbf{u}}^{-1}.$$

As the matrix $\mathbf{u}$ is unitary and $\tilde{\mathbf{A}}^{(1)}$ is Hermitian

$$\widetilde{\mathbf{A}^{(2)}} = (\mathbf{u}^{-1}\mathbf{A}^{(1)}\mathbf{u})^* = \mathbf{A}^{*(2)}$$

and so $\mathbf{A}^{(2)}$ is also Hermitian.

*Theorem 4*

The eigenvalues of a Hermitian matrix remain invariant under a similarity transformation.

*Proof*

Suppose the operator $\alpha$ is represented by the matrix $\mathbf{A}^{(1)}$ in the original representation. The eigenvalue $a$ satisfies

$$\mathbf{A}^{(1)}\mathbf{c} = a\,\mathbf{c} \tag{6.57}$$

where $\mathbf{c}$ is the representative of the eigenfunction. If the basis functions are transformed using (6.44) the operator matrix undergoes a similarity transformation (6.56), so that

$$\mathbf{A}^{(1)} = \mathbf{u}\mathbf{A}^{(2)}\mathbf{u}^{-1} \tag{6.58}$$

and the eigenvalue equation (6.57) may be written

$$\mathbf{u}\mathbf{A}^{(2)}\mathbf{u}^{-1}\mathbf{c} = a\mathbf{c}.$$

On premultiplying by $\mathbf{u}^{-1}$

$$\mathbf{A}^{(2)}(\mathbf{u}^{-1}\mathbf{c}) = a(\mathbf{u}^{-1}\mathbf{c}).$$

From (6.54) this becomes

$$\mathbf{A}^{(2)}\mathbf{f} = a\,\mathbf{f} \tag{6.59}$$

where $\mathbf{f}$ is the eigenfunction representative in the new basis and the theorem is proved.

*Theorem 5*

The matrix of the product of two operators is the product of the separate matrices representing the operators.

*Proof*

Suppose $\gamma$ and $\lambda$ are two quantum differential operators and that $\{\phi_i\}$ is a complete set of orthonormal wave functions. Clearly

$$\lambda\phi_i = \sum_j \lambda_{ji}\phi_j. \tag{6.60}$$

$\lambda_{ji}$ is an element of the matrix $\mathbf{\Lambda}$ representing $\lambda$ in the basis $\{\phi_i\}$ and

$$\lambda_{ji} = \int \phi_j^* \lambda \phi_i \, d\tau. \tag{6.61}$$

Then

$$\begin{aligned} \gamma\lambda\phi_i &= \sum_i \lambda_{ji}\gamma\phi_j \\ &= \sum_j \lambda_{ji} \sum_l \gamma_{lj}\phi_l \\ &= \sum_j \sum_l \gamma_{lj}\lambda_{ji}\phi_l \end{aligned} \tag{6.62}$$

where

$$\gamma_{lj} = \int \phi_l^* \gamma \phi_j \, d\tau$$

and is a matrix element of $\mathbf{\Gamma}$.

Premultiply (6.62) by $\phi_m^*$ and integrate over all the space. Then

$$(\gamma\lambda)_{mi} = \int \phi_m^* \gamma\lambda\phi_i \, d\tau = \sum_j \gamma_{mj}\lambda_{ji}. \tag{6.63}$$

and the theorem is proved.

The matrices satisfy the same commutation relationships as the differential operators. In particular the matrices representing the coordinates $q_j$ and the conjugate momentum $p_j$ satisfy

$$\mathbf{p}_j\mathbf{q}_k - \mathbf{q}_k\mathbf{p}_j = \frac{\hbar}{i}\mathbf{I}\delta_{kj}. \tag{6.64}$$

As diagonal matrices (of the same order) commute, then (6.64) implies that there is no representation in which $\mathbf{p}_j$ and $\mathbf{q}_j$ are simultaneously of diagonal form. This of course is an expression of the uncertainty principle in that the operators $\hat{p}_j$ and $q_j$ do not possess a simultaneous set of eigenfunctions.

## *Angular momentum matrices*

Simultaneous eigenfunctions can be found for the angular momentum operators $\mathscr{L}^2$ and $\mathscr{L}_z$ with eigenvalues $l(l+1)\hbar^2$ and $\hbar m_l$ respectively. These simultaneous eigenfunctions $\psi(l, m_l)$ form a complete set and an arbitrary wave-function can be expanded in terms of them.

$$\Psi(\mathbf{r}) = \sum_l \sum_{m_l} c(l, m_l)\psi(l, m_l). \tag{6.65}$$

One choice for the basis functions is the orthonormal hydrogen type functions (5.41).

$$\psi(l, m_l) = R_{n_l}(r) P_l^{m_l}(\cos\theta)\, e^{im_l\phi} \qquad m_l \leqslant l. \tag{6.66}$$

The matrix elements for the angular momentum operators in this representation are

$$(\mathscr{L}^2)_{lm_l, l'm_{l'}} = \hbar^2 l(l+1)\delta_{ll'}\delta_{m_l m_{l'}} \tag{6.67}$$

and

$$(\mathscr{L}_z)_{lm_l, l'm_{l'}} = \hbar m_l \delta_{ll'}\delta_{m_l m_{l'}}. \tag{6.68}$$

The matrices are both diagonal

$$\mathbf{L}^2 = \hbar^2 \begin{pmatrix} 0 & & \\ & \begin{array}{|ccc|} \hline 2 & 0 & 0 \\ 0 & 2 & 0 \\ 0 & 0 & 2 \\ \hline \end{array} & 0 \\ 0 & & \begin{array}{|ccccc|} \hline 6 & 0 & 0 & 0 & 0 \\ 0 & 6 & 0 & 0 & 0 \\ 0 & 0 & 6 & 0 & 0 \\ 0 & 0 & 0 & 6 & 0 \\ 0 & 0 & 0 & 0 & 6 \\ \hline \end{array} \end{pmatrix} \tag{6.69}$$

$$\mathbf{L}_z = \hbar \begin{pmatrix} 0 & & \\ & \begin{array}{|ccc|} \hline 1 & 0 & 0 \\ 0 & 0 & 0 \\ 0 & 0 & -1 \\ \hline \end{array} & 0 \\ 0 & & \begin{array}{|ccccc|} \hline 2 & 0 & 0 & 0 & 0 \\ 0 & 1 & 0 & 0 & 0 \\ 0 & 0 & 0 & 0 & 0 \\ 0 & 0 & 0 & -1 & 0 \\ 0 & 0 & 0 & 0 & -2 \\ \hline \end{array} \end{pmatrix}. \tag{6.70}$$

The matrices representing $\mathscr{L}_x, \mathscr{L}_y$ are not diagonal.

*The Schrödinger picture*

In Chapter 3 the basic postulates of the Schrödinger theory were given. The differential operators representing observables are in general time independent and the time-dependence of a state is included in the wave-function which undergoes a rotation in Hilbert space as time passes.

The Schrödinger equation has an analogy in matrix mechanics. Let $\{\phi_j\}$ be a complete set of time-independent orthonormal basis functions which can be used to expand an arbitrary state wave-function.

$$\Psi(\mathbf{r}, t) = \sum_j c_j(t)\phi_j(\mathbf{r}). \tag{6.71}$$

As $\Psi(\mathbf{r}, t)$ must satisfy the Schrödinger equation

$$\sum_j \mathscr{H} c_j(t)\phi_j(\mathbf{r}) = i\hbar \sum_j \frac{\mathrm{d}c_j}{\mathrm{d}t}(t)\phi_j(\mathbf{r}).$$

Premultiply this equation by $\phi_i^*(\mathbf{r})$ and integrate over all space.

$$\sum_j \mathscr{H}_{ij} c_j(t) = i\hbar \dot{c}_i(t)$$

where

$$\mathscr{H}_{ij} = \int \phi_i^* \mathscr{H} \phi_j \, \mathrm{d}\tau.$$

In terms of matrices

$$\mathbf{H}\mathbf{c}(t) = i\hbar \dot{\mathbf{c}}(t). \tag{6.72}$$

The time-dependence of the wave-function $\Psi(\mathbf{r}, t)$ is reflected in the time-dependence of the column vector.

In the 'Schrödinger' picture where the expansion (6.71) is employed any time-independent differential operator is represented by a time-independent matrix.

When the basis functions used in (6.71) are the energy eigenfunctions, the matrix $\mathbf{H}$ is diagonal and the time variation of the vector $\mathbf{c}$ is given by

$$c_j(t) = e^{-iE_j t/\hbar} c_j(0).$$

*The Heisenberg picture*

There is another important way of discussing the time variation of a state. This is the Heisenberg picture in which the basis functions used to expand an arbitrary function are time-dependent. Suppose $\Phi_i(\mathbf{r}, t)$ and $\Phi_j(\mathbf{r}, t)$ are solutions of the system Schödinger equation corresponding to the initial states $\Phi_i(\mathbf{r}, 0)$ and $\Phi_j(\mathbf{r}, 0)$ respectively. The Hermitian character of the Hamiltonian $\mathscr{H}$ determines that the 'inner product' of the two functions remains constant in time.

$$\frac{\mathrm{d}}{\mathrm{d}t}\int \Phi_i^*(\mathbf{r}, t)\Phi_j(\mathbf{r}, t)\,\mathrm{d}\tau = \int \left[\frac{\partial \Phi_i^*}{\partial t}\Phi_j + \Phi_i^* \frac{\partial \Phi_j}{\partial t}\right] \mathrm{d}\tau = 0.$$

In particular if the complete set $\{\Phi_i\}$ is orthonormal at $t = 0$ then it will remain so for all time if

$$\mathscr{H}\Phi_j = i\hbar \frac{\partial \Phi_j}{\partial t} \qquad \text{for all } j. \tag{6.73}$$

Since the set is complete it can be used to expand an arbitrary solution of the Schrödinger equation (6.73)

$$\Psi(\mathbf{r}, t) = \sum_j c_j \Phi_j(\mathbf{r}, t) \tag{6.74}$$

Since the set is orthonormal for all time

$$c_j = \int \Phi_j^*(\mathbf{r}, t)\Psi(\mathbf{r}, t)\, \mathrm{d}\tau$$

and from (6.73) the coefficients $c_j$ are constants.

The column vector representing the wave-function is now time-independent and the time variation is included in the operator matrix.

Suppose $\alpha$ is a Schrödinger differential operator representing the observable $A$. The elements of the operator matrix $\mathbf{A}$ are given by

$$\alpha_{kj} = \int \Phi_k^*(\mathbf{r}, t)\alpha\Phi_j(\mathbf{r}, t)\, \mathrm{d}\tau. \tag{6.75}$$

From Chapter 5 (equation (5.47) and question 4) the elements of $\dot{\mathbf{A}}$ are given by

$$\frac{\mathrm{d}}{\mathrm{d}t}\alpha_{kj} = \int \Phi_k^*(\mathbf{r}, t)\left[\frac{\partial \alpha}{\partial t} + \frac{i}{\hbar}(\mathscr{H}\alpha - \alpha\mathscr{H})\right]\Phi_j(\mathbf{r}, t)\, \mathrm{d}\tau \tag{6.76}$$

and in matrix notation

$$\dot{\mathbf{A}} = \frac{\partial \mathbf{A}}{\partial t} + \frac{i}{\hbar}(\mathbf{HA} - \mathbf{AH}) \tag{6.77}$$

where the elements of the matrix $\partial\mathbf{A}/\partial t$ are

$$\left(\frac{\partial \mathbf{A}}{\partial t}\right)_{kj} = \int \Phi_k^*(\mathbf{r}, t)\frac{\partial \alpha}{\partial t}\Phi_j(\mathbf{r}, t)\, \mathrm{d}\tau.$$

If the differential operator is time-independent then Heisenberg's equation (6.77) becomes

$$\dot{\mathbf{A}} = \frac{i}{\hbar}(\mathbf{HA} - \mathbf{AH}). \tag{6.78}$$

As an example consider the matices $\mathbf{q}_i$ and $\mathbf{p}_i$ representing the $i$th components of position and momentum for a free particle. From (6.64)

$$\dot{\mathbf{q}}_i = \frac{i}{\hbar}(\mathbf{Hq}_i - \mathbf{q}_i\mathbf{H}) = \frac{\mathbf{p}_i}{m}$$

$$i = 1, 2, 3.$$

$$\dot{\mathbf{p}}_i = \frac{i}{\hbar}(\mathbf{Hp}_i - \mathbf{p}_i\mathbf{H}) = 0$$

since

$$\mathbf{H} = \frac{1}{2m}(\mathbf{p}_1^2 + \mathbf{p}_2^2 + \mathbf{p}_3^2).$$

A useful representation is obtained when the basis functions are the energy eigenfunctions so diagonalizing the Hamiltonian matrix. Then the elements of the time derivative matrix $\dot{\mathrm{A}}$ are (from (6.78))

$$\dot{\alpha}_{kj} = \frac{i}{\hbar}\,[E_k - E_j]\,\alpha_{kj} \tag{6.79}$$

where $E_k$ and $E_j$ are energy eigenvalues. Integration of (6.79) gives

$$\alpha_{kj}(t) = \alpha_{kj}(0)\exp\left[i\frac{(E_k - E_j)}{\hbar}t\right]. \tag{6.80}$$

In this representation, off-diagonal elements oscillate in time with frequencies $\omega_{kj} = (E_k - E_j)/\hbar$.

*The density matrix*

The quantum mechanical theory so far developed deals with 'pure states' for which the state function is completely known. Even in this case the value of an observable is not well defined unless the state function is an eigenfunction of the observable operator. It is often the case that the state-function is not known completely and then additional uncertainties must appear. There may be many wave-functions compatible with the incomplete information known about the system and the effects of these must be suitably averaged. This second statistical aspect, due to the fact that not all the information that can be known is known, is dealt with by statistical mechanics and occurs in both quantum and classical mechanics.

Such a quantum-mechanical system which is not known as completely as is possible, is said to be in a 'mixed state'. It is useful to introduce the concept of a representative ensemble. This is a collection of similar systems each described by different wave-functions but with the property that the values of the known observables are the same in all systems. The values of the remaining observables (of a complete set) form a suitable chosen distribution.

Consider a representative ensemble composed of $N$ systems described by the normalized wave-functions $\Psi_1, \Psi_2, \ldots, \Psi_N$. Let $\{\Phi_j\}$ be a complete orthonormal set which will be assumed to be discrete. Then

$$\Psi_i = \sum_j c_{ji}\Phi_j \qquad \text{with} \qquad c_{ji} = \int \Phi_j^* \Psi_i \,\mathrm{d}\tau \tag{6.81}$$

A 'density operator' $\rho$ is defined by specifying its matrix elements in this $\{\Phi_j\}$ representation.

$$\rho_{kj} = \frac{1}{N}\sum_{i=1}^{N} c_{ji}^* c_{ki}. \tag{6.82}$$

It can be shown that the density matrix $\boldsymbol{\rho}$ is Hermitian and transforms like any other operator (6.56).

In the special case that the wave-function is known completely ($\Psi_i$ say), the density matrix takes the form

$$\rho_{kj} = c_{ji}^* c_{ki}.$$

The square of this matrix is given by

$$(\boldsymbol{\rho}^2)_{kj} = \sum_l (c_{li}^* c_{ki})(c_{ji}^* c_{li}) = \rho_{kj}$$

since the wave-function is normalized and

$$\sum_l c_{li}^* c_{li} = 1.$$

The density matrix is equal to its own square when the ensemble is described by a single wave-function.

In the general case, the density matrix contains all the physical information that is known about the system and can be used to find the mean value of any observable. Suppose $A$ is an observable represented by the Schrödinger operator $\alpha$. The expectation value for $A$ for the $i$th system in the ensemble is

$$\langle A \rangle = \int \Psi_i^* \alpha \Psi_i \, d\tau = \sum_j \sum_k c_{ji}^* c_{ki} \alpha_{jk} \tag{6.83}$$

where

$$\alpha_{jk} = \int \Phi_j^* \alpha \Phi_k \, d\tau.$$

The elements $\alpha_{jk}$ form the matrix representing $A$ in the representation defined by the set $\{\Phi_j\}$. The mean value for this observable for the system represented by the ensemble is from (6.83) and (6.82)

$$[A] = \frac{1}{N} \sum_{i=1}^{N} \left( \sum_j \sum_k c_{ji}^* c_{ki} \alpha_{jk} \right) = \text{Trace}\,(\boldsymbol{\rho\alpha}). \tag{6.84}$$

This equation is valid in all representations.

In classical mechanics the observable $A$ is a function of the co-ordinates and momenta. The classical average for this observable for the ensemble is

$$[A] = \int f(p, q, t) A(p, q) \, dp \, dq$$

where the integration is over phase space and $f(p, q, t)$ is a suitably chosen distribution function. The density matrix in quantum mechanics is analogous to the classical distribution function.

In the Schrödinger picture the expansion coefficients are in general time-dependent. A simple expression can readily be obtained for $\dot{\rho}$. The expansion coefficients satisfy (6.72).

$$\sum_j \mathscr{H}_{kj} c_{ji} = i\hbar \dot{c}_{ki}.$$

From this equation together with (6.82) and remembering that $\mathscr{H}$ is Hermitian it is easily shown that

$$i\hbar\dot{\rho}_{kl} = \sum_j [\mathscr{H}_{kj}\rho_{jl} - \rho_{kj}\mathscr{H}_{jl}]$$

i.e.

$$\dot{\boldsymbol{\rho}} = -\frac{i}{\hbar}(\mathbf{H}\boldsymbol{\rho} - \boldsymbol{\rho}\mathbf{H}). \tag{6.85}$$

Note carefully that this equation differs in sign from the equation of motion (6.78) of an observable in the Heisenberg picture.

Equation (6.85) does not hold in the Heisenberg picture as the time dependence is then in the matrices representing the observable and in this case $\boldsymbol{\rho}$ is a constant matrix.

For a more complete discussion of the density matrix the reader is referred to the review paper *Theory and Application of the Density Matrix* by D. Ter Haar.

## PROBLEMS

1 Explain how any real square matrix can be written as the sum of a symmetric matrix and a skew-symmetric matrix.

2 Let $\boldsymbol{\Delta}$ be a square matrix with $n$ rows and let $\mathbf{c}_i$ be an eigenvector with eigenvalue $d_i$ so that

$$(\boldsymbol{\Delta} - d_i\mathbf{I})\mathbf{c}_i = 0.$$

An arbitrary column vector $\mathbf{x}$ with $n$ rows can be expressed as a sum of the $n$ eigenvectors, i.e.

$$\mathbf{x} = \sum_{i=1}^{n} \alpha_i \mathbf{c}_i.$$

Show that $\boldsymbol{\Delta}$ satisfies the Hamilton-Cayley equation

$$(\boldsymbol{\Delta} - d_1\mathbf{I})(\boldsymbol{\Delta} - d_2\mathbf{I})\ldots(\boldsymbol{\Delta} - d_n\mathbf{I}) = \mathbf{0}.$$

(Note that is the characteristic polynomial of (6.34).)

3 Prove that the eigenvalues of a skew-symmetry matrix are purely imaginary or zero.

4 Prove that eigenvalues of a unitary matrix have the absolute value one and so lie on the unit circle in the complex plane.

5 Prove that if the matrix $\boldsymbol{\Gamma}$ has eigenvalues $d_1, d_2, \ldots$ then the matrix $\boldsymbol{\Gamma}^2$ has eigenvalues $d_1^2, d_2^2, \ldots$ etc.

6 Consider the three matrices

$$\boldsymbol{\tau}_1 = \begin{pmatrix} 0 & 1 \\ 1 & 0 \end{pmatrix} \quad \boldsymbol{\tau}_2 = \begin{pmatrix} 0 & -i \\ i & 0 \end{pmatrix} \quad \boldsymbol{\tau}_3 = \begin{pmatrix} 1 & 0 \\ 0 & -1 \end{pmatrix}.$$

Show that

(a) $\boldsymbol{\tau}_1^2 = \boldsymbol{\tau}_2^2 = \boldsymbol{\tau}_3^2 = \mathbf{I}$

and

(b) $\boldsymbol{\tau}_1\boldsymbol{\tau}_2 - \boldsymbol{\tau}_2\boldsymbol{\tau}_1 = 2i\boldsymbol{\tau}_3$.

Show also that $\boldsymbol{\tau}_1, \boldsymbol{\tau}_2$ and $\boldsymbol{\tau}_3$ have the common eigenvalues $\pm 1$. (These are the Pauli spin matrices.)

Construct the matrix

$$\frac{e}{2}(\mathbf{I} + \boldsymbol{\tau}_3)$$

and confirm that it has eigenvalues $+e$ and zero. (This result is used in the isotopic spin formalism to distinguish between protons and neutrons.)

7 Verify that the eigenfunction $\psi_n = \sqrt{2}\sin(n\pi x)$ are orthonormal over the range $0 \leqslant x \leqslant 1$ and satisfy the Schrödinger equation for a free particle in the one-dimensional box $0 \leqslant x \leqslant 1$.

Construct the matrix representing the linear momentum

$$p \equiv \frac{\hbar}{i}\frac{\mathrm{d}}{\mathrm{d}x}$$

obtaining

$$p_{nm} = -i4\hbar nm/(n^2 - m^2) \qquad (n \pm m) \text{ odd}$$

$$= 0 \qquad (n \pm m) \text{ even.}$$

Confirm that the matrix $\mathbf{p}$ is Hermitian. Use the relation

$$\mathbf{H} = \frac{1}{2m}\mathbf{p}^2$$

to obtain the elements $\mathscr{H}_{11}$ and $\mathscr{H}_{22}$ of the diagonal matrix $\mathbf{H}$ and show that

$$\mathscr{H}_{11} = \frac{8\hbar^2}{m}\sum_{k=1}^{\infty}\frac{4k^2}{(4k^2-1)^2} = \frac{\hbar^2\pi^2}{2m}$$

$$\mathscr{H}_{12} = 0.$$

8 In the text it was argued that as the quantum mechanical differential operators are Hermitian then the representative matrices must also be Hermitian.

This final result may be obtained without reference to the Hermitian character of the operators.

Clearly, any physical observable must have a representation in which it is represented by a diagonal matrix with real elements. Using the result of theorem three show that an observable matrix is Hermitian in all representations.

9 The adjoint of a matrix $\mathbf{\Gamma}$ is denoted by $\mathbf{\Gamma}^+$ and is the transpose of the complex conjugate, i.e.

$$\mathbf{\Gamma}^+ = \tilde{\mathbf{\Gamma}}^*.$$

Show that a Hermitian matrix is equal to its adjoint (Self-adjoint) and that the inverse of a unitary matrix is equal to its adjoint.

10 Suppose the Hamiltonian operator $\mathscr{H}$ for a system can be written

$$\mathscr{H} \equiv \mathscr{H}_0 + \mathscr{H}'.$$

Let the orthonormal set $\{\Psi_n\}$ satisfy the equation

$$\mathscr{H}_0 \Psi_n = i\hbar \frac{\partial \Psi_n}{\partial t}.$$

A wave-function $\Psi$ satisfying the Schrödinger equation for the total Hamiltonian $\mathscr{H}$ can be expanded as

$$\Psi = \sum_n c_n(t) \Psi_n(\mathbf{r}, t).$$

(Observe that as the functions $\Psi(\mathbf{r}, t)$ and $\Psi_n(\mathbf{r}, t)$ satisfy different Schrödinger equations the expansion coefficients are functions of time. Compare equation (6.74) for the Heisenberg picture.)

Show that

$$\mathscr{H}' \sum_n c_n \Psi_n = i\hbar \sum_n \frac{\partial c_n}{\partial t} \Psi_n$$

and derive the matrix equation $\mathbf{H}'\mathbf{c} = i\hbar\dot{\mathbf{c}}$. (This representation is called the interaction representation. $\mathscr{H}_0$ represents two non-interacting systems and $\mathscr{H}'$ represents their mutual interaction. It reduces to the Heisenberg picture when $\mathbf{H}' = 0$.)

Show that the equation of motion of a matrix $\mathbf{\Gamma}$ representing a time independent operator is

$$\dot{\mathbf{\Gamma}} = \frac{i}{\hbar}(\mathbf{H}_0\mathbf{\Gamma} - \mathbf{\Gamma}\mathbf{H}_0).$$

Compare equation (6.78).

11 Using equation (6.23) show that the trace of a matrix is invariant under a similarity transformation. Hence deduce that the trace is equal to the sum of the eigenvalues.

12 Consider the definition of the density matrix (6.82). Show that under a change of basis functions (6.52–6.54) the density matrix transforms like an operator (6.56), i.e.

$$\boldsymbol{\rho}^{(2)} = \mathbf{u}^{-1}\boldsymbol{\rho}^{(1)}\mathbf{u}.$$

Hence show that

$$\text{Trace}\,(\boldsymbol{\rho}^{(2)}\boldsymbol{\alpha}^{(2)}) = \text{Trace}\,(\boldsymbol{\rho}^{(1)}\boldsymbol{\alpha}^{(1)}).$$

confirming that the mean value $[A]$ (6.84) is unchanged by the transformation.

13 It has been shown in the text that the density matrix of a pure state is idempotent, i.e.

$$\boldsymbol{\rho}^2 = \boldsymbol{\rho}.$$

This equation can be written

$$\boldsymbol{\rho}(\boldsymbol{\rho} - \mathbf{I}) = \mathbf{0}.$$

Explain why the eigenvalues of $\boldsymbol{\rho}$ are zero and one.
Show that

$$\text{Trace}\,\boldsymbol{\rho} = 1.$$

Hence deduce that the eigenvalue one is non-degenerate.

14 In Chapter 4 it was shown that the eigenfunctions of the one-dimensional harmonic oscillator are the Hermite functions

$$\psi_n(y) = e^{-y^2/2}H_n(y) \qquad (4.30)$$

where $H_n(y)$ is a Hermite polynomial.
The Hermite functions are orthogonal and satisfy

$$\int_{-\infty}^{\infty} \psi_n\psi_n \, dy = 2^n n!\sqrt{\pi}.$$

Given the recurrence relations

$$2n\psi_{n-1} = y\psi_n + \psi_n'$$

$$2y\psi_n = 2n\psi_{n-1} + \psi_{n+1}$$

show that

$$\int_{-\infty}^{\infty} \psi_m y \psi_n \, dy = \begin{cases} 0 & m \neq n \pm 1 \\ 2^n (n+1)! \sqrt{\pi} & m = n+1 \\ 2^{n-1} n! \sqrt{\pi} & m = n-1 \end{cases}$$

and

$$\int_{-\infty}^{\infty} \psi_m \psi_n' \, dy = \begin{cases} 0 & m \neq n \pm 1 \\ -2^n (n+1)! \sqrt{\pi} & m = n+1 \\ 2^{n-1} n! \sqrt{\pi} & m = n-1. \end{cases}$$

Multiply $\psi_n(y)$ by a suitable constant to normalize the wave-functions $\psi(x)$ (solution of (4.27)) to unity and confirm the following matrix representations for the position co-ordinate and momentum.

$$\mathbf{x} = \sqrt{\frac{\hbar}{2\omega m}} \begin{bmatrix} 0 & \sqrt{1} & 0 & 0 & 0 \ldots \\ \sqrt{1} & 0 & \sqrt{2} & 0 & 0 \\ 0 & \sqrt{2} & 0 & \sqrt{3} & 0 \\ \vdots & & & & \end{bmatrix}$$

$$\mathbf{p} = i\sqrt{\frac{\hbar\omega m}{2}} \begin{bmatrix} 0 & -\sqrt{1} & 0 & 0 & 0 \ldots \\ \sqrt{1} & 0 & -\sqrt{2} & 0 & 0 \\ 0 & \sqrt{2} & 0 & -\sqrt{2} & 0 \\ \vdots & & & & \end{bmatrix}.$$

Using the relation

$$\mathbf{H} = \frac{1}{2m}\mathbf{p}^2 + \frac{1}{2}\omega^2 m \mathbf{x}^2$$

show that **H** is diagonal.

## References

Neumann, J. von. *Mathematical foundations of Quantum Mechanics.* Princeton University Press (1955).

Sneddon, I. N. *Special functions of Mathematical Physics and Chemistry.* Oliver & Boyd, Edinburgh (2nd edition, 1961).

Ter Haar, D. Theory and applications of the Density Matrix. *Reports on Progress In Physics.* Vol. XXIV (1961).

CHAPTER 7

# Approximate Methods of Solution

## 7.1 Introduction

Most problems in quantum mechanics do not possess exact analytic solutions. Some do of course, and a few of these have been mentioned in Chapter 4. However, approximate methods of solution play an important part in the theory and the present chapter, together with Chapter 8, will be devoted to a discussion of them. This chapter will deal with stationary state problems where the aim is to evaluate the energies and eigenfunctions of some time-independent Hamiltonian operator.

Three methods will be discussed, the variational method, the method of linear combinations including the perturbation method and the Wentzel–Kramers–Brillouin–Jeffreys (W.K.B.J.) approximation. The relationship between the variational and perturbation methods will be emphazised. (See also Slater.)

## 7.2 The Variational Method

*Theorem 1. The Rayleigh–Ritz variational principle*

Suppose $\psi$ is a wave-function that is not necessarily normalized. Then the quotient

$$\langle\mathcal{H}\rangle = \int \psi^*\mathcal{H}\psi \, \mathrm{d}\tau \Big/ \int \psi^*\psi \, \mathrm{d}\tau \tag{7.1}$$

is an upper bound to the least eigenvalue of the quantum system described by the time-independent Hamiltonian $\mathcal{H}$.

*Proof*

As the Hamiltonian is time-independent it has a complete set of orthonormal eigenfunctions $\{\phi_n\}$ such that

$$\mathscr{H}\phi_n = E_n\phi_n. \tag{7.2}$$

An arbitrary wave-function can be expanded in terms of this set.

$$\psi = \sum_{n=1} c_n\phi_n, \qquad c_n \text{ constant.} \tag{7.3}$$

Consider the integral

$$\int \psi^*\mathscr{H}\psi \, d\tau = \int \sum_n c_n^*\phi_n^* \mathscr{H} \sum_m c_m\phi_m \, d\tau$$

$$= \sum_n E_n |c_n|^2.$$

If $E_1$ is the ground state of the system then $E_1 \leqslant E_n$ and clearly

$$\int \psi^*\mathscr{H}\psi \, d\tau \geqslant E_1 \sum_n |c_n|^2. \tag{7.4}$$

But

$$\sum_n |c_n|^2 = \int \psi^*\psi \, d\tau$$

and so

$$\frac{\int \psi^*\mathscr{H}\psi \, d\tau}{\int \psi^*\psi \, d\tau} \geqslant E_1 \tag{7.5}$$

as required. The equality applies when $\psi$ is the exact wave-function. This is a most important result. In practice an informed guess is made for the ground state wave-function which contains some parameters $\alpha_1, \alpha_2, \ldots$ say. The quotient (7.1) is evaluated and the parameters given the values that satisfy

$$\frac{\partial \langle\mathscr{H}\rangle}{\partial\alpha_i} = 0 \qquad i = 1, 2, \ldots \tag{7.6}$$

so that the expectation value $\langle\mathscr{H}\rangle$ is minimized. In general the energy eigenvalue obtained is considerably more accurate than the wave-function.

The method can be extended to find upper bounds to the excited states. Suppose it is desired to find an approximation to the first excited level $E_2$. The choice for $\psi$ is then of the form

$$\psi = \chi - \psi_1 \int \psi_1^*\chi \, d\tau \tag{7.7}$$

where $\chi$ contains the free parameters. In this case $\psi$ is orthogonal to the normalized determined ground state wave-function $\psi_1$. If $\psi_1$ is a good approximation to the true ground state function $\phi_1$ then the expansion for $\psi$ (analogous to (7.3)) will exclude $\phi_1$, i.e.

$$\psi = \sum_{n=2} c_n \phi_n. \tag{7.8}$$

It is left to the reader to verify that

$$\langle \mathscr{H} \rangle \geqslant E_2. \tag{7.9}$$

*Theorem 2. The virial theorem*

The virial theorem states that when a system is in a stationary state for which the potential is a homogeneous function of the co-ordinates of degree $s$ the expectation value of the potential energy is

$$\langle V \rangle = \frac{2}{s} \langle T \rangle \tag{7.10}$$

where $\langle T \rangle$ is the average kinetic energy. This theorem holds in both quantum and classical mechanics.

*Proof*

Consider a quantum mechanical system composed of $n$ particles of masses $m_i$. If they are at positions $\mathbf{r}_i$ the Hamiltonian operator is

$$\mathscr{H} \equiv -\sum_{i=1}^{n} \frac{\hbar^2}{2m_i} \nabla_i^2 + V(\mathbf{r}_1, \ldots, \mathbf{r}_n) \tag{7.11}$$

where $\nabla_i^2$ operates on the co-ordinates of the $i$th particle only and $V(\mathbf{r}_1, \ldots, \mathbf{r}_n)$ is the potential energy of the system. Since $V$ is homogeneous

$$V(\alpha \mathbf{r}_1, \ldots, \alpha \mathbf{r}_n) = \alpha^s V(\mathbf{r}_1, \ldots, \mathbf{r}_n). \tag{7.12}$$

Let $\chi(\mathbf{r}_1, \ldots, \mathbf{r}_n)$ be an arbitrary function and define the quotients

$$A = -\frac{\displaystyle\int \chi^*(\mathbf{r}_1, \ldots, \mathbf{r}_n) \sum_i \frac{\hbar^2}{2m_i} \nabla_i^2 \chi(\mathbf{r}_1, \ldots, \mathbf{r}_n)\, d\tau}{\displaystyle\int \chi^*(\mathbf{r}_1, \ldots, \mathbf{r}_n) \chi(\mathbf{r}_1, \ldots, \mathbf{r}_n)\, d\tau}$$

$$B = \frac{\displaystyle\int \chi^*(\mathbf{r}_1, \ldots, \mathbf{r}_n) V(\mathbf{r}_1, \ldots, \mathbf{r}_n) \chi(\mathbf{r}_1, \ldots, \mathbf{r}_n)\, d\tau}{\displaystyle\int \chi^*(\mathbf{r}_1, \ldots, \mathbf{r}_n) \chi(\mathbf{r}_1, \ldots, \mathbf{r}_n)\, d\tau}$$

the integrations being over all values of the co-ordinates.

Consider now the trial wave function

$$\psi(\mathbf{r}_1, \ldots, \mathbf{r}_n) = \chi(\alpha \mathbf{r}_1, \ldots, \alpha \mathbf{r}_n) \tag{7.13}$$

where $\alpha$ is the variational parameter. The expectation values of the kinetic and potential energies are respectively

$$\langle T\rangle = \alpha^2 A \qquad \text{and} \qquad \langle V\rangle = \alpha^{-s} B.$$

(To show this carry out the transformation $\mathbf{r}_i' = \alpha \mathbf{r}_i$, $i = 1, \ldots, n$ and note that the Jacobian is the same constant for all integrals and cancels.)

The expectation value of the Hamiltonian is

$$\langle \mathcal{H}\rangle = \alpha^2 A + \alpha^{-s} B$$

and the best choice for $\alpha$ is that which satisfies

$$\frac{\partial\langle\mathcal{H}\rangle}{\partial\alpha} = 0 = 2\alpha A - s\alpha^{-s-1} B$$

or

$$\frac{2}{s}(\alpha^2 A) = (\alpha^{-s} B) \tag{7.14}$$

then

$$\langle V\rangle = \frac{2}{s}\langle T\rangle.$$

In the case that the particles have a charge $e_i$ respectively and interact through a Coulombic potential

$$V(\mathbf{r}_1, \ldots, r_n) = \sum_{i=1}^{n} \sum_{j>i} \frac{e_i e_j}{4\pi\epsilon_0 |\mathbf{r}_i - \mathbf{r}_j|} \tag{7.15}$$

and $s = -1$. Then

$$\langle V\rangle = -2\langle T\rangle = +2\langle\mathcal{H}\rangle.$$

This result must hold in the case that $\psi$ is an exact eigenfunction of the Hamiltonian with energy $E$ and then

$$E = -\langle T\rangle = \tfrac{1}{2}\langle V\rangle. \tag{7.16}$$

Note that the kinetic energy has a maximum value in the lowest energy state.

*Ground state of the helium atom*

The Hamiltonian operator for the helium atom may be written

$$\mathcal{H} \equiv -\frac{\hbar^2}{2m}(\nabla_1^2 + \nabla_2^2) - \frac{2e^2}{4\pi\epsilon_0 r_1} - \frac{2e^2}{4\pi\epsilon_0 r_2} + \frac{e^2}{4\pi\epsilon_0|\mathbf{r}_1 - \mathbf{r}_2|} \tag{7.17}$$

where the two electrons are at $\mathbf{r}_1$ and $\mathbf{r}_2$ respectively. The kinetic energy of the nucleus and the spin-interaction terms are ignored.

If the electron–electron interaction term $e^2/(4\pi\epsilon_0|\mathbf{r}_1 - \mathbf{r}_2|)$ is ignored, the Schrödinger equation is separable and the helium wave-function is simply a product of two ground-state hydrogen type wave-functions (see (4.73)), i.e.

$$\psi(\mathbf{r}_1, \mathbf{r}_2) = \frac{1}{\pi}\left(\frac{2}{a_0}\right)^3 \exp\left[-\frac{2}{a_0}(r_1 + r_2)\right]. \tag{7.18}$$

In fact the two electrons will screen each other from the nucleus and so a reasonable simple trial function is to write

$$\psi(\mathbf{r}_1, \mathbf{r}_2) = \frac{1}{\pi}\left(\frac{\alpha}{a_0}\right)^3 \exp\left[-\frac{\alpha}{a_0}(r_1 + r_2)\right] \tag{7.19}$$

where $\alpha$ is a parameter. This wave-function is already normalized.

The expectation value of the kinetic energy is clearly

$$\langle T\rangle = 2\langle T_H\rangle \tag{7.20}$$

where $\langle T_H\rangle$ is the expectation value of the hydrogen type atom ($Z = \alpha$) in its ground state. From the virial theorem $\langle T_H\rangle$ is equal to the binding energy of the hydrogen type atom in its ground state and so from (4.71)

$$\langle T\rangle = 2\left(\frac{\alpha^2 e^4 m}{2\hbar^2}\right)\frac{1}{(4\pi\epsilon_0)^2}$$

$$= \frac{\alpha^2 e^2}{a_0(4\pi\epsilon_0)}. \tag{7.21}$$

Similarly, the expectation value of the potential energy is

$$\langle V\rangle = \frac{4}{\alpha}\langle V_H\rangle + \left\langle\frac{e^2}{4\pi\epsilon_0|\mathbf{r}_1 - \mathbf{r}_2|}\right\rangle, \tag{7.22}$$

where $\langle V_H\rangle$ is the potential energy of the hydrogen type atom ($z = \alpha$) and from the virial theorem

$$\langle V_H\rangle = \frac{-\alpha^2 e^2}{a_0(4\pi\epsilon_0)}. \tag{7.23}$$

The remaining term in the potential energy (7.22) represents the interaction of two superimposed spherical symmetric charge distributions and it can be shown that

$$\left\langle\frac{e^2}{4\pi\epsilon_0|\mathbf{r}_1 - \mathbf{r}_2|}\right\rangle = \frac{5}{8}\alpha\frac{e^2}{a_0(4\pi\epsilon_0)}. \tag{7.24}$$

From equations (7.21)–(7.24) the expectation value of the Hamiltonian is

$$\langle\mathscr{H}\rangle = \left(\alpha^2 - 4\alpha + \frac{5}{8}\alpha\right)\frac{e^2}{4\pi\epsilon_0 a_0}. \tag{7.25}$$

The best value for the parameter is obtained from

$$\frac{\partial\langle\mathscr{H}\rangle}{\partial\alpha} = \frac{e^2}{4\pi\epsilon_0 a_0}\left(2\alpha - \frac{27}{8}\right) = 0$$

i.e.

$$\alpha = \tfrac{27}{16}. \tag{7.26}$$

The least upper bound for the ground state energy that can be obtained from the trial function (7.19) is

$$\langle\mathscr{H}\rangle = -\left(\frac{27}{16}\right)^2 \frac{e^2}{4\pi\epsilon_0 a_0} \simeq -2{\cdot}848\,\frac{e^2}{4\pi\epsilon_0 a_0}. \tag{7.27}$$

The corresponding binding energy, that is, the energy required to remove both of the electrons, is $+2{\cdot}848\ e^2/4\pi\epsilon_0 a_0$. The experimental value is $+2{\cdot}904\ e^2/4\pi\epsilon_0 a_0$.

## 7.3 The Method of Linear Combinations

As has already been mentioned in the introduction, this chapter deals with stationary perturbation theory in which the Hamiltonian is time-independent. The variation principle may be used to derive the appropriate expansions.

The object is to find the eigensolutions of some given Hamiltonian $\mathscr{H}$. That is, to solve

$$\mathscr{H}\psi = E\psi. \tag{7.28}$$

Let $\{\chi_j\}$ be a set of $n$ linearly-independent functions which for the moment will not be assumed to be orthogonal. Observe that this is a finite set of functions and so is not complete. Suppose this set is used to give an approximate expansion for an eigenfunction of $\mathscr{H}$, i.e.

$$\psi = \sum_{j=1}^{n} c_j\chi_j. \tag{7.29}$$

The wave-function (7.29) may be used in a variation problem where the expansion coefficients are regarded as parameters. The expectation value for the Hamiltonian (7.1) is

$$\langle \mathscr{H} \rangle = \frac{\int \sum_{i=1}^{n} c_i^* \chi_i^* \mathscr{H} \sum_{j=1}^{n} c_j \chi_j \, d\tau}{\int \sum_{i=1}^{n} c_i^* \chi_i^* \sum_{j=1}^{n} c_j \chi_j \, d\tau}$$

$$= \frac{\sum_{i=1}^{n} \sum_{j=1}^{n} c_i^* \mathscr{H}_{ij} c_j}{\sum_{i=1}^{n} \sum_{j=1}^{n} c_i^* S_{ij} c_j} \tag{7.30}$$

where

$$\mathscr{H}_{ij} = \int \chi_i^* \mathscr{H} \chi_j \, d\tau$$

and

$$S_{ij} = \int \chi_i^* \chi_j \, d\tau. \tag{7.31}$$

In matrix notation

$$\langle \mathscr{H} \rangle = \frac{\tilde{\mathbf{c}}^* \mathbf{H} \mathbf{c}}{\tilde{\mathbf{c}}^* \mathbf{S} \mathbf{c}}. \tag{7.32}$$

**H** and **S** are both Hermitian and **c** is a column matrix. A very simple argument shows that this expectation value is real. As a scalar quantity is equal to its own transpose

$$\langle \mathscr{H} \rangle = \frac{\tilde{\mathbf{c}} \tilde{\mathbf{H}} \mathbf{c}^*}{\tilde{\mathbf{c}} \tilde{\mathbf{S}} \mathbf{c}^*}.$$

However as **H** and **S** are both Hermitian then

$$\langle \mathscr{H} \rangle = \frac{\tilde{\mathbf{c}} \mathbf{H}^* \mathbf{c}^*}{\tilde{\mathbf{c}} \mathbf{S}^* \mathbf{c}^*}.$$

This final result shows that $\langle \mathscr{H} \rangle$ is equal to its own complex conjugate and so is real.

The best trial solution of the type (7.29) is that which satisfies

$$\frac{\partial \langle \mathscr{H} \rangle}{\partial c_l^*} = 0 \qquad l = 1, 2, \ldots, n \tag{7.33}$$

$$\therefore \quad \left( \sum_{j=1}^{n} \mathscr{H}_{lj} c_j - \frac{\sum_i^n \sum_j^n c_i^* \mathscr{H}_{ij} c_j}{\sum_i^n \sum_j^n c_i^* S_{ij} c_j} \sum_j^n S_{lj} c_j \right) = 0 \qquad l = 1, 2, \ldots, n \tag{7.34}$$

Writing $E_A$ for the real approximate value of the energy, these $n$ equations are in matrix form

$$\mathbf{Hc} - E_A \mathbf{Sc} = 0. \tag{7.35}$$

If $\langle \mathscr{H} \rangle$ is differentiated with respect to $c_l$ instead of $c_l^*$ an equivalent set of equations is obtained.

The equation (7.35) can be obtained directly by substituting (7.29) in (7.28), integrating and replacing $E$ by $E_A$. The condition for the set of $n$ equation (7.35) to have a non-trivial solution is that the characteristic determinant be zero, i.e.

$$\det(\mathbf{H} - E_A \mathbf{S}) = 0. \tag{7.36}$$

The determinant (7.36) is a polynomial of degree $n$ in $E_A$ and the lowest root is the best approximation to the ground state of the system.

### *Hydrogen-molecular ion $H_2^+$*

This is the simplest *molecular* structure and may be approximated by a one-electron problem, the two nuclei being in fixed positions. The electron Hamiltonian is

$$\mathscr{H} \equiv -\frac{\hbar^2}{2m} \nabla^2 - \frac{1}{4\pi\epsilon_0} \left( \frac{e^2}{r_1} + \frac{e^2}{r_2} \right) + \frac{e^2}{4\pi\epsilon_0 R}$$

where $r_1, r_2$ are the distances of the electron from the protons which are a distance $R$ apart. This problem can be solved exactly in elliptic coordinates and the energy found for different values of $R$. The minimum energy occurs when $R = 1{\cdot}06$ Å in agreement with experiment. The corresponding binding energy is 2·8 eV.

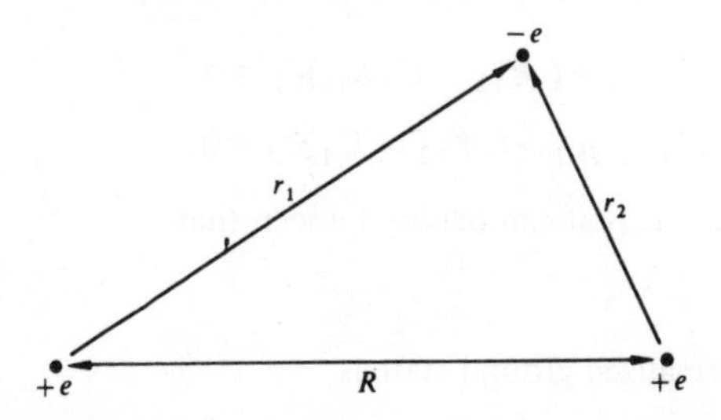


Figure 7.1 The hydrogen molecular ion.

However, as an illustration, this problem is solved by expressing the molecular orbital as a linear combination of atomic orbitals. The simplest approximation to the ground state is

$$\psi = c_1\psi_1 + c_2\psi_2$$

where $\psi_1$, $\psi_2$ are the real normalized hydrogen (1s) wave functions centred on the two protons and $c_1, c_2$ are constants.

The best solution of this type is obtained by solving the characteristic determinant

$$\begin{vmatrix} \mathscr{H}_{11} - E_A & \mathscr{H}_{12} - E_A S_{12} \\ \mathscr{H}_{21} - E_A S_{21} & \mathscr{H}_{22} - E_A \end{vmatrix} = 0$$

where

$$S_{12} = S_{21} = \int \psi_1 \psi_2 \, d\tau$$

as $\psi_1$ and $\psi_2$ are real functions, and also

$$\mathscr{H}_{12} = \mathscr{H}_{21} = \int \psi_1 \mathscr{H} \psi_2 \, d\tau$$

$$\mathscr{H}_{11} = \mathscr{H}_{22} = \int \psi_1 \mathscr{H} \psi_1 \, d\tau.$$

The two roots of the determinant are

$$E_A = \frac{\mathscr{H}_{11} \pm \mathscr{H}_{12}}{1 \pm S_{12}}.$$

As $\psi_1$ and $\psi_2$ are both positive in the region of overlap then it can be shown that $\mathscr{H}_{12} < 0$ and so the lowest root is

$$E_{A1} = \frac{\mathscr{H}_{11} + \mathscr{H}_{12}}{1 + S_{12}}.$$

This corresponds to a binding energy of $\sim 1{\cdot}8$ eV which involves an error of 1 eV.

The values of $c_1$ and $c_2$ are obtained by solving the two equations

$$(\mathscr{H}_{11} - E_A)c_1 + (\mathscr{H}_{12} - E_A S_{12})c_2 = 0$$

$$(\mathscr{H}_{21} - E_A S_{21})c_1 + (\mathscr{H}_{22} - E_A)c_2 = 0.$$

By putting $E_A = E_{A1}$ it can readily be seen that

$$c_1 = c_2$$

and so the normalized ground state is

$$\psi_g = (\psi_1 + \psi_2)/\sqrt{2(1 + S_{12})}.$$

This is an even function, (gerade or $g$-state) and represents a bonding state in which charge is transferred to the bond, i.e.

$$|\psi_g|^2 = \frac{1}{2(1+S_{12})}(\psi_1^2 + 2\psi_1\psi_2 + \psi_2^2).$$

The other root of the determinant corresponds to the odd function (ungerade or $u$-state), i.e.

$$\psi_u = (\psi_1 - \psi_2)/\sqrt{2(1-S_{12})}$$

and represents an antibonding state in which charge is removed from the bond.

*Theorem 3*

Let $E_r$ be the $r$th root of (7.36) and $c_{rj}, j = 1, \ldots, n$, the corresponding set of expansion coefficients. The $n$ approximate eigenfunctions are

$$\psi_r = \sum_{j=1}^{n} c_{rj}\chi_j \qquad r = 1, 2, \ldots, n. \tag{7.37}$$

These approximate eigenfunctions are orthogonal.

*Proof*

Let $E_q$ and $E_r$ be two different approximate eigenvalues with corresponding eigenfunctions $\psi_q, \psi_r$

$$\int \psi_q^*\psi_r \,\mathrm{d}\tau = \sum_{i=1}^{n}\sum_{j=1}^{n} c_{qi}^* c_{rj}\int \chi_i^*\chi_j \,\mathrm{d}\tau = \sum_i\sum_j c_{qi}^* S_{ij} c_{rj}$$
$$= \tilde{\mathbf{c}}_q^* \mathbf{S}\mathbf{c}_r. \tag{7.38}$$

From (7.35)

$$\mathbf{H}\mathbf{c}_q = E_q\mathbf{S}\mathbf{c}_q \quad \text{(a)} \qquad \mathbf{H}\mathbf{c}_r = E_r\mathbf{S}\mathbf{c}_r \quad \text{(b)}.$$

Transpose (a) and take the complex conjugate

$$\tilde{\mathbf{c}}_q^*\tilde{\mathbf{H}}^* = E_q\tilde{\mathbf{c}}_q^*\tilde{\mathbf{S}}^*.$$

As the matrices **H** and **S** are Hermitian

$$\tilde{\mathbf{c}}_q^*\mathbf{H} = E_q\tilde{\mathbf{c}}_q^*\mathbf{S}.$$

Post multiply this equation by $\mathbf{c}_r$ and premultiply (b) by $\tilde{\mathbf{c}}_q^*$

$$\tilde{\mathbf{c}}_q^*\mathbf{H}\mathbf{c}_r = E_q\tilde{\mathbf{c}}_q^*\mathbf{S}\mathbf{c}_r \qquad \tilde{\mathbf{c}}_q^*\mathbf{H}\mathbf{c}_r = E_r\tilde{\mathbf{c}}_q^*\mathbf{S}\mathbf{c}_r \tag{7.39}$$

On subtraction these equations give

$$(E_q - E_r)\tilde{\mathbf{c}}_q^*\mathbf{S}\mathbf{c}_r = 0.$$

If the eigenvalues are not degenerate, this equation implies

$$\tilde{\mathbf{c}}_q^* \mathbf{S} \mathbf{c}_r = 0.$$

From (7.38), the approximate eigenfunctions are orthogonal. If these functions are normalized then

$$\int \psi_q^* \psi_r \, d\tau = \delta_{qr}. \tag{7.40}$$

*Discussion*

The linearly-independent functions $\{\chi_j\}$, $j = 1, \ldots, n$ define an $n$-dimensional sub-space of the system function space. In the Rayleigh–Ritz variational method the best approximation to the ground state is obtained by minimizing the quotient (7.1). The best approximation to the $r$th excited state is obtained by minimizing (7.1) subject to the condition that the trial function is orthogonal to the $(r-1)$ approximate eigenfunctions for the lower eigenvalues. The results of the Rayleigh–Ritz method are upper bounds to the true eigenvalues.

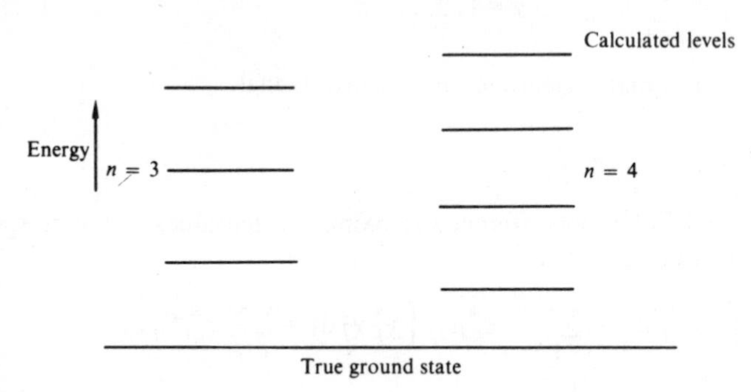


Figure 7.2

The method of linear combination is essentially the Rayleigh–Ritz method with the trial functions confined to the $n$-dimensional sub-space $\{\chi_j\}$. The $n$ roots of the characteristic determinant

$$D_n(E_A) \equiv \det(\mathbf{H} - E_A \mathbf{S}) = 0 \tag{7.41}$$

give the corresponding best possible approximations to the lowest eigenvalues. Of course the approximation is best for the deepest levels.

Consider now the determinant $D_{n+1}(E_A)$ constructed from the $(n+1)$ independent functions $\chi_1, \ldots, \chi_n, \chi_{n+1}$. This problem is equivalent to Rayleigh–Ritz minimization method with the trial function $\psi$ restricted to the $(n+1)$-dimensional sub-space spanned by $\{\chi_j\}$, $j = 1, \ldots, n+1$, i.e.

$$\psi = \sum_{j=1}^{n+1} c_j \chi_j. \tag{7.42}$$

The greater flexibility of the form of $\psi$ implies that the roots of $D_{n+1}(E_A)$ will be less than or equal to the corresponding roots of $D_n(E_A)$. If the $r$th roots are written $E_r^{(n+1)}$ and $E_r^{(n)}$, the superscript denoting the dimensionality of the associated function space, then

$$E_r^{(n+1)} \leqslant E_r^{(n)}. \tag{7.43}$$

The roots of the new determinant $D_{n+1}(E_A)$ give a better approximation to the previous $n$ eigenvalues as well as introducing an approximation to another eigenvalue.

*Perturbation of non-degenerate levels*

The method of linear combinations can be used to obtain the well-known perturbation expansions. Consider the case where the Hamiltonian can be written

$$\mathscr{H} \equiv \mathscr{H}^0 + \hbar \tag{7.44}$$

where the energy associated with $\mathscr{H}^0$ is large compared with that associated with $\hbar$. $\mathscr{H}^0$ must be chosen so that its eigenvalues and eigenfunctions are known. Let the functions $\{\chi_j\}$ be these eigenfunctions

$$\mathscr{H}^0\chi_j = E_j^0\chi_j.$$

It is convenient to consider the Hamiltonian

$$\mathscr{H} \equiv \mathscr{H}^0 + \lambda\hbar \qquad 0 \leqslant \lambda < 1. \tag{7.45}$$

The parametric value $\lambda = 1$ corresponds to 'complete' perturbation and $\lambda = 0$ to zero perturbation. The Hamiltonian (7.45) is solved and then $\lambda$ put equal to unity in the results to obtain the required solution of (7.44).

The unperturbed Hamiltonian eigenfunctions $\{\chi_j\}$ are orthogonal and may be chosen to be normalized. Following (7.29), the first $n$ of these eigenfunctions, corresponding to the $n$ lowest energies, are used as an expansion set. The overlap integrals (7.31) are

$$S_{ij} = \delta_{ij}$$

and the elements of the Hamiltonian matrix are

$$\mathscr{H}_{ij} = E_j^0\delta_{ij} + \lambda h_{ij}$$

where

$$h_{ij} = \int \chi_i^* \hbar \chi_j \, d\tau.$$

The $n$ equation (7.35) become

$$\sum_{j=1}^{n} (E_j^0\delta_{lj} + \lambda h_{lj} - E_A\delta_{lj})c_j = 0 \qquad l = 1, 2, \ldots, n. \tag{7.46}$$

In the unperturbed case $\lambda = 0$, the determinant of (7.46) is diagonal and

$$\prod_{j=1}^{n} (E_j^0 - E_A) = 0 \tag{7.47}$$

giving the $n$ solutions

$$E_A = E_1^0, E_2^0, \ldots, E_n^0$$

as expected.

The basic assumption of perturbation theory is that it is possible to expand the perturbed energies and eigenfunctions as power series in $\lambda$. There are some cases when this assumption is invalid. For example if $\mathscr{H}^0$ has a discrete spectrum and $\mathscr{H}$ has a continuous spectrum. For the ground state, the unperturbed eigenvalue and eigenfunction are $E_1^0$ and $\chi_1$, respectively. The perturbed ground state energy is written

$$E_{1A} = E_1^0 + \lambda E_1^{(1)} + \lambda^2 E_1^{(2)} + \ldots \tag{7.48}$$

and the perturbed ground state wave function

$$\psi_1 = \chi_1 + \lambda \psi_1^{(1)} + \lambda^2 \psi_2^{(2)} + \ldots. \tag{7.49}$$

$E_1^{(1)}$ is of the same order as $\hbar$ and $E_1^{(2)}$ is of the second order of smallness. (The reader is warned not to confuse $E_1^{(1)}$ etc. with the entirely different meaning given to $E_1^{(n)}$ in (7.43).)

With the energy expansion (7.48), the characteristic determinant of the equations (7.46) becomes

$$\begin{vmatrix} \lambda h_{11} - \lambda E_1^{(1)} - \lambda^2 E_1^{(2)}, & \lambda h_{12}, \ldots & & \lambda h_{1n} \\ \lambda h_{21}, & \lambda h_{22} + E_2^0 - E_1^0 - \lambda E_1^{(1)} - \lambda^2 E_1^{(2)}, \ldots & & \\ \cdot & & & \\ \cdot & & & \\ \cdot & & & \\ \lambda h_{n1}, & & & \end{vmatrix} = 0 \tag{7.50}$$

This determinant is a polynomial in $\lambda$ of degree $n$ and is identically zero for $0 \leqslant \lambda \leqslant 1$. The coefficient of each power of $\lambda$ must vanish. The term in $\lambda$, corresponding to the first order in smallness, is the diagonal determinant

$$\begin{vmatrix} \lambda(h_{11} - E_1^{(1)}), & 0, & 0, \ldots, 0 & \\ 0, & E_2^0 - E_1^0, & & \\ \cdot & & E_3^0 - E_1^0, & \\ \cdot & & & \\ \cdot & & & \\ 0 & & & E_n^0 - E_1^0 \end{vmatrix} = 0$$

i.e.

$$(h_{11} - E_1^{(1)})(E_2^0 - E_1^0)\dots(E_n^0 - E_1^0) = 0. \tag{7.51}$$

If the ground state eigenvalue $E_1^0$ of the unperturbed Hamiltonian is non-degenerate

$$E_1^{(1)} = h_{11} = \int \chi_1^* h \chi_1 \, d\tau. \tag{7.52}$$

This is the first-order perturbation of the ground state energy. By taking

$$E_{jA} = E_j^0 + \lambda E_j^{(1)} + \lambda^2 E_j^{(2)} + \dots \tag{7.53}$$

it can similarly be shown that the first order perturbation of the $j$th state, (assumed non-degenerate), is

$$E_j^{(1)} = h_{jj} = \int \chi_j^* h \chi_j \, d\tau$$

and so

$$E_{jA} = E_j^0 + \lambda \int \chi_j^* h \chi_j \, d\tau. \tag{7.54}$$

For complete perturbation $\lambda = 1$ and so to first order, the energy of the perturbed $j$th state is

$$E_{jA} = E_j^0 + \int \chi_j^* h \chi_j \, d\tau. \tag{7.55}$$

(Strictly $E_{jA}$ is a power series in $\lambda$ and ought to be investigated for convergence to decide if it is valid to put $\lambda = 1$.) The energy (7.55) is the expectation value of the Hamiltonian (7.44) for the unperturbed wave function $\chi_j$.

The first-order change in the ground state wave function is expanded in terms of the first $n$ functions of the set $\{\chi_j\}$ so that

$$\psi_1^{(1)} = \sum_{i=1}^{n} c_i^{(1)} \chi_i$$

and

$$\psi_1 = (1 + \lambda c_1^{(1)})\chi_1 + \lambda \sum_{i=2}^{n} c_i^{(1)} \chi_i.$$

When $E_{1A}$ is given its value (7.54), the $n$ equations (7.46) are, keeping first-order terms only,

$$h_{l1} + (E_l^0 - E_1^0)c_l^{(1)} = 0 \qquad l = 2, 3, \dots, n$$

$$c_l^{(1)} = \frac{h_{l1}}{E_1^0 - E_l^0} \qquad l = 2, 3, \dots, n.$$

To first order in $\lambda$, the first of the equations (7.46) is the trivial identity zero equals zero and so $c_1^{(1)}$ is not obtained by this method. The value of $c_1^{(1)}$ is chosen so that $\psi_1$ is normalized and by considering first-order terms it is readily shown that the choice $c_1^{(1)} = 0$ is satisfactory.

So, the coefficients in the expansion (7.29), for the ground state wave function, are to first order

$$c_1 = 1 \qquad c_l^{(1)} = \frac{h_{l1}}{E_1^0 - E_l^0} \qquad l = 2, 3, \ldots, n. \tag{7.56}$$

By an exactly similar argument the coefficients for the $j$th state wave function are, to first order

$$c_j = 1 \qquad c_l^{(1)} = \frac{h_{lj}}{E_j^0 - E_l^0} \qquad l \neq j \tag{7.57}$$

and for complete perturbation with $\lambda = 1$, the $j$th perturbed wave function is

$$\psi_j = \chi_j + \sum_{l \neq j}^{n} \frac{h_{lj}}{E_j^0 - E_l^0} \chi_l. \tag{7.58}$$

The second-order changes due to the perturbation can also be calculated. To find the second-order change in the energy of the ground state the determinant (7.50) is expanded by the first row keeping only those terms in $\lambda^2$. Then

$$\begin{aligned}\lambda^2\{&-E_1^{(2)}(E_2^0 - E_1^0)(E_3^0 - E_1^0)\ldots(E_n^0 - E_1^0)\\ &-|h_{12}|^2(E_3^0 - E_1^0)\ldots(E_n^0 - E_1^0) - |h_{13}|^2(E_2^0 - E_1^0)(E_4^0 - E_1^0)\\ &\ldots(E_n^0 - E_1^0)\ldots - |h_{1n}|^2(E_2^0 - E_1^0)\ldots(E_{n-1}^0 - E_1^0)\} = 0.\end{aligned}$$

This is equivalent to the bordered diagonal determinant

$$\begin{vmatrix} -\lambda^2 E_1^{(2)} & \lambda h_{12} & \lambda h_{13} & \ldots & \lambda h_{1n} \\ \lambda h_{21} & E_2^0 - E_1^0 & 0 & \ldots & 0 \\ \lambda h_{31} & 0 & E_3^0 - E_1^0 & & \\ \vdots & \vdots & & \ddots & \\ \lambda h_{n1} & 0 & & & E_n^0 - E_1^0 \end{vmatrix} = 0.$$

On the assumption that the ground state energy is non-degenerate, the second-order change in the energy of the ground state is

$$E_1^{(2)} = -\sum_{l=2}^{n} \frac{|h_{1l}|^2}{E_l^0 - E_1^0}.$$

This result may be obtained directly by substituting (7.56) into the first of the equations (7.46) and picking the second-order terms. Observe that the second-order change in the energy of the ground state is always negative as $E_l^0 > E_1^0$.

By a similar argument, the second-order change of the energy of the $j$th state, (assumed non-degenerate), is

$$E_j^{(2)} = -\sum_{l \neq j}^{n} \frac{|h_{jl}|^2}{E_l^0 - E_j^0}.$$

So to second order, the perturbed energy of the $j$th state with $\lambda = 1$ is

$$E_{jA} = E_j^0 + h_{jj} - \sum_{l \neq j}^{n} \frac{|h_{jl}|^2}{E_l^0 - E_j^0}. \tag{7.59}$$

Note that if the wave function is known to first order (7.58), enough information is known to calculate the energy to second order.

If the number of basis functions is increased the calculated value will be improved. In the case that the complete set of eigenfunctions of $\mathscr{H}^0$ is used, the summations (7.58) and (7.59) include an infinite number of terms. These are obtained directly in the more usual method of obtaining the perturbation expansions but this does not show the relationship between the variation and perturbation methods.

The perturbation method is applicable when although the true Schrödinger equation cannot be solved, solutions can be found for a derived equation, which differs from the true one by terms that are small. Often this cannot be done or perhaps the convergence is poor. Then the variation method is often useful.

*Van der Waals interaction*

The attractive forces that hold most atoms together to form molecules are very short range and can be ignored for atomic separations greater than a few angström units. Even so there are weak attractive forces that act between neutral atoms at separations large compared with the Bohr radius. These are the van der Waals forces and they vary as $1/R^6$ where $R$ is the atomic separation. They are due to a mutual polarization of the two atoms and help to determine the properties of gases.

Consider two hydrogen atoms at a fixed distance $R$ apart. This is a two-electron problem and the Hamiltonian is

$$\mathscr{H} \equiv -\frac{\hbar^2}{2m}(\nabla_1^2 + \nabla_2^2) - \frac{1}{4\pi\epsilon_0}\left(\frac{e^2}{r_{1a}} + \frac{e^2}{r_{1b}} + \frac{e^2}{r_{2a}} + \frac{e^2}{r_{2b}}\right) + \frac{e^2}{4\pi\epsilon_0 r_{12}} + \frac{e^2}{4\pi\epsilon_0 R}$$

where $r_{1a}, r_{1b}$ are the distances of the first electron from the first and second protons at '$a$' and '$b$' and similarly $r_{2a}, r_{2b}$ are the distances of the second electron from the protons. The electron separation is $r_{12}$ (see Figs. 7.1 and 11.2).

The Hamiltonian

$$\mathscr{H}^0 \equiv -\frac{\hbar^2}{2m}(\nabla_1^2 + \nabla_2^2) - \frac{1}{4\pi\epsilon_0}\left(\frac{e^2}{r_{1a}} + \frac{e^2}{r_{2b}}\right)$$

represents the separated hydrogen atoms and has known exact solutions. The 'atom-atom' interaction Hamiltonian

$$h \equiv -\frac{e^2}{4\pi\epsilon_0}\left(\frac{1}{r_{1b}} + \frac{1}{r_{2a}}\right) + \frac{e^2}{4\pi\epsilon_0}\left(\frac{1}{r_{12}} + \frac{1}{R}\right)$$

can be regarded as a perturbation.

For the ground state both electrons are in the $(1s)$ state (about different protons) and the wave function is

$$\exp \psi_1(\mathbf{r}_1, \mathbf{r}_2) = \frac{1}{\pi}\left(\frac{1}{a_0}\right)^3 \exp\left[-\frac{1}{a_0}(r_{1a} + r_{2b})\right].$$

It is convenient to choose a co-ordinate system with the $z$-axis along the line of atoms and an expansion of $h$ in powers of $1/R$, keeping the first term only, gives

$$h \simeq \frac{e^2}{4\pi\epsilon_0 R^3}(x_{1a}x_{2b} + y_{1a}y_{2b} - 2z_{1a}z_{2b})$$

where $(x_{1a}, y_{1a}, z_{1a})$ and $(x_{2b}, y_{2b}, z_{2b})$ are the co-ordinates of each electron relative to their respective protons. This can be written

$$h \simeq \frac{1}{4\pi\epsilon_0 R^3}\left(\mathbf{p}_1 \cdot \mathbf{p}_2 - \frac{3(\mathbf{p}_1 \cdot \mathbf{R})(\mathbf{p}_2 \cdot \mathbf{R})}{R^2}\right)$$

where

$$\mathbf{p}_1 = -e(x_{1a}\mathbf{i} + y_{1a}\mathbf{j} + z_{1a}\mathbf{k})$$

and similarly for $\mathbf{p}_2$. The perturbation represents the mutual interaction energy of the two atomic dipoles at any instant.

Since $h$ is an odd function the first-order perturbation integral vanishes. The second-order energy change of the ground state is

$$E_1^{(2)} = -\sum_{l>1} \frac{|h_{1l}|^2}{E_l^0 - E_1^0}$$

where $h_{1l}$ is the matrix element between the ground state and the $l$th excited state. Since $E_l^0 > E_1^0$ the energy change is negative and represents an attraction between the atoms. The form of the perturbation indicates that $E_1^{(2)} \alpha - 1/R^6$.

Eisenchitz and London (*Z. f. Phys.* **60**, 491, 1930) carried out this calculation and obtained the result

$$E_1^{(2)} = \mp \frac{6{\cdot}47\, e^2 a_0^5}{(4\pi\epsilon_0)^2 R^6}.$$

*The second-order Stark effect*

Consider the effect of an electric field $\mathscr{E}$, acting along the $z$-axis, on the ground state of the hydrogen atom. The unperturbed Hamiltonian is

$$\mathscr{H}^0 \equiv -\frac{\hbar^2}{2\mu}\nabla^2 - \frac{e^2}{4\pi\epsilon_0 r}$$

where $\mu$ is the reduced electron mass. The perturbation $h$ is the extra energy of the system due to the electric field and is equal to $-\mathbf{p}\,.\,\mathscr{E}$ where $\mathbf{p}$ is the dipole moment of the atom, i.e.

$$h = e\,\mathscr{E}\,z.$$

(The electron change is $-e$).

The unperturbed ground state wave-function is (4.73),

$$\psi_{100} = \left(\frac{1}{a_0^3\pi}\right)^{1/2} e^{-r/a_0}.$$

As this function is spherically symmetric then clearly the first-order perturbation to the ground state energy is zero.

However there is a second-order change in the energy. From (7.59) this is

$$E_1^{(2)} = -e\,\mathscr{E}^2 \sum_{q=2}^{\infty} \frac{|z_{1q}|^2}{E_q^0 - E_1^0} \tag{7.60}$$

where $z_{1q} = \int \psi_{100} z \psi_q \,d\tau$ and $\psi_q$ is an unperturbed excited hydrogen wave-function. From the form of the hydrogen atom wave-functions (4.72), it is clear that $z_{1q}$ is non-zero only for those excited states for which the quantum number $m = 0$. By further symmetry arguments it can also be shown that $z_{1q}$ is non-zero only when the excited state $l$ quantum number is one.

The change in the ground state energy is proportional to the square of the field strength and a polarizability $\alpha$ may be defined so that this change is $-\alpha\mathscr{E}^2/2$. From (7.60)

$$\alpha = 2\,e^2 \sum_{q=2}^{\infty} \frac{|z_{1q}|^2}{E_q^0 - E_1^0}. \tag{7.61}$$

The dominant term in (7.61) is clearly that obtained from the $(2p_z)$ state when $n = 2$. A fairly short calculation then shows that

$$\alpha > 2{\cdot}9(a_0^3 4\pi\epsilon_0).$$

The experimental value is $\alpha = 4{\cdot}5\ (a_0^3 4\pi\epsilon_0)$.

*Perturbation of a degenerate level*

The perturbation theory developed up to this point has been concerned only with non-degenerate levels. However, degenerate levels occur very frequently and this work must be extended. This is not in fact very difficult.

It is clear that the results already obtained do not apply for a degenerate level as there would then be singularities in both the first-order expansion coefficients (7.58), and in the second-order change in energy (7.59). Examination suggests that these difficulties could possibly be removed if the perturbation interactions $h_{lj}$ between degenerate levels, could be chosen to be zero.

As a simple example consider the case of a two-fold degenerate level. In particular, in the previous treatment, (7.44-7.59), let $E_1^0 = E_2^0$. The first- and second-order corrections to these degenerate levels are given by the solutions of (compare (7.50).)

$$\begin{vmatrix} \lambda h_{11} - \lambda E_1^{(1)} - \lambda^2 E_1^{(2)}, \ \lambda h_{12} & ,\ldots, & h_{1n} \\ \lambda h_{21}, & \lambda h_{22} - \lambda E_1^{(1)} - \lambda^2 E_1^{(2)}, & \\ \cdot & \lambda h_{33} + E_3^0 - E_1^0 - \lambda E_1^{(1)} - \lambda^2 E_1^{(2)}, \ldots & \\ \cdot & \cdot & \\ \cdot & \cdot & \\ \lambda h_{n1} & & \cdot \end{vmatrix} = 0. \tag{7.62}$$

This determinant is again a polynomial in $\lambda$. The lowest power is $\lambda^2$ and this term is the determinant

$$\begin{vmatrix} \lambda h_{11} - \lambda E_1^{(1)}, & \lambda h_{12}, & 0,\ldots, & 0 \\ \lambda h_{21}, & \lambda h_{22} - \lambda E_1^{(1)}, & 0 & \cdot \\ 0 & 0, & E_3^0 - E_1^0 & \\ \cdot & & \cdot & \\ \cdot & & & \cdot \\ \cdot & & & \cdot \\ 0 & 0 & & E_n^0 - E_1^0 \end{vmatrix} = 0$$

i.e.

$$(h_{11} - E_1^{(1)})(h_{22} - E_1^{(1)}) - |h_{12}|^2 = 0. \tag{7.63}$$

The two roots of this quadratic give the first-order changes in the degenerate levels. Generally these roots are distinct and the perturbation removes the degeneracy.

The degenerate basis functions can always be chosen so that the sub-matrix referring to the degenerate levels is diagonal. That is, in the present case, it is possible to obtain from $\chi_1$ and $\chi_2$, two new basis functions.

$$\begin{aligned} \chi_1' &= c_{11}\chi_1 + c_{12}\chi_2 \\ \chi_2' &= c_{21}\chi_1 + c_{22}\chi_2 \end{aligned} \tag{7.64}$$

so that the off-diagonal elements

$$h_{12}' = \int \chi_1'^* \mathscr{h} \chi_2' \, d\tau = 0, \qquad h_{21}' = \int \chi_2'^* \mathscr{h} \chi_1' \, d\tau = 0.$$

In this case the two solutions of (7.63) are

$$E_1^{(1)} = h_{11}' = \int \chi_1'^* \mathscr{h} \chi_1' \, d\tau \qquad \text{and} \qquad E_1^{(1)} = h_{22}' = \int \chi_2'^* \mathscr{h} \chi_2' \, d\tau.$$

The first-order effects of the perturbation has split the originally doubly-degenerate level $E_1^0$, into the separate levels

$$E_1^0 + h_{11}' \qquad \text{and} \qquad E_1^0 + h_{22}'. \tag{7.65}$$

The functions $\chi_1'$ and $\chi_2'$ are the zero-order wave-functions in that the perturbed wave-functions must approach them as $\lambda \to 0$. For the perturbed level $E_1^0 + h_{11}'$, the first-order wave-function is

$$\psi_1 = \chi_1' + \lambda \left[ c_2^{(1)} \chi_2' + \sum_{i=3}^{n} c_i^{(1)} \chi_i \right] + \ldots. \tag{7.66}$$

Keeping first-order terms only, equations (7.46) give (after diagonalizing the sub-matrix),

$$h_{l1}' + (E_l^0 - E_1^0) c_l^{(1)} = 0 \qquad l = 3, 4, \ldots, n$$

where

$$h_{l1}' = \int \chi_l^* \mathscr{h} \chi_1' \, d\tau.$$

There are no first-order terms in the first two equations. The second-order terms in the second equation give

$$(h_{22}' - h_{11}') c_2^{(1)} + \sum_{j=3}^{n} h_{2j}' c_j^{(1)} = 0 \qquad \text{where} \qquad h_{2j}' = \int \chi_2'^* \mathscr{h} \chi_j \, d\tau.$$

So the first-order coefficients for the wave-function with perturbed energy $E_1^0 + h'_{11}$ are

$$c_1 = 1 \qquad c_2^{(1)} = -\sum_{j=3}^{n} \frac{h'_{2j} h'_{j1}}{(h'_{22} - h'_{11})(E_1^0 - E_j^0)},$$

$$c_l^{(1)} = \frac{h'_{l1}}{(E_1^0 - E_l^0)} \quad l > 2. \tag{7.67}$$

It is left to the reader to find the first-order wave-function corresponding to the level $E_1^0 + h'_{22}$.

This procedure can obviously be generalized to a degenerate level of any order. The problem is really to find the correct 'zero-order' (unperturbed) wave-functions that diagonalize the appropriate submatrix. With this choice of basis functions the first-order change in the energy is given by the same formula as in the non-degenerate case (7.52).

The second-order changes in the energy can also be calculated. In the doubly-degenerate example discussed above the expansion of the determinant (7.62), keeping only those terms in $\lambda^3$ gives (with $E_1^{(1)} = h'_{11}$ and $h'_{12} = h'_{21} = 0$)

$$-\lambda^3 E_1^{(2)}(h'_{22} - E_1^{(1)})(E_3^0 - E_1^0) \dots (E_n^0 - E_1^0)$$
$$-\lambda^3 |h'_{13}|^2 (h'_{22} - E_1^{(1)})(E_4^0 - E_1^0) \dots (E_n^0 - E_1^0) \dots$$
$$-\lambda^3 |h'_{1n}|^2 (h'_{22} - E_1^{(1)})(E_3^0 - E_1^0) \dots (E_{n-1}^0 - E_1^0) = 0$$

i.e.

$$\begin{vmatrix} -\lambda^2 E_1^{(2)} & 0 & \lambda h'_{13} & \lambda h'_{1n} \\ 0 & \lambda(h'_{22} - E_1^{(1)}) & 0 & \\ \lambda h'_{31} & 0 & (E_3^0 - E_1^0) & \\ \cdot & & \cdot & \\ \cdot & & & \cdot \\ \cdot & & & \\ \lambda h'_{n1} & 0 & & (E_n^0 - E_1^0) \end{vmatrix} = 0. \tag{7.68}$$

(This is of course the determinant (7.62) with the correct zero-order basis functions.)

The second-order change in the level $E_1^0 + h'_{11}$ is

$$E_1^{(2)} = -\sum_{m=3}^{n} \frac{|h'_{1m}|^2}{E_m^0 - E_1^0}. \tag{7.69}$$

The second-order change in the level $E_1^0 + h'_{22}$ is

$$E_2^{(2)} = -\sum_{m=3}^{n} \frac{|h'_{2m}|^2}{E_m^0 - E_1^0}. \tag{7.70}$$

These results can be extended to a degenerate level of any order. When the correct set of basis functions is used the non-degenerate formula (7.59) applies except that the sum excludes all unperturbed levels degenerate with the level considered. In this way the singularities are avoided.

### *The first-order Zeeman effect*

It is well known that degenerate atomic levels may be split when a magnetic field is applied. This is the Zeeman effect, and the theory for the hydrogen atom is outlined below.

The classical Hamiltonian for a charged particle in an electromagnetic field was derived in Chapter 1 (1.45) and it was shown in Chapter 3 (question two) that to obtain a Hermitian operator on quantization it should be written (for an electron of charge $-e$ and mass $\mu$)

$$H = \frac{p^2}{2\mu} + \frac{e}{2\mu}(\mathbf{p}\,.\,\mathbf{A} + \mathbf{A}\,.\,\mathbf{p}) + \frac{e^2}{2\mu}A^2 - \mathrm{eV} \tag{7.71}$$

where the magnetic induction and electric vectors are given by

$$\mathbf{B} = \text{curl}\,\mathbf{A} \qquad \text{and} \qquad \mathscr{E} = -\text{grad V} - \frac{\partial \mathbf{A}}{\partial t}. \tag{7.72}$$

The quantum mechanical Hamiltonian is

$$\mathscr{H} \equiv -\frac{\hbar^2}{2\mu}\nabla^2 - \frac{e}{\mu}i\hbar\,\mathbf{A}\,.\,\text{grad} - \frac{e}{2\mu}i\hbar\,\text{div}\,\mathbf{A} + \frac{e^2A^2}{2\mu} - \mathrm{eV}. \tag{7.73}$$

For a hydrogen atom in a uniform induction field of magnitude $B$ parallel to the $z$-axis the vector and scalar potentials may be chosen so that

$$A_x = -\frac{1}{2}By \qquad A_y = \frac{1}{2}Bx \qquad A_z = 0 \qquad V = \frac{e}{4\pi\epsilon_0 r}. \tag{7.74}$$

As div $\mathbf{A} = 0$ and

$$\mathbf{A}\,.\,\text{grad} \equiv \frac{1}{2}B\left(-y\frac{\partial}{\partial x} + x\frac{\partial}{\partial y}\right) \equiv \frac{1}{2}B\frac{\partial}{\partial \phi} \tag{7.75}$$

(where $\phi$ is the polar angle) the quantum Hamiltonian is

$$\mathscr{H} \equiv -\frac{\hbar^2}{2\mu}\nabla^2 - \frac{e^2}{4\pi\epsilon_0 r} - i\hbar\frac{e}{\mu}\frac{B}{2}\frac{\partial}{\partial\phi} + \frac{e^2}{8\mu}B^2(x^2 + y^2). \tag{7.76}$$

For the fields normally met in the laboratory the term in $B^2$ is much smaller than the electric field term $-e^2/4\pi\epsilon_0 r$ and can be ignored. The unperturbed Hamiltonian is that of the free hydrogen atom and the perturbation term is linear in $B$.

$$\mathscr{H}^0 \equiv -\frac{\hbar^2}{2\mu}\nabla^2 - \frac{e^2}{4\pi\epsilon_0 r}, \qquad \mathscr{h} = -\frac{i\hbar e}{\mu}\frac{B}{2}\frac{\partial}{\partial\phi}. \tag{7.77}$$

The eigenfunctions of the unperturbed Hamiltonian are

$$R_{nl}(r)P_l^{m_l}(\cos\theta)e^{im_l\phi}. \tag{4.72}$$

The quantum number $n$ determines the energy and there are degenerate eigenfunctions for

$$l = 0, 1, \ldots, (n-1), \qquad m_l = -l, -(l-1), \ldots, 0, \ldots, +l.$$

The functions (4.72) are already eigenfunctions of $\mathscr{h}$ with eigenvalues $e\hbar Bm_l/2\mu$ and so the perturbation matrix is diagonal. When the magnetic field is applied the degenerate levels with the same quantum number $l$ and $n$ are split into the $(2l + 1)$ levels with energies

$$E_n + \frac{e\hbar B}{2\mu}m_l \qquad m_l = -l, \ldots, +l. \tag{7.78}$$

These energy levels explain the spectrum of the normal Zeeman effect. However there are additional effects due to the spin magnetic moment of the electron and these will be discussed in Chapter 10.

## 7.4 The W.K.B.J. Approximation

This method carries the names of G. Wentzel, H. A. Kramers, L. Brillouin, and H. Jeffreys and it can be used to deduce the Bohr–Sommerfeld quantum conditions from quantum mechanics. The W.K.B.J. approximation is applicable to ordinary differential equations. That is, one-dimensional eigenvalue problems or three-dimensional problems that can be reduced to one-dimensional form by separation of variables. It is useful when the potential is a slowly varying function of position in the sense that the fractional change of potential is small over several de Broglie wave-lengths. This will arise in the case of large quantum numbers when the wave-function has many nodes and the distance between them is small, corresponding to a short de Broglie wave-length. This method has been termed 'quasi-classical' in that it is most appropriate in the almost classical limit of large quantum numbers. Consider the one-dimensional, time-independent Schrödinger equation,

$$\frac{d^2\psi}{dx^2} + \frac{2m}{\hbar^2}(E - V(x))\psi = 0. \tag{7.79}$$

In Chapter 4 it was shown that the classical limiting form of the wave-function is essentially $\exp(iS/\hbar)$ (4.9) where $S$ is Hamilton's characteristic function.

Look now for solutions of (7.79) of the form

$$\psi(x) = \exp(iA/\hbar) \tag{7.80}$$

where $A(x)$ is some function of position. For $\psi$ to satisfy (7.79) $A$ must satisfy the non-linear equation

$$\left(\frac{\mathrm{d}A}{\mathrm{d}x}\right)^2 - i\hbar\frac{\mathrm{d}^2A}{\mathrm{d}x^2} - 2m(E-V) = 0. \tag{7.81}$$

Assume that it is now permissible to expand $A(x)$ in series in $h$.

$$A(x) = A_0(x) + \hbar A_1(x) + \frac{\hbar^2}{2}A_2(x) + \ldots. \tag{7.82}$$

The classical limit corresponds to $\hbar \to 0$ and as expected $A_0(x)$ is related to Hamilton's function $S(x)$. Quantum effects are included in the higher terms. If (7.82) is substituted into (7.81) and terms in $\hbar$ and higher powers ignored then, remember $\hbar$ is small

$$\left(\frac{\mathrm{d}A_0}{\mathrm{d}x}\right)^2 = 2m(E-V), \tag{7.83a}$$

i.e.

$$A_0 = \pm\int\{2m(E-V)\}^{1/2}\,\mathrm{d}x. \tag{7.83b}$$

This is the expected result since the classical momentum

$$p(x) = \{2m(E-V)\}^{1/2}$$

and Hamilton's function

$$S(x,t) = -Et \pm \int p\,\mathrm{d}x.$$

A wave-number $k(x)$ may be introduced with

$$2m(E-V) = p^2(x) = \hbar^2k^2 \tag{7.84}$$

and then

$$A_0(x) = \pm\hbar\int k(x)\,\mathrm{d}x. \tag{7.85}$$

If the first two terms of (7.82) are now substituted into (7.81) and the terms up to and including first order in $\hbar$ retained

$$\left(\frac{\mathrm{d}A_0}{\mathrm{d}x}\right)^2 + 2\hbar\left(\frac{\mathrm{d}A_0}{\mathrm{d}x}\right)\left(\frac{\mathrm{d}A_1}{\mathrm{d}x}\right) - i\hbar\frac{\mathrm{d}^2A_0}{\mathrm{d}x^2} - 2m(E-V) = 0.$$

From (7.83) this becomes

$$\frac{dA_1}{dx} = \frac{i}{2} \frac{\dfrac{d^2A_0}{dx^2}}{\dfrac{dA_0}{dx}}.$$

In terms of $k$,

$$A_1(x) = \frac{i}{2} \ln k(x). \tag{7.86}$$

The approximate solution of Schrödinger's equation (7.79) obtained by using the first two terms in the expansion (7.82) is

$$\psi(x) = \frac{1}{k^{1/2}}\left[K_1 \exp\left(i \int k \, dx\right) + K_2 \exp\left(-i \int k \, dx\right)\right] \tag{7.87}$$

where $K_1$ and $K_2$ are constants chosen to satisfy the boundary conditions. If $E > V$, $k$ is real but in the classically forbidden regions where $V > E$, $k$ is imaginary.

The validity of the approximation requires (in (7.81))

$$\left|\hbar \frac{d^2A}{dx^2}\right| \ll \left|\left(\frac{dA}{dx}\right)\right|^2. \tag{7.88}$$

If this is evaluated using $A_0$, and writing $\lambda(x) = 2\pi/k(x)$

$$\left|\left(\frac{\dfrac{dV}{dx}}{E - V}\right) \frac{\lambda}{4\pi}\right| \ll 1.$$

That is, the approximation will be good if the fractional change in $(E - V)$ over the de Broglie wave-length $\lambda$ is small. Clearly the approximation breaks down near those points where $E = V(x)$. Classically these are the turning points of the motion where the kinetic energy and hence the momentum is instantaneously zero and the particle changes direction.

At these turning points $k = 0$ corresponding to an infinite de Broglie wave-length. Then the approximate form (7.87) is singular although the exact wave-function will not be. In fact the expansion for $A(x)$ (7.82) is not a convergent series but is asymptotically correct and can only be used several wave-lengths away from a turning point.

Turning points generally occur in problems and a method must be developed to deal with them. Suppose there is a single turning point at $x = a$ and

$$E < V(x) \qquad x < a,$$
$$E > V(x) \qquad x > a.$$

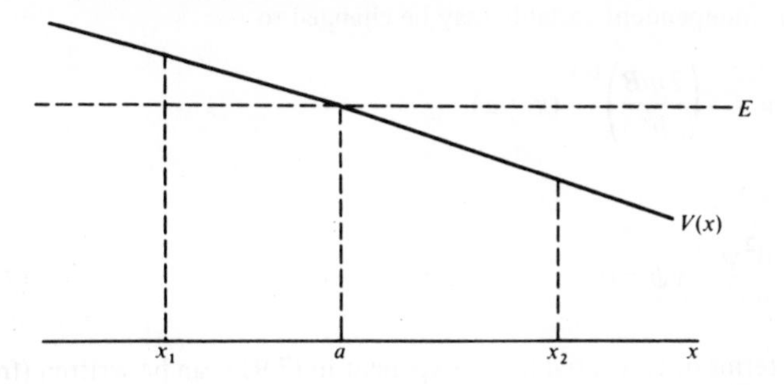


Figure 7.3.

In some region about '$a$' say, $x_1 < x < x_2$ the approximate wave-function (7.87) is not valid.

Outside this region (7.87) may be used with $k$ given as below.

(a) For $x > x_2$

$$k = \frac{(2m)^{1/2}}{\hbar}(E - V)^{1/2} > 0 \tag{7.89}$$

(b) For $x < x_1$

$$k = i\gamma$$

and

$$\gamma = \frac{(2m)^{1/2}}{\hbar}(V - E)^{1/2} > 0 \tag{7.90}$$

The physically acceptable solution must vanish as $x \to -\infty$ and is

$$\psi(x) \sim \frac{1}{\gamma^{1/2}} \exp\left[-\int_x^a \gamma \, \mathrm{d}x\right] \quad x < x_1 \tag{7.91}$$

Observe that the exponent decreases as $x$ moves away from '$a$'. The problem is to find the corresponding solution of the form (7.87) valid for $x > x_2$.

As $V(x)$ is analytic it can be expanded in Taylor series and in the linear approximation it is represented by two terms

$$V(x) = E - B(x - a) \qquad B > 0. \tag{7.92}$$

In the region $x_1 < x < x_2$ where (7.87) does not apply, Schrödinger's equation is solved exactly using the potential (7.92), i.e.

$$\frac{\mathrm{d}^2\psi}{\mathrm{d}x^2} + \frac{2m}{\hbar^2}B(x - a)\psi = 0 \qquad x_1 < x < x_2.$$

The independent variable may be changed to

$$y = -\left(\frac{2mB}{\hbar^2}\right)^{1/3} (x - a)$$

and then

$$\frac{d^2\psi}{dy^2} - y\psi = 0. \tag{7.93}$$

In terms of this variable the exponent in (7.91) can be written (from (7.90), (7.92))

$$\int_x^a \gamma \, dx = \frac{2}{3} y^{3/2} \tag{7.94}$$

and similarly (7.89), (7.92)

$$\int_a^x k \, dx = \frac{2}{3} |y|^{3/2}. \tag{7.95}$$

Equation (7.93) is Airey's differential equation. One of the two solutions is finite for all $y$ and the asymptotic form of this for large $|y|$ is

$$\psi(y) \sim \frac{D}{2} \frac{1}{y^{1/4}} \exp\left(-\frac{2}{3} y^{3/2}\right) \qquad y > 0 \qquad (x < a)$$

$$\psi(y) \sim D \frac{1}{|y|^{1/4}} \cos\left(\frac{2}{3} |y|^{3/2} - \frac{\pi}{4}\right) \qquad y < 0 \qquad (x > a). \tag{7.96}$$

(The second solution is also oscillatory for $y < 0$ but increases exponentially for $y > 0$.)

From (7.94), (7.95) and (7.96) it is clear that the solution of the form (7.87) that is valid for $x > x_2$ and connects with (7.91) is

$$\psi(x) \sim \frac{2}{k^{1/2}} \cos\left[\int_a^x k \, dx - \frac{\pi}{4}\right] \qquad x > x_2.$$

This can be written

$$\frac{1}{\gamma^{1/2}} \exp\left[-\int_x^a \gamma \, dx\right] \rightarrow \frac{2}{k^{1/2}} \cos\left[\int_a^x k \, dx - \frac{\pi}{4}\right]. \tag{7.97}$$

The direction of the connection arrow is important. For further details see Kembles's Quantum Mechanics.

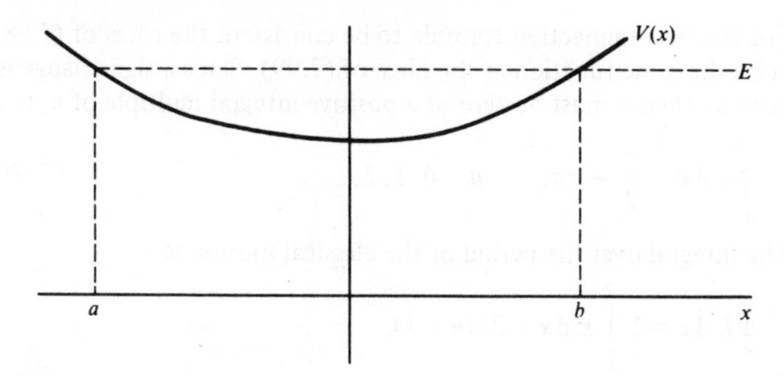


Figure 7.4

*The Bohr–Sommerfeld conditions*

The problem here is to find the energy levels for a particle moving in a one-dimensional potential well.

Consider the diagram. For any energy $E$, there are two classical turning points, say $x = a$ and $x = b$.

The asymptotic W.K.B.J. solution valid outside the region $a < x < b$ must converge as $|x| \to \infty$ and so (from (7.87))

$$\psi(x) \sim \frac{1}{\gamma^{1/2}} \exp\left[-\int_x^a \gamma \, \mathrm{d}x\right] \qquad x < a$$

$$\psi(x) \sim \frac{1}{\gamma^{1/2}} \exp\left[-\int_b^x \gamma \, \mathrm{d}x\right] \qquad x > b.$$

The connection formula at $x = a$ is (from (7.97))

$$\frac{1}{\gamma^{1/2}} \exp\left[-\int_x^a \gamma \, \mathrm{d}x\right] \to \frac{2}{k^{1/2}} \cos\left[\int_a^x k \, \mathrm{d}x - \frac{\pi}{4}\right]. \tag{7.98}$$

Similarly the connection formula at $x = b$ is

$$\frac{2}{k^{1/2}} \cos\left[\int_x^b k \, \mathrm{d}x - \frac{\pi}{4}\right] \leftarrow \frac{1}{\gamma^{1/2}} \exp\left[-\int_b^x \gamma \, \mathrm{d}x\right] \tag{7.99}$$

The l.h.s of (7.99) can be written

$$\frac{2}{k^{1/2}} \cos\left[\int_a^x k \, \mathrm{d}x - \frac{\pi}{4} - \alpha\right] \tag{7.100}$$

where

$$\alpha = \int_a^b k \, \mathrm{d}x - \frac{\pi}{2}.$$

For the two connection formula to be consistent the r.h.s. of (7.98) must be the same function as the l.h.s. of (7.99). Since a sign change is immaterial then $\alpha$ must be zero or a positive integral multiple of $\pi$, i.e.

$$\int_a^b k\,\mathrm{d}x - \frac{\pi}{2} = n\pi, \qquad n = 0, 1, 2, \ldots. \tag{7.101}$$

The integral over the period of the classical motion is

$$\oint k\,\mathrm{d}x = 2\int_a^b k\,\mathrm{d}x = 2\pi(n + \tfrac{1}{2}).$$

Writing $p = \hbar k$ this condition becomes

$$\oint p\,\mathrm{d}x = (n + \tfrac{1}{2})h, \qquad n = 0, 1, 2, \ldots. \tag{7.102}$$

The equation (7.101) (or (7.102)) determines the eigenvalues of the problem. The last equation is the Bohr-Sommerfeld quantization rule with half-integer quantum numbers.

## PROBLEMS

1 Suppose $\psi$ is some normalized approximate wave-function for the ground state of a system. Define the integrals

$$I = \int \psi^*(\mathscr{H} - K_1)^2\psi\,\mathrm{d}\tau \qquad \text{and} \qquad K_n = \int \psi^*\mathscr{H}^n\psi\,\mathrm{d}\tau,$$

where $\mathscr{H}$ is the Hamiltonian operator. By expanding $\psi$ in terms of the exact eigenfunctions of $\mathscr{H}$ show that

$$I \geqslant (E_1 - K_1)^2$$

where $E_1$ is the true ground state energy, if $K_1$ is closer to $E_1$ than to the energy of any higher state.

Express $I$ in terms of $K_1$ and $K_2$ and show that

$$E_1 \geqslant K_1 - (K_2 - K_1^2)^{1/2}.$$

This formula, due to Weinstein, provides a lower bound to the ground state energy.

2 Equation (4.26) gives the Hamiltonian for a linear harmonic oscillator. Using the normalized function

$$\psi(x) = \sqrt{\frac{2\alpha^3}{\pi}}\,\frac{1}{(\alpha^2 + x^2)}$$

obtain the expectation value of the Hamiltonian. Use the variational principle to show that the best choice for $\alpha$ for the ground state is $(\hbar^2/2mk)^{1/4}$ and that the best value for $\langle\mathscr{H}\rangle$ is $\hbar(k/2m)^{1/2}$.

3 Consider the one-dimensional particle whose Hamiltonian is

$$\mathscr{H} \equiv -\frac{\hbar^2}{2m}\frac{d^2}{dx^2} + kx^4.$$

Using the non-normalized trial function

$$\psi(x) = e^{-\alpha^2 x^2/2}$$

show that the expectation value for the energy is

$$\langle \mathscr{H} \rangle = \frac{\hbar^2\alpha^2}{4m} + \frac{3k}{4\alpha^4}$$

Using the variational principle show that the best value for $\alpha$ is $(6km/\hbar^2)^{1/6}$ and obtain the corresponding value for the ground state energy.

You may assume the integral

$$\int_{-\infty}^{\infty} e^{-\alpha^2 x^2}\,dx = \sqrt{\pi}/\alpha.$$

4 Suppose a system of classical particles is in motion. Newton's second law states that $\chi_i = m_i\ddot{x}_i$ where $\chi_i$ is the force in the $x$-direction acting on the $i$th particle of mass $m_i$. Confirm the expression

$$\frac{x_i}{2}\chi_i = -\frac{m_i}{2}\dot{x}_i^2 + \frac{1}{2}\frac{d}{dt}(m_i x_i \dot{x}_i).$$

Show that provided the particle is confined to a finite region of space, the time average as $t \to \infty$ of the last term on the right-hand side is zero. For conservative systems the forces acting on the particles may be derived from a scalar potential.

$$\chi_i = -\frac{\partial}{\partial x_i} V(x_1, y_1, z_1, \ldots).$$

If $V$ is a homogeneous function of the co-ordinates of degree $n$ show that

$$\frac{n}{2}\bar{V} = \bar{T}$$

where $\bar{V}$ and $\bar{T}$ are the time averages of the potential and kinetic energies of the system as $t \to \infty$.

(You may use Euler's theorem

$$\sum_i \left( x_i \frac{\partial V}{\partial x_i} + y_i \frac{\partial V}{\partial y_i} + z_i \frac{\partial V}{\partial z_i} \right) = nV.)$$

This is the virial theorem of classical mechanics.

5 Let $\mathbf{r}_1$ and $\mathbf{r}_2$ be the position vectors of two points in space relative to some origin. A well-known result is that

$$\frac{1}{|\mathbf{r}_1 - \mathbf{r}_2|} = \frac{1}{r_1} \sum_{n=0}^{\infty} \left(\frac{r_2}{r_1}\right)^n P_n(\cos\theta) \qquad r_1 > r_2$$

$$= \frac{1}{r_2} \sum_{n=0}^{\infty} \left(\frac{r_1}{r_2}\right)^n P_n(\cos\theta) \qquad r_2 > r_1$$

where $\theta$ is the angle between $\mathbf{r}_1$ and $\mathbf{r}_2$ and $P_n$ is the Legendre polynomial of degree $n$. Another result is that

$$P_n(\cos\theta) = P_n(\cos\theta_1) P_n(\cos\theta_2) + \sum_{m=1}^{n} (-1)^m 2 \frac{(n-m)!}{(n+m)!} P_n^m(\cos\theta_1) P_n^m(\cos\theta_2) \times \cos m(\phi_1 - \phi_2)$$

where $(\theta_1, \phi_1)$ and $(\theta_2, \phi_2)$ are the polar angles corresponding to $\mathbf{r}_1$ and $\mathbf{r}_2$ respectively.

Hence show that if

$$\psi(\mathbf{r}_1, \mathbf{r}_2) = \frac{1}{\pi}\left(\frac{\alpha}{a_0}\right)^3 \exp\left[-\frac{\alpha}{a_0}(\mathbf{r}_1 + \mathbf{r}_2)\right]$$

then

$$\frac{e^2}{4\pi\epsilon_0} \int \psi^* \frac{1}{|\mathbf{r}_1 - \mathbf{r}_2|} \psi \, d\tau_1 \, d\tau_2 = \frac{5}{8} \alpha \frac{e^2}{a_0 4\pi\epsilon_0}.$$

This is the result (7.24). You may assume that

$$\int_{-1}^{+1} P_n(\mu) \, d\mu = 0 \qquad n > 0$$

and that

$$P_0(\mu) = 1.$$

6 Consider the eigenvalue equation (7.28). Substitute for $\psi$ using (7.29). Premultiply the equation by $\chi_l^*$ and integrate and confirm that (7.35) is obtained.

7 Confirm that the perturbed wave functions (7.58) obtained by choosing $c_j = 1$ are orthonormal to first order.

8 Suppose the Hamiltonian $\mathscr{H}$ is a function of the parameter $\lambda$ (e.g. (7.45)). Let $\psi_n$ be the normalized eigenfunction with eigenvalue

$$E_n(\lambda) = \int \psi_n^* \mathscr{H} \psi_n \, d\tau.$$

Show that

$$\frac{dE_n}{d\lambda} = \int \psi_n^* \frac{\partial \mathscr{H}}{\partial \lambda} \psi_n \, d\tau \quad \text{(i)}$$

This is the Feynman theorem. Note that $\psi_n$ will depend upon $\lambda$ and that

$$\frac{\partial}{\partial \lambda} \int \psi_n^* \psi_n \, d\tau = 0.$$

In the case of perturbation theory $\partial \mathscr{H}/\partial \lambda = \mathscr{h}$ (7.45). By expanding both sides of (i) in powers of $\lambda$ show that this equation is consistent with the perturbation theory results (7.54), (7.58), (7.59).

9 The Hamiltonian operator describing the Helium atom is given by (7.17). This may be regarded as an unperturbed system with Hamiltonian

$$\mathscr{H}^0 \equiv -\frac{\hbar^2}{2m}(\nabla_1^2 + \nabla_2^2) - \frac{2e^2}{4\pi\epsilon_0}\left(\frac{1}{r_1} + \frac{1}{r_2}\right)$$

and a perturbation

$$\mathscr{h} = \frac{e^2}{4\pi\epsilon_0} \frac{1}{|\mathbf{r}_1 - \mathbf{r}_2|}.$$

Show that to first order, the energy of the ground state of the perturbed system is

$$E_1 = \frac{-e^2}{4\pi\epsilon_0 a_0}\left(\frac{11}{4}\right).$$

(Compare this with (7.27).)

10 A perturbation $\alpha x^4$ is applied to a one-dimensional simple harmonic oscillator (4.26). Show that the first-order change in the energy is

$$\tfrac{3}{4}(2n^2 + 2n + 1)\hbar^2\alpha/m^2\omega^2.$$

You may use the recurrence relation

$$2yH_n(y) = 2nH_{n-1}(y) + H_{n+1}(y)$$

where $H_n(y)$ is the Hermite polynomial of degree $n$, and also the orthogonality relation

$$\int_{-\infty}^{\infty} e^{-y^2} H_n(y) H_m(y) \, dy = 2^n n! \sqrt{\pi}\,\delta_{mn}.$$

11 The Schrödinger equation for a plane rigid rotator with a uniform electric field $\mathscr{E}$ applied in the plane is

$$-\frac{\hbar^2}{2I}\frac{d^2\psi}{d\phi^2} - (E + \mu\,\mathscr{E}\cos\phi)\psi = 0$$

where $I$ is the moment of inertia, $\mu$ is the electric dipole moment and $\phi$ is the angle of rotation. Regarding the term $-\mu\,\mathscr{E}\cos\phi$ as a perturbation, show that the energy, correct to second order is

$$\frac{\hbar^2}{2I}m^2 + \frac{I\mu^2\mathscr{E}^2}{\hbar^2(4m^2-1)} \qquad m = 0, \pm 1, \pm 2, \ldots$$

Find an expression for the polarizability and deduce that the dipole is parallel to the field when $m = 0$. Comment on the results for the states $m = \pm 1$. (Hint: remember this is strictly a degenerate problem.)

12 Show that the eigenfunctions for a free-state electron moving along the $x$-axis, and satisfying the periodic boundary conditions

$$\psi(x + L) = \psi(x)$$

are

$$\psi(x) = \frac{1}{\sqrt{L}}\exp\left[il\frac{2\pi}{L}x\right] \qquad l \text{ integer}$$

with energies

$$E^0(l) = \frac{h^2 l^2}{2mL}.$$

Suppose a periodic potential

$$V(x) = V(x + a)$$

with

$$L = Na \qquad N \text{ integer}$$

is introduced as a perturbation. If the mean value is zero

$$V(x) = \sum_{n \neq 0} V_n e^{-in\frac{2\pi}{a}x} \qquad \text{with} \qquad V_n^* = -V_{-n}.$$

Show that to second-order terms, the perturbed energy is

$$E(l) = E^0(l) + \sum_{n \neq 0} \frac{|V_n|^2}{E^0(l) - E^0(l - nN)}.$$

Explain why this result is not valid if $l = nN/2$.

In this case show, that to first order

$$E\left(\frac{nN}{2}\right) = E^0\left(\frac{nN}{2}\right) \pm |V_n|.$$

13 The 'phase' of the wave-function (7.98) is

$$\phi = \int_a^x k \, \mathrm{d}x - \pi/4.$$

Explain why the quantization condition (7.101) implies that the wave-function has $n$ and only $n$ zeros, (nodes).

By considering the quantization rule (7.102) for adjacent levels of large $n$ and using the classical relation

$$V = \frac{\partial E}{\partial p}$$

where $V$ is the velocity and $E$ the energy, show that the difference between the neighbouring levels is

$$\delta E = \hbar\omega$$

where $\omega = 2\pi/T$ and $T$ is the 'period' of the motion.

## References

Kemble, E. C. *The Fundamental Principles of Quantum Mechanics with Elementary Applications.* Dover, New York (1958).

Landau, L. D. and Lifshitz, E. M. *Quantum Mechanics.* Pergamon, London (1959).

Macdonald, J. K. L. *Phys. Rev.* **43**, 830–33 (1933).

Marcus, M. and Min, H. *Introduction to Linear Algebra.* Collier-Macmillan London (1965).

Pauling, L. and Wilson, E. B. *Introduction to Quantum Mechanics.* McGraw-Hill, New York (1935).

Slater, T. C. *Quantum Theory of Matter.* McGraw-Hill, New York (1968).

CHAPTER 8

# Time-Dependent Perturbation Theory

## 8.1 Variation of Constants

The time-independent perturbation theory was discussed in the previous chapter. Then the object was to find the changes in the eigenvalues and eigenfunctions of the stationary states of a system when a constant perturbation was applied. Suppose now that the perturbation is a function of the time. In this case the perturbed Hamiltonian operator is time-dependent and stationary states do not exist. The problem is now to find the time-dependent state function.

Consider a system where the Hamiltonian can be written

$$\mathscr{H} \equiv \mathscr{H}^0 + \lambda \hbar(t) \qquad 0 \leqslant \lambda \leqslant 1 \tag{8.1}$$

where $\mathscr{H}^0$ is independent of time and $\hbar(t)$ is a function of time. It is assumed that the energy associated with $\hbar(t)$ is small compared with that associated with $\mathscr{H}^0$. As before the parametric value $\lambda = 1$ corresponds to 'complete' perturbation and $\lambda = 0$ to zero perturbation.

$\mathscr{H}^0$ must be chosen so that its eigenfunctions are known. Let the orthonormal set $\{\chi_k\}$ be these time-independent eigenfunctions.

$$\mathscr{H}^0 \chi_k = E_k^0 \chi_k. \tag{8.2}$$

$\mathscr{H}^0$ is assumed to possess a discrete spectrum.

The general solution of the unperturbed time-dependent wave equation

$$\mathscr{H}^0 \Psi^0 = -\frac{\hbar}{i} \frac{\partial \Psi^0}{\partial t} \tag{8.3}$$

can be expressed as a linear combination of the variable separable solutions (3.15), i.e.

$$\Psi^0 = \sum_{k=1}^{\infty} a_k \exp\left[-iE_k^0 t/\hbar\right] \chi_k. \tag{8.4}$$

The $a_k$'s are constants.

The perturbed time-dependent wave equation is

$$(\mathcal{H}^0 + \lambda \mathscr{h})\Psi = -\frac{\hbar}{i}\frac{\partial \Psi}{\partial t}. \tag{8.5}$$

At any instant, $\Psi$ can be expanded as in (8.4) but the values of the expansion coefficients will no longer be constants in time.

$$\Psi = \sum_k a_k(t) \exp\left[-iE_k^0 t/\hbar\right] \chi_k. \tag{8.6}$$

Substitute (8.6) into (8.5) and with the help of (8.2) it is easily shown that

$$\lambda \sum_k a_k \mathscr{h}\, e^{-iE_k^0 t/\hbar} \chi_k = -\frac{\hbar}{i} \sum_k \frac{\mathrm{d}a_k}{\mathrm{d}t} e^{-iE_k^0 t/\hbar} \chi_k.$$

Multiply this last equation by $e^{iE_l^0 t/\hbar}\, \chi_l^*$ and integrate over the space co-ordinates to obtain

$$\lambda \sum_k a_k h_{lk} e^{i(E_l^0 - E_k^0)t/\hbar} = -\frac{\hbar}{i}\frac{\mathrm{d}a_l}{\mathrm{d}t} \qquad l = 1, 2, \ldots \tag{8.7}$$

where

$$h_{lk}(t) = \int \chi_l^* \mathscr{h} \chi_k \,\mathrm{d}\tau.$$

The equations (8.7) are the basic equations of the method of variation of constants and determine the expansion coefficients. However to solve these equations it is necessary to use perturbation theory. As in the previous chapter (7.49) it is assumed that the wave-function is a continuous function of $\lambda$ so that it is possible to expand $a_k$ as a power series in $\lambda$

$$a_k = a_k^0 + \lambda a_k^{(1)} + \ldots. \tag{8.8}$$

Equation (8.8) may be substituted into (8.7) and the resulting equation regarded as a power series in $\lambda$. The zero-order term gives

$$i\hbar \frac{\mathrm{d}a_l^{(0)}}{\mathrm{d}t} = 0 \qquad l = 1, 2, \ldots \tag{8.9}$$

showing that the zero-order expansion coefficients are constants. The first-order term gives

$$i\hbar \frac{da_l^{(1)}}{dt} = \sum_k a_k^{(0)} h_{lk} \, e^{i(E_l^0 - E_k^0)t/\hbar} \qquad l = 1, 2, \ldots. \tag{8.10}$$

The general power $\lambda^n$ yields the higher order perturbation coefficients

$$i\hbar \frac{da_l^{(n)}}{dt} = \sum_k a_k^{(n-1)} h_{lk} \, e^{i(E_l^0 - E_k^0)t/\hbar} \qquad l = 1, 2, \ldots. \tag{8.11}$$

The zero-order coefficients, (8.9) do not change with time and their values are specified by the initial state of the system. Suppose the system is in the $i$th stationary state of the unperturbed Hamiltonian when the perturbation is introduced at $t = 0$. Then

$$a_i^{(0)} = 1 \qquad a_f^{(0)} = 0 \qquad f \neq i. \tag{8.12}$$

The first-order changes in the expansion coefficients are then obtained by solving the equations

$$i\hbar \frac{da_f^{(1)}}{dt} = h_{fi} \, e^{i(E_f^0 - E_i^0)t/\hbar} \qquad f = 1, 2, \ldots. \tag{8.13}$$

The solutions satisfying the initial conditions $a_f^{(1)} = 0$ are

$$a_f^{(1)}(t) = -\frac{i}{\hbar} \int_0^t h_{fi}(t) \, e^{i(E_f^0 - E_i^0)t/\hbar} \, dt \qquad f = 1, 2, \ldots. \tag{8.14}$$

In the simple case that the perturbation does not depend explicitly on the time but is switched on at time $t = 0$ and switched off at $t = t'$, the solutions of (8.14) are

$$a_f^{(1)}(t') = h_{fi} \frac{1 - e^{i(E_f^0 - E_i^0)t'/\hbar}}{E_f^0 - E_i^0} \qquad f \neq i$$

$$a_i^{(1)}(t') = -i \frac{h_{ii}}{\hbar} t'. \tag{8.15}$$

These values remain unchanged for $t > t'$ after the perturbation is switched off and the wave-function which for $t < 0$ was $\chi_i \exp[-iE_i^0 t/\hbar]$ is

$$\Psi = [1 + a_i^{(1)}(t')] e^{-iE_i^0 t/\hbar} \chi_i + \sum_{f \neq i} a_f^{(1)}(t') e^{-iE_f^0 t/\hbar} \chi_f \qquad t > t'$$

where the coefficients are given by (8.15). If the set $\{\chi_f\}$ is orthonormal then this wave-function is normalized to first order.

The product $a_f^{(1)*}(t)a_f^{(1)}(t)$ gives the probability of finding the system in the state $\chi_f \exp(-iE_f^0 t/\hbar)$ after a time $t$.

$$|a_f^{(1)}(t)|^2 = |h_{fi}|^2 \frac{\sin^2 (E_f^0 - E_i^0)t/2\hbar}{\hbar^2\left(\dfrac{E_f^0 - E_i^0}{2\hbar}\right)^2} \qquad \begin{matrix} f \neq i \\ 0 < t < t'. \end{matrix} \tag{8.16}$$

The system may be considered to make transitions into all possible states. The expression (8.16) is periodic in time and the time $\Delta t$ of a transition is of the order of $\hbar/(E_f^0 - E_i^0)$. Strictly $E_f^0$ is not an eigenvalue of the perturbed Hamiltonian but if the perturbation is small enough it may be taken as a system eigenvalue. If the energy change $(E_f^0 - E_i^0)$ is significantly different from zero the time spent in the upper state is small. This is to be expected as in such transitions energy is not conserved and the change in energy $\Delta E = (E_f^0 - E_i^0)$ and the time available for measurement must be consistent with the uncertainty principle,

$$(\Delta E)(\Delta t) \sim \hbar. \tag{8.17}$$

Energy is conserved within the limits set by the uncertainty principle.

At time $t \geqslant t'$ after the perturbation is switched off the probability of finding the system in the exact eigenstate $\chi_f \exp(-iE_f^0 t/\hbar)$ is simply (8.16) with $t$ replaced by $t'$. If the matrix element $h_{fi}$ is zero the transition from the $i$th state to the $f$th state cannot take place by a 'first-order' process. However, if $h_{ji}$ and $h_{fj}$ are both non-zero then this transition may take place by a 'second-order' process through the 'virtual' state $j$. The details of this depend upon the second-order change in the wave-function and will not be dealt with here.

The expression for $|a_f^{(1)}(t')|^2$ may be written

$$|a_f^{(1)}(t')|^2 = \frac{|h_{fi}|^2}{\hbar^2}\frac{\sin^2(xt')}{x^2} \quad \text{with} \quad x = \frac{(E_f^0 - E_i^0)}{2\hbar}.$$

The function $\sin^2 (xt')/x^2$ is sharply peaked about $x = 0$ with zeros at $xt' = \pm n\pi, n = 1, 2, \ldots$. This is another expression of the result that energy is conserved in transitions, the most probable transition being those for which $E_f^0 \sim E_i^0$. If the perturbation is applied for a very small time $t'$ then $|a_f^{(1)}(t')|^2$ is proportional to $t'^2$ and the transition probability per unit time is not a constant. This undesirable result is avoided if it is assumed that there is a continuous distribution of states.

Suppose it is desired to find the probability that a system originally in the $i$th state will finally be found in the continuous range

$$f - \frac{\delta f}{2} < f < f + \frac{\delta f}{2}.$$

The probability of finding the system in one of these states after the perturbation is switched off is

$$\int_{\delta f} |a_f^{(1)}(t')|^2 \, \mathrm{d}f \tag{8.18}$$

$f$ is a continuous variable (eigenvalue) that defines the state uniquely. This can be written

$$\int_{\delta E} \rho(E_f^0)\, |a_f^{(1)}(t')|^2 \, \mathrm{d}E_f \tag{8.19}$$

where $\rho(E_f^0)$ is defined so that $\rho(E_f^0)\,\mathrm{d}E_f$ is the number of states in the energy range $E_f^0 + \mathrm{d}E_f$ and the range of integration $\delta E$ corresponds to $\delta f$.

For $t'$ sufficiently large (so that $\delta E \gg \hbar/t'$) the transition probability has the properties of a delta function (see equation (3.36)) and (8.19) may be written

$$\frac{2\pi t'}{\hbar} \int_{\delta E} |h_{fi}|^2 \rho(E_f^0) \delta(E_f^0 - E_i^0)\, \mathrm{d}E_f = \frac{2\pi}{\hbar} t' \, |h_{fi}|^2 \rho(E_i^0). \tag{8.20}$$

This result assumes that the energy range $\delta E$ includes $E_i^0$ and that $\rho(E_f^0)$ is a slowly-varying function. This probability is linear in time and the transition probability per unit time $P$, that the system undergoes a transition from the initial state $i$ to one of the final states is

$$P = \frac{2\pi}{\hbar} \rho(E_i^0)\, |h_{fi}|^2 \tag{8.21}$$

$P$ is a constant as expected and equation (8.21) is Fermi's golden rule.

## 8.2 Semi-Classical Treatment of Radiation

In this section the interaction between a material particle and electromagnetic radiation will be discussed. The particle will be assumed to obey quantum mechanics and the radiation to satisfy the classical Maxwell equations. Clearly such a semi-classical hybrid of quantum and classical mechanics cannot be completely satisfactory and the radiation field should be quantized and described in terms of photons. Such a treatment is beyond the scope of this book and the interested reader is referred to *Introductory Quantum Electrodynamics*, by E. A. Power.

Classically, an electromagnetic field can be described by the scalar and vector potentials $\phi$ and $\mathbf{A}$ with

$$\mathbf{B} = \text{curl}\,\mathbf{A} \qquad \text{and} \qquad \mathscr{E} = -\text{grad}\,\phi - \frac{\partial \mathbf{A}}{\partial t} \tag{8.22}$$

where $\mathbf{B}$ and $\mathscr{E}$ are the magnetic induction and electric field vectors respectively. The vector field $\mathbf{A}$ can be resolved into an irrotational and a solenoidal part. The first of equations (8.22) defines the solenoidal part only and in free space the vector **A** can always be chosen to satisfy

$$\operatorname{div} \mathbf{A} + \frac{1}{c^2}\frac{\partial \phi}{\partial t} = 0 \tag{8.23}$$

$c$ is the velocity of light.

It has already been shown (7.73) that the Schrödinger equation for an electron of charge $-e$ and mass $m$ in an electromagnetic field is

$$\left[-\frac{\hbar^2}{2m}\nabla^2 - \frac{e}{m} i\hbar \mathbf{A} . \operatorname{grad} - \frac{e}{2m} i\hbar \operatorname{div} \mathbf{A} + \frac{e^2 A^2}{2m} - e\phi\right]\Psi$$
$$= i\hbar \frac{\partial \Psi}{\partial t}. \tag{8.24}$$

More than one set of scalar and vector potentials describe the same electromagnetic field. If $\chi$ is a solution of the wave equation

$$\nabla^2 \chi - \frac{1}{c^2}\frac{\partial^2 \chi}{\partial t^2} = 0 \tag{8.25}$$

then new potentials

$$\mathbf{A}' = \mathbf{A} - \operatorname{grad} \chi \qquad \phi' = \phi + \frac{\partial \chi}{\partial t} \tag{8.26}$$

can be defined which still satisfy (8.22) and (8.23). Such a transformation is called a gauge transformation. In regions where there is no free charge the scalar potential must also be a solution of the wave equation (8.25) and in this case $\chi$ can always be chosen so that the scalar potential is zero and from (8.23) div $\mathbf{A} = 0$.

With this choice for **A** and $\phi$ and if the field is weak so that the term in $A^2$ can be neglected, the Schrödinger Hamiltonian becomes

$$-\frac{\hbar^2}{2m}\nabla^2 - \frac{e}{m} i\hbar \mathbf{A} . \operatorname{grad}.$$

If the electron is bound to an atom it is necessary to introduce an additional potential energy term $V$ to represent the central field and it is then appropriate to use the reduced mass $\mu$ for the electron. The Hamiltonian is then

$$\mathscr{H} \equiv -\frac{\hbar^2}{2\mu}\nabla^2 + V - \frac{e}{\mu} i\hbar \mathbf{A} . \operatorname{grad}. \tag{8.27}$$

This Hamiltonian can be written

$$\mathscr{H} \equiv \mathscr{H}^0 + h$$

where $\mathscr{H}^0$ is the Hamiltonian for an electron in a central field and $\hbar$ is the perturbation operator

$$\hbar \equiv -\frac{e}{\mu} i\hbar \mathbf{A} . \text{grad} \equiv +\frac{e}{\mu} \mathbf{A} . \hat{\mathbf{p}} \tag{8.28}$$

where $\hat{\mathbf{p}}$ is the operator representing the linear momentum of the electron. For a monochromatic plane wave, the real vector potential is

$$\mathbf{A} = (\mathbf{A}_0 e^{i\alpha}) e^{i(\mathbf{k}.\mathbf{r}-\omega t)} + (\mathbf{A}_0 e^{-i\alpha}) e^{-i(\mathbf{k}.\mathbf{r}-\omega t)} \tag{8.29}$$

where $\mathbf{A}_0$ is a constant real vector and $\omega$ is the angular frequency. To satisfy div $\mathbf{A} = 0$ it is necessary for

$$\mathbf{A}_0 . \mathbf{k} = 0.$$

That is, the plane of polarization of the wave is perpendicular to the direction of propagation.

The expansion coefficients can be found using the result (8.14). If the perturbation is switched on at $t = 0$ when the system is in the unperturbed atomic state $i$ then to first order

$$a_f^{(1)}(t) = -\left[h'_{fi} \frac{e^{i(\omega_{fi}-\omega)t} - 1}{(\omega_{fi} - \omega)} + h''_{fi} \frac{e^{i(\omega_{fi}+\omega)t} - 1}{(\omega_{fi} + \omega)}\right] \tag{8.30}$$

where

$$h'_{fi} = \frac{e}{\mu\hbar} e^{i\alpha} \int \chi_f^* e^{i\mathbf{k}.\mathbf{r}} \mathbf{A}_0 . \hat{\mathbf{p}} \chi_i \, d\tau$$

$$h''_{fi} = \frac{e}{\mu\hbar} e^{-i\alpha} \int \chi_f^* e^{-i\mathbf{k}.\mathbf{r}} \mathbf{A}_0 . \hat{\mathbf{p}} \chi_i \, d\tau$$

and

$$\omega_{fi} = (E_f^0 - E_i^0)/\hbar.$$

If $(\omega_{fi} \pm \omega)$ is not close to zero then $a_f^{(1)}(t)$ oscillates very rapidly and the time spent in the state $f$ is very small. Following the argument from (8.14) to (8.16) the probability of finding the system in the state $f$ at a time $t$ is

$$|a_f^{(1)}(t)|^2 = |h'_{fi}|^2 \frac{\sin^2 (\omega_{fi} - \omega)t/2}{\left(\frac{\omega_{fi} - \omega}{2}\right)^2} + |h''_{fi}|^2 \frac{\sin^2 (\omega_{fi} + \omega)t/2}{\left(\frac{\omega_{fi} + \omega}{2}\right)^2}$$

$$+ \text{(product terms).} \tag{8.31}$$

In all cases the product terms in the expansion for $|a_f^{(1)}(t)|^2$ are small and can be neglected.

The probability of finding the system in the state $f$ at the time $t$ is appreciable only if the frequency of the incident radiation is such that

$$\omega_{fi} - \omega \simeq 0 \qquad \text{or} \qquad \omega_{fi} + \omega \simeq 0.$$

These are the cases of resonance. At most one of these conditions is satisfied for a given frequency and two given energy levels. Transitions between the states $i$ and $f$ are probable only if

$$E_f^0 - E_i^0 \simeq \pm\hbar\omega. \tag{8.32}$$

The positive sign applies when the particle absorbs the energy $\hbar\omega$ from the field. If (8.32) is satisfied with the negative sign the particle loses energy $\hbar\omega$ to the field.

In atomic spectra the initial and final states are discrete and the transition probability per unit time would appear to be time-dependent (c.f. (8.16)). The difficulty is overcome by remembering that for a light source the radiation is produced by the random vibration of atoms that are widely-spaced. In practice the radiation is not strictly monochromatic but covers a small but finite range of frequencies

$$\omega - \frac{\delta\omega}{2} < \omega < \omega + \frac{\delta\omega}{2},$$

the frequencies having random relative phases. In this case interference effects can be neglected and the contributions from the different frequencies are additive.

Consider the case of absorption where energy is absorbed from the field. In this case $\omega_{fi} \simeq \omega$ and only the first term in (8.31) need be considered. The total probability is obtained by integrating this term over the frequency range $\delta\omega$. The delta function properties of (8.31) give the probability for absorption (upward transition) to be for $t$ sufficiently large,

$$\int_{\delta\omega} |a_f^{(1)}(t)|^2 \, d\omega = 2\pi t \int_{\delta\omega} |h'_{fi}|^2 \delta(\omega_{fi} - \omega) \, d\omega.$$

The integral is independent of time and so the transition probability per unit time is a constant.

$$P = 2\pi \int_{\delta\omega} |h'_{fi}|^2 \delta(\omega_{fi} - \omega) \, d\omega.$$

The magnitude of the vector potential $A_0$ is a function of the frequency and this integral can be written

$$P = \frac{2\pi e^2}{\mu^2\hbar^2} A_0^2(\omega_{fi}) \left| \int \chi_f^* \, e^{i\mathbf{k}\cdot\mathbf{r}} \hat{p}_0 \chi_i \, d\tau \right|^2 \tag{8.33}$$

where $\hat{p}_0$ represents the momentum component along the direction $\mathbf{A}_0$. The use of the delta function is equivalent to assuming that $A_0^2(\omega)$ is a slowly varying function of $\omega$.

The transition probability depends upon the intensity of the radiation at the 'resonance' frequency. At any time instant the rate of flow of energy per unit area is given by the Poynting vector

$$\mathbf{N} = \mathscr{E} \wedge \mathbf{H}. \tag{8.34}$$

For a plane wave in free space

$$|\mathbf{N}| = \sqrt{\frac{\epsilon}{\mu_0}}\, \mathscr{E}^2$$

where $\mathscr{E}$ is the magnitude of the electric field. The value of $|\mathbf{N}|$ associated with the frequency $\omega$ and averaged over one oscillation of period $2\pi/\omega$ is (8.22), (8.29), (8.34)

$$\frac{2\omega^2}{\mu_0 c} A_0^2(\omega) = E(\omega)$$

where $E(\omega)$ is intensity of radiation at frequency $\omega$ and $\mu_0$ is the permeability of free space. From (8.33)

$$P = \frac{\pi \mu_0 c e^2}{\mu^2 \hbar^2 \omega_{fi}^2} E(\omega_{fi}) \left| \int \chi_f^* e^{i\mathbf{k}\cdot\mathbf{r}} \hat{p}_0 \chi_i \, \mathrm{d}\tau \right|^2 \tag{8.35}$$

where $E_f^0 \simeq E_i^0 + \hbar\omega$.

Consider now the case of emission where the energies are such that $\omega_{fi} \simeq -\omega$, i.e.

$$E_f^0 \simeq E_i^0 - \hbar\omega.$$

The second term in (8.30) is now dominant and the probability for the downward transition from $i$ to $f$ is simply (8.35) with $e^{i\mathbf{k}\cdot\mathbf{r}}$ replaced by $e^{-i\mathbf{k}\cdot\mathbf{r}}$. This is the result for emission induced by the presence of a field. Note that not only is the absorption probability proportional to the field intensity as expected but that the stimulated emission probability is also proportional to the intensity. Also the probabilities for induced upward or downward transitions between a given pair of states in a given field are equal and proportional to the integral

$$\left| \int \psi_f^* e^{i\mathbf{k}\cdot\mathbf{r}} \hat{p}_0 \chi_i \, \mathrm{d}\tau \right|^2. \tag{8.36}$$

This is an example of the principle of detailed balancing. In an upward transition a photon (a quantum of radiation energy) is absorbed from the field and in a downward transition a photon is emitted by the particle.

*The electric-dipole approximation*

The exponential in (8.35) can be expanded

$$e^{i\mathbf{k}.\mathbf{r}} = 1 + i\mathbf{k}.\mathbf{r} - \frac{(\mathbf{k}.\mathbf{r})^2}{2} + \ldots. \tag{8.37}$$

The wave-length of visible radiation is much greater than atomic dimensions and in the vicinity of the atomic system where the wave-functions $\chi_f$ and $\chi_i$ are appreciable, $\mathbf{k}.\mathbf{r}$ is much smaller than one.

In the dipole approximation (8.35) is evaluated with $e^{i\mathbf{k}.\mathbf{r}}$ replaced by unity and the induced probability per unit time is proportional to the square of the integral,

$$\int \chi_f^* \hat{p}_0 \chi_i \, d\tau.$$

This approximation will not be valid for X-rays. Following an argument similar to that used in Chapter 5 (5.70) it can be shown that

$$\int \chi_f^* \hat{p}_0 \chi_i \, d\tau = \mu \frac{d}{dt} \int \chi_f^* r_0 \chi_i \, d\tau$$

where $r_0$ is the component of the electron displacement $\mathbf{r}$, relative to the origin, parallel to the direction of polarization. $(-e\mathbf{r})$ is the electric dipole moment of the electron. From the generalization of (5.47) (question 4, Chapter 5)

$$\frac{d}{dt} \int \chi_f^* r_0 \chi_i \, d\tau = \frac{i}{\hbar} (E_f^0 - E_i^0) \int \chi_f^* r_0 \chi_i \, d\tau.$$

Finally, the integral in (8.35) becomes

$$\int \chi_f^* \hat{p}_0 \chi_i \, d\tau = i\mu\omega_{fi} \int \chi_f^* r_0 \chi_i \, d\tau.$$

The transition probabilities per unit time for absorption and induced emission are now

$$\frac{\pi\mu_0 c e^2}{\hbar^2} E(\omega_{fi}) \left| \int \chi_f^* r_0 \chi_i \, d\tau \right|^2.$$

If the radiation field is isotropic this becomes

$$\frac{\pi\mu_0 c e^2}{3\hbar^2} E(\omega_{fi}) \left| \int \chi_f^* \mathbf{r} \chi_i \, d\tau \right|^2 \tag{8.38}$$

where $\mathbf{r} = \mathbf{i}x + \mathbf{j}y + \mathbf{k}z$ and $E(\omega_{fi})/3$ is the intensity in the $x$, $y$, $z$ directions. The coefficient of the intensity in this formula is the Einstein B-coefficient of absorption or induced emission.

If the dipole integral $\int \chi_f^* \mathbf{r} \chi_i \, d\tau$ is zero for a pair of states the transition probability vanishes in this approximation and the transition is said to be forbidden. In fact the transition probability may not be exactly zero as there may be contributions from the other terms in the expansion (8.37).

If the expression (8.35) vanishes the transition is said to be strictly forbidden. Even in this case, however, the transition may take place by a second-order transition. The reader is referred to *Quantum Mechanics*, by Schiff, p. 254, for a deeper treatment.

*Spontaneous emission*

Even in the absence of an external field there is a chance that an atom in an excited state with energy $E_i$ will make a transition to a lower energy state $E_f$ with a corresponding emission of radiation energy. It can be shown (Schiff, Power) that the transition probability per unit time for such a spontaneous emission is again proportional to the square of the modulus of the dipole integral, i.e.

$$\left| \int \chi_f^* \mathbf{r} \chi_i \, d\tau \right|^2 .$$

*Selection rules*

The conditions which the functions $\chi_i$ and $\chi_f$ must satisfy to prevent the dipole integral (8.38) from vanishing are called 'selection rules'. Note that if both functions are spherically symmetric (atomic-states with $l = 0$) the integral vanishes and the transition is forbidden. Similarly, using the hydrogen-type wave-functions it can be shown that the electron transition is 'forbidden' unless the changes in the $l$ and $m_l$ quantum numbers satisfy

$$\delta l = \pm 1$$

and

$$\delta m_l = 0 \qquad \text{or} \qquad \pm 1.$$

This is dealt with in more detail in Chapter 11.

In Chapter 6, problem 14 it is shown that the matrix elements of $x_{nm}$ with respect to the harmonic oscillator eigenfunctions are zero unless $n = m \pm 1$. This implies the selection rule

$$\delta n = \pm 1$$

so that the spectrum should consist of a single frequency equal to that of the oscillator. This has already been mentioned in Chapter 4 with reference to the vibrations of diatomic molecules.

*Line breadth*

The uncertainty principle can be expressed in the form

$$(\Delta t)(\Delta E) \geqslant \hbar/2$$

where $(\Delta t)$ is the uncertainty in the time measurement and $(\Delta E)$ is the uncertainty in the energy.

The probability per unit time $\lambda$ for a transition from a state $i$ to a lower state $f$ of a system is independent of the time. Such a time-independent process obeys an exponential law of decay and $1/\lambda$ is of the order of magnitude that the system stays in the upper-state. Clearly a determination of the energy of this upper state cannot take longer than the lifetime $1/\lambda$ of the state and so the uncertainty in the energy is

$$(\Delta E) \geqslant \hbar\lambda/2.$$

This uncertainty in the energy corresponds to a broadening of the emitted spectrum line (uncertainty in the angular frequency of $\lambda/2$) and in general, the longer the lifetime of the state the sharper the spectrum line. This broadening imposed by the uncertainty principle is very small and is of the order of that observed although there are other factors such as radiation damping which affect the observed line width.

## 8.3 Elastic Collisions

*Scattering cross-section*

Suppose a parallel stream of non-interacting particles impinges upon a fixed target. Let the density of incident particles be such that $n$ particles

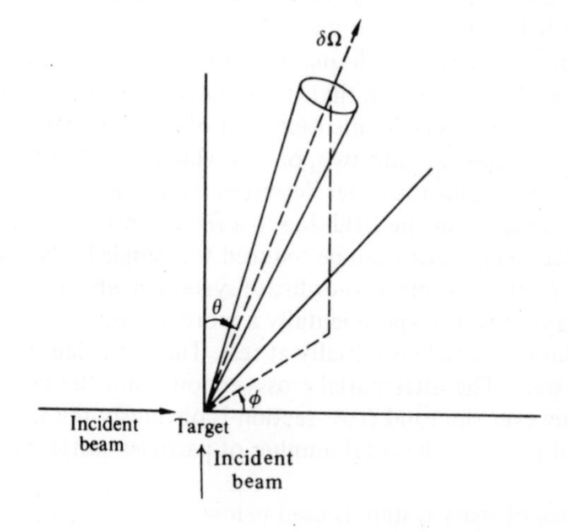


Figure 8.1

cross a unit area perpendicular to the flux in unit time. The incident particles will be scattered due to the interaction with the target and the number of particles scattered into an element of solid angle $\delta\Omega$ will be proportional to $n\delta\Omega$. The constant of proportionality $\sigma$ has the dimensions of area and is called the differential cross-section. It is not a constant but depends upon the polar co-ordinates $(\theta, \phi)$ of $\delta\Omega$ relative to the incident flux.

$\sigma(\theta, \phi)$ is the probability per unit time for a particle to be scattered into unit solid angle about $(\theta, \phi)$ when a beam of unit flux bombards the target. The total cross-section is defined to be the integral over the solid angle

$$\int \sigma(\theta, \phi)\, d\Omega = \int_0^{2\pi} \int_0^{\pi} \sigma(\theta, \phi) \sin\theta \, d\theta \, d\phi.$$

*Quantum theory of scattering*

The result of a collision in classical mechanics is determined by

(a) the initial particle velocities;
(b) the distance at which the particles would pass if there was no interaction;
(c) the interaction potential between the particles.

Since the concept of particle trajectory loses its meaning in quantum mechanics a different approach is required.

The quantum mechanical Hamiltonian representing a system of two particles whose interaction depends only on their relative displacement commutes with the operator representing the total linear momentum of the system. So the expectation value of the momentum in any state that satisfies the Schrödinger equation is a constant. Linear momentum is conserved in collisions in both quantum and classical mechanics.

In Chapter 4 it was shown that if the potential in a two-body problem depends only on the relative displacement of the particles the Schrödinger equation can be split up into two, one of which describes the motion of the centre of mass and the other describes the motion of a single particle with reduced mass $\mu$ in the field due to a fixed centre of force.

The scattering problem can be reduced to a single-body problem by choosing the centre of mass co-ordinate system in which the centre of mass is always at rest. Experimentally a more obvious system is one in which the target particle is initially at rest. This is the laboratory co-ordinate system. The differential cross-section is not the same in both systems. However the total cross-section is obviously the same in both systems as it refers to the total number of particles scattered in all directions.

The centre of mass system is used below.

Consider a collision in which an incident particle is scattered by a target. If the 'internal energy' of the two particle is left unchanged, the collision is said to be elastic. In the centre of mass system, the problem is to solve the equation

$$-\frac{\hbar^2}{2\mu}\nabla^2\psi(\mathbf{r}) + V(\mathbf{r})\psi(\mathbf{r}) = E\psi(\mathbf{r})$$

where $E$ is the energy of the relative motion. The precise form of the scattering potential $V(\mathbf{r})$ may be quite complicated and consequently most collision problems cannot be solved analytically. However it is not difficult to show that at large distances from the scatterer the wave-function will take the form

$$e^{ikz} + f(\theta, \phi)\frac{e^{ikr}}{r}.$$

The first term represents a particle incident along the positive $z$-axis whereas the second term represents the scattered wave. The problem is to find $f(\theta, \phi)$.

Collision problems can be solved by time-dependent perturbation theory. In this treatment the short-range interaction is regarded as a weak perturbation which is zero initially and decreases to zero after the collision. Equation (8.21) will now be used to evaluate the transition probability when the initial and final states of the scattered particles are free-particle momentum eigenfunctions.

The time-independent part of the wave-function of the incident particle is, when normalized to give unit incident flux

$$\chi_i = \sqrt{\frac{\mu}{\hbar k_0}}\exp(i\mathbf{k}_0 . \mathbf{r}) \qquad (8.39)$$

This represents particles with momentum $\hbar\mathbf{k}_0$. The wave-function of a scattered state is, using the $\delta$-function normalization for a continuous spectrum,

$$\chi_k = \frac{1}{\sqrt{(2\pi)^3}}\exp(i\mathbf{k} . \mathbf{r}). \qquad (8.40)$$

For elastic collisions the internal energy of the particle is left unchanged and

$$k_0 = k.$$

The change of momentum of the incident particle is

$$\hbar(\mathbf{k} - \mathbf{k}_0) = -\hbar\mathbf{K}.$$

The perturbation matrix element corresponding to the initial and final states (8.39) and (8.40) is

$$h_{ki} = \sqrt{\frac{\mu}{\hbar k_0 (2\pi)^3}} \int V(\mathbf{r}) \exp(i\mathbf{K}.\mathbf{r})\, d\tau \tag{8.41}$$

where the potential $V(\mathbf{r})$ is due to the interaction between the incident particle and the scatterer. This integral depends in general on the direction of scattering. The probability of finding the scattered particle with wave vector in the range $k + \delta k$ and in the solid angle $\delta\Omega = \sin\theta\delta\theta\delta\phi$ after time $t'$ is (8.18)

$$\int_{\delta k} |a_k^{(1)}(t')|^2 k^2 \, dk \, d\Omega. \tag{8.42}$$

Since $E_k = \hbar^2 k^2/2\mu$ then (8.20) this can be written

$$(|h_{ki}|^2 \mu k 2\pi/\hbar^3)\delta\Omega t'. \tag{8.43}$$

Hence the number of particles scattered into the solid angle $\delta\Omega$ per unit time is ((8.41), (8.43))

$$\frac{\mu^2}{(2\pi)^2\hbar^4} \left| \int V(\mathbf{r}) e^{i\mathbf{K}.\mathbf{r}} \, d\tau \right|^2 \delta\Omega$$

and so the differential cross-section is

$$\sigma(\theta, \phi) = \left(\frac{\mu}{2\pi\hbar^2}\right)^2 \left| \int V(\mathbf{r}) e^{i\mathbf{K}.\mathbf{r}} \, d\tau \right|^2. \tag{8.44}$$

This is the result of the first Born approximation and will be valid if the interaction potential $V(\mathbf{r})$ is small compared with the kinetic energy of the incident particles.

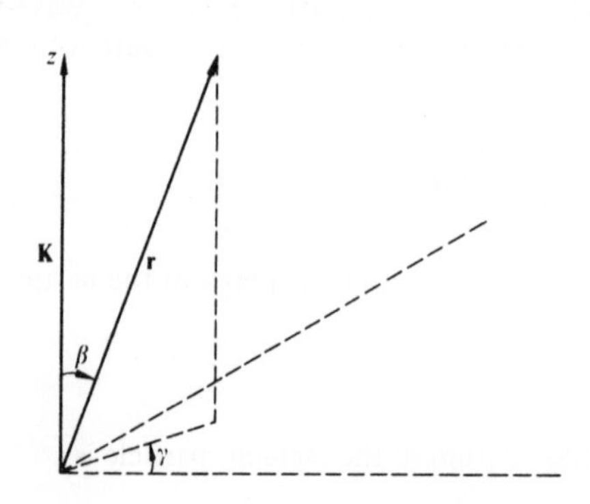


Figure 8.2

The simplest scattering potential is that of the spherical well

$$V(\mathbf{r}) = -V_0 \qquad r \leqslant \alpha$$
$$= 0 \qquad r > \alpha.$$

In this case the polar axis can be conveniently taken along the momentum transfer direction **K**.

$$\int V(\mathbf{r}) e^{i\mathbf{K}\cdot\mathbf{r}}\,\mathrm{d}\tau = V_0 \int_0^\alpha \int_0^\pi \int_0^{2\pi} e^{iKr\cos\beta} r^2 \sin\beta\,\mathrm{d}\gamma\,\mathrm{d}\beta\,\mathrm{d}r$$

$$= -\frac{4\pi}{K} V_0 \int_0^\alpha \sin(Kr) r\,\mathrm{d}r$$

$$= +\frac{4\pi}{K} V_0 \left[\frac{\alpha}{K}\cos K\alpha - \frac{1}{K^2}\sin K\alpha\right].$$

$K$ can be expressed in terms of the scattering angle

$$K = 2k_0 \sin\theta/2.$$

The differential scattering cross-section is

$$\sigma(u) = \left(\frac{2\mu V_0 \alpha^3}{\hbar^2}\right)^2 \left(\frac{\sin u - u\cos u}{u^3}\right)^2$$

where

$$u = K\alpha = 2k_0\alpha \sin\frac{\theta}{2}.$$

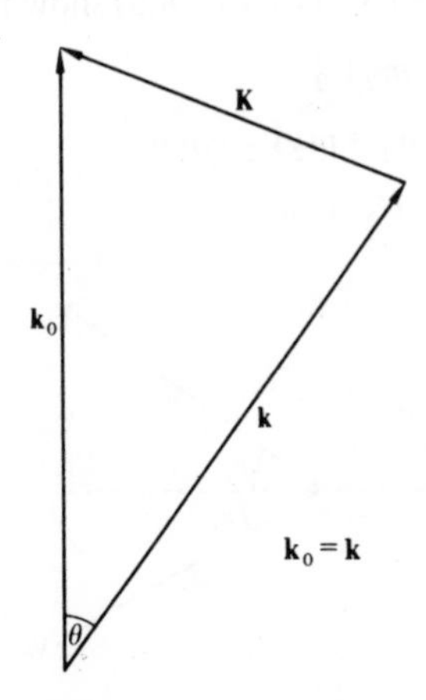


Figure 8.3

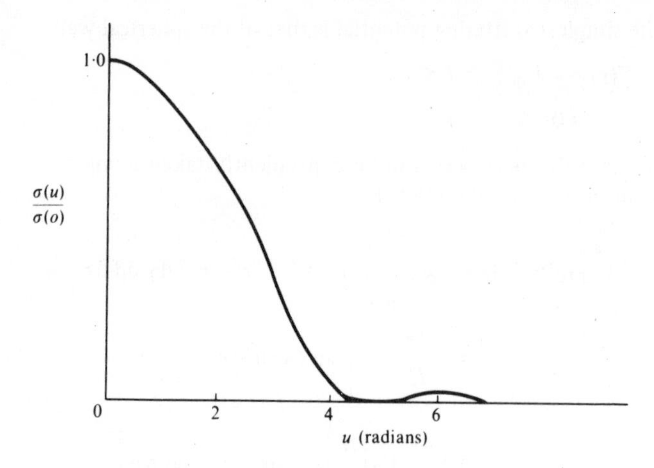


Figure 8.4 Differential scattering cross-section for a spherical potential-well.

This extremely brief treatment of the theory of collisions has been given primarily to illustrate the time-dependent perturbation theory. The reader is referred to Burhop's review article 'Theory of Collisions' in *Quantum Theory, Elements,* edited by D. R. Bates, for a fuller treatment.

## PROBLEMS

1 In the laboratory co-ordinate system the stationary target mass $m_2$ is initially at 0 and the projected mass $m_1$ has a velocity $(0, 0, V)$. If after the collision, assumed elastic, the particles are scattered through $\theta_1$ and $\theta_2$ with velocities $\mathbf{V}_1$ and $\mathbf{V}_2$ (see Fig. 8.5) show that

$$m_1 V^2 = m_1 V_1^2 + m_2 V_2^2$$

$$m_1 V = m_1 V_1 \cos\theta_1 + m_2 V_2 \cos\theta_2$$

$$m_1 V_1 \sin\theta_1 = m_2 V_2 \sin\theta_2.$$

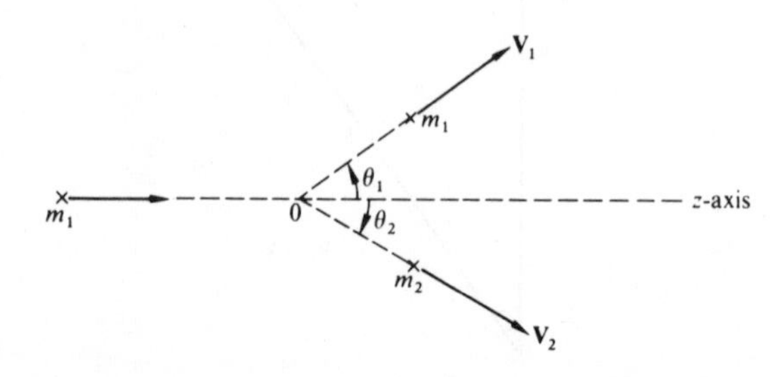


Figure 8.5

Show also that the centre of mass moves with a velocity $(0, 0, V_0)$ where $V_0 = m_1 V/(m_1 + m_2)$.

In the centre of mass system the total linear momentum is zero and the momenta of the particles are the same in magnitude but opposite in

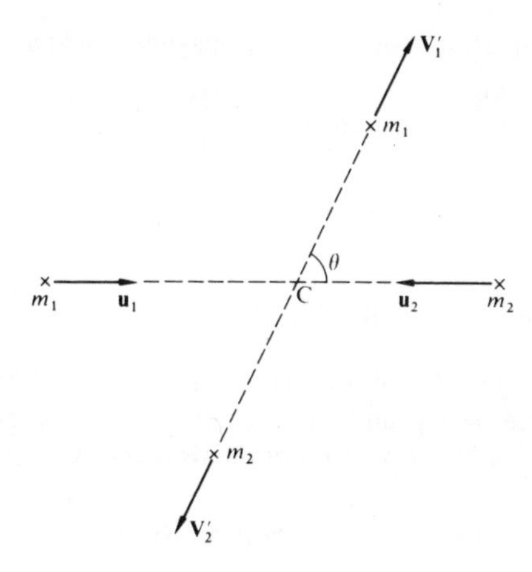


Figure 8.6

sign. In Fig. 8.6, C is the fixed centre of the mass, $\mathbf{u}_1$, $\mathbf{u}_2$ are the initial velocities and $\mathbf{V}'_1$, $\mathbf{V}'_2$ are the final velocities. Explain why for elastic collisions $V'_1 = u_1$ and $V'_2 = u_2$ and show that

$$V_1 \sin \theta_1 = u_1 \sin \theta$$

$$V_1 \cos \theta_1 = u_1 \cos \theta + V_0.$$

Hence show that

$$\tan \theta_1 = \frac{m_2 \sin \theta}{m_1 + m_2 \cos \theta} \qquad \text{and} \qquad \theta_2 = \tfrac{1}{2}(\pi - \theta).$$

(Hint: observe that $\mathbf{V}_1$ is the resultant of $\mathbf{V}'_1$ and the centre of mass velocity $(0, 0, V_0)$.)

2 Let $\sigma$ and $\sigma'$ be the differential cross-sections in the laboratory and centre of mass co-ordinate systems respectively. Then

$$\sigma(\theta_1)\, \mathrm{d}\Omega = \sigma'(\theta)\, \mathrm{d}\Omega'$$

where $\mathrm{d}\Omega$ and $\mathrm{d}\Omega'$ are the corresponding respective solid angles. Show that (see problem one)

$$\sin \theta_1\, \mathrm{d}\theta_1 = \frac{(1 + \mu \cos \theta)}{(1 + \mu^2 + 2\mu \cos \theta)^{3/2}} \sin \theta\, \mathrm{d}\theta$$

where $\mu = m_1/m_2$ and hence show that

$$\sigma(\theta_1) = \sigma'(\theta)\,\frac{(1+\mu^2+2\mu\cos\theta)^{3/2}}{|1+\mu\cos\theta|}.$$

3 Maxwell's equations for the electromagnetic field are

$$\text{curl}\,\mathcal{E} = -\frac{\partial \mathbf{B}}{\partial t} \qquad \text{curl}\,\mathbf{H} = \mathbf{j} + \frac{\partial \mathbf{D}}{\partial t}$$

$$\text{div}\,\mathbf{D} = \rho \qquad \text{div}\,\mathbf{B} = 0$$

with

$$\mathbf{D} = \epsilon_r\epsilon_0\mathcal{E} \qquad \mathbf{B} = \mu_r\mu_0\mathbf{H}$$

where $\epsilon_r$, $\mu_r$ are the relative permittivity and permeability of the medium.

Show that the electric and magnetic induction vectors can be expressed in terms of the scalar and vector potentials $\phi$ and $\mathbf{A}$ such that

$$\mathcal{E} = -\frac{\partial \mathbf{A}}{\partial t} - \text{grad}\,\phi, \qquad \mathbf{B} = \text{curl}\,\mathbf{A} \quad \text{(i)}$$

with

$$\nabla^2\phi - \frac{\mu_r\epsilon_r}{c^2}\frac{\partial^2\phi}{\partial t^2} = -\frac{\rho}{\epsilon_r\epsilon_0} \quad \text{(ii)}$$

and

$$\nabla^2\mathbf{A} - \frac{\mu_r\epsilon_r}{c^2}\frac{\partial^2\mathbf{A}}{\partial t^2} = -\mu_r\mu_0\mathbf{j} \quad \text{(iii)}$$

provided

$$\text{div}\,\mathbf{A} = -\frac{\mu_r\epsilon_r}{c^2}\frac{\partial\phi}{\partial t}.$$

More than one set of potentials specify the fields. Show that equations (i), (ii) and (iii), are invariant under a change to the new potentials

$$\mathbf{A}' = \mathbf{A} - \text{grad}\,\chi, \qquad \phi' = \phi + \frac{\partial\chi}{\partial t}$$

provided

$$\nabla^2\chi - \frac{\mu_r\epsilon_r}{c^2}\frac{\partial^2\chi}{\partial t^2} = 0.$$

This is called a gauge transformation.

4 (i) Prove the vector identity

$$\text{div}\,(\mathscr{E}\wedge\mathbf{H}) = \mathbf{H}\,.\,\text{curl}\,\mathscr{E} - \mathscr{E}\,.\,\text{curl}\,\mathbf{H}.$$

(ii) Write down Maxwell's equations for free space (problem 3) and show that

$$\mathbf{H}\,.\,\text{curl}\,\mathscr{E} - \mathscr{E}\,.\,\text{curl}\,\mathbf{H} = -\frac{\partial}{\partial t}(\tfrac{1}{2}\mu_0 H^2 + \tfrac{1}{2}\epsilon_0\mathscr{E}^2).$$

The energy density of the electromagnetic field is

$$\tfrac{1}{2}\mu_0 H^2 + \tfrac{1}{2}\epsilon_0\mathscr{E}^2.$$

Use the divergence theorem to show that the flux of energy out of a closed surface $s$ is

$$\oint_s \mathbf{N}\,.\,\mathbf{n}\,\mathrm{d}s$$

where **n** is the unit outward normal and **N** is the Poynting vector

$$\mathbf{N} = \mathscr{E}\wedge\mathbf{H}.$$

5 The commutator for two quantum operators $\hat{\mathbf{p}}$, $\hat{\mathbf{A}}$ is related to the corresponding Poisson bracket of the two dynamical variables **p**, **A** by (3.6)

$$[\hat{\mathbf{p}}, \hat{\mathbf{A}}] \equiv i\hbar\{\mathbf{p}, \mathbf{A}\}.$$

Show that if $\hat{\mathbf{p}}$ is the momentum operator and **A** is a function of the position only then

$$[\hat{\mathbf{p}}, \hat{\mathbf{A}}] \equiv -i\hbar\ \text{div}\,\mathbf{A}. \quad \text{(i)}$$

(Hint: use equation (1.28).)

The classical Hamiltonian for an electron in an electromagnetic field is

$$H = \frac{p^2}{2m} + \frac{e}{2m}(\mathbf{p}\,.\,\mathbf{A} + \mathbf{A}\,.\,\mathbf{p}) + \frac{e^2}{2m}A^2 - e\phi.$$

(Chapter 3, problem 2). Explain how equation (i) can be used to obtain the correct quantum Hamiltonian of (8.24).

## References

Bates, D. R. *Quantum Theory I. Elements.* Academic Press, London (1961).

Power, E. A. *Introductory Quantum Electrodynamics.* Longmans, London (1964).

Schiff, L. I. *Quantum Mechanics.* McGraw-Hill, New York (1955).

CHAPTER 9

# Group Theory

## 9.1 Symmetry Operations

Many physical systems possess symmetry properties which have important consequences for the occurrence of energy levels. For example, crystalline solids have a periodic structure and this translational symmetry enables the energy eigenfunctions to be classified by a wave vector **k**. These functions are the Bloch waves and will be referred to again later in the chapter.

Although finite molecules cannot possess a periodic structure they can have other symmetry properties, such as reflection planes and rotation axes.

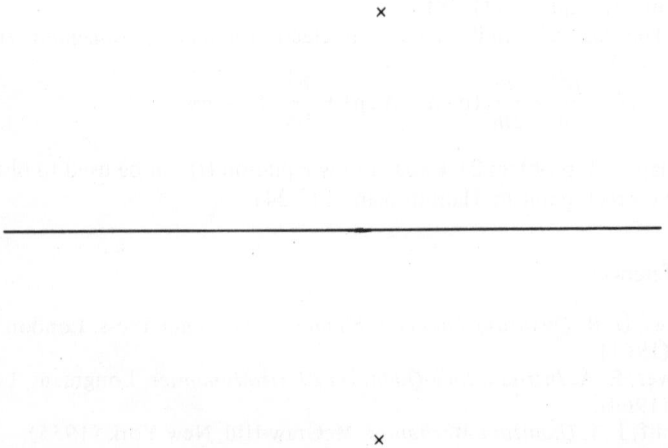

Figure 9.1 Plane of symmetry.

A plane of symmetry is a plane in the system such that each point on one side of the plane is the mirror image of a corresponding point on the other.

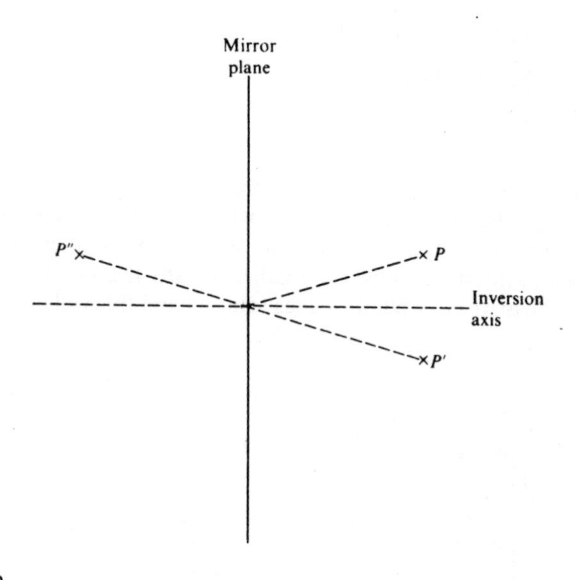


Figure 9.2

An axis of symmetry is an axis such that the system takes up an identical position on rotation through a suitable angle about it. The angle of rotation must be of the form $2\pi/n$ where $n$ is an integer and such an axis is called an $n$-fold axis.

A centre of inversion is a point such that the system is invariant under the operation $\mathbf{r} \rightarrow -\mathbf{r}$ where $\mathbf{r}$ is the vector displacement of any point in the system referred to the inversion centre.

A system possesses a rotation-inversion axis if it is brought into self-coincidence by a rotation followed by an inversion.

Some symmetry operations are equivalent. For example, a two-fold inversion axis is equivalent to a mirror plane perpendicular to the axis (Fig. 9.2).

*Symmetry operators*

Mathematically, a symmetry operation such as a rotation or reflection involves a linear transformation of the co-ordinates. There are two distinct ways of thinking about such a rotation. A physical rotation of a body through an angle $\alpha$ in the clockwise direction about an axis, say the $x_3$ axis, is equivalent to a rotation of the complete axes system through the same angle but in the anti-clockwise sense in that the relation between the new and old co-ordinates of any point in the body is the same in both cases (Fig. 9.3). In each case

$$
\begin{aligned}
x_1 &= x_1' \cos\alpha - x_2' \sin\alpha \\
x_2 &= x_1' \sin\alpha + x_2' \cos\alpha \\
x_3 &= x_3'.
\end{aligned}
\tag{9.1}
$$

where $x_1, x_2, x_3$ and $x_1', x_2', x_3'$ represent the old and new co-ordinates respectively. The first point of view in which the system is imagined to undergo an actual rotation is called the 'active' viewpoint and is perhaps the easiest to interpret. However, mathematically it is far more convenient to use the 'passive' viewpoint of axes rotation and this will be done here.

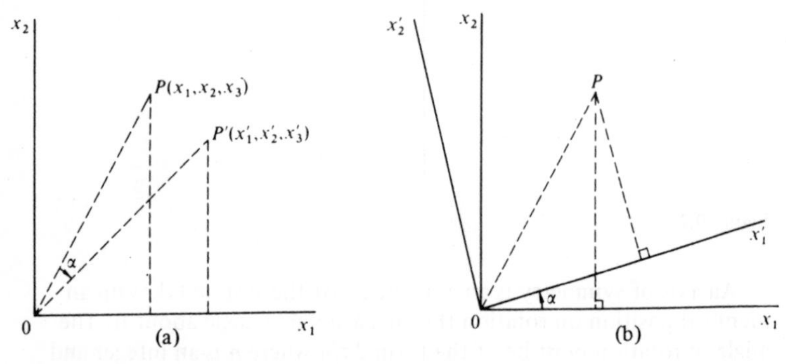


Figure 9.3 (a) Rotation of $P$ to $P'$ in a clockwise sense.
(b) Rotation of axis in anti-clockwise sense.

A general rotation, reflection or inversion can be represented by a linear relationship of the form,

$$
\begin{aligned}
x_1' &= \alpha_{11}x_1 + \alpha_{12}x_2 + \alpha_{13}x_3 \\
x_2' &= \alpha_{21}x_1 + \alpha_{22}x_2 + \alpha_{23}x_3 \\
x_3' &= \alpha_{31}x_1 + \alpha_{32}x_2 + \alpha_{33}x_3
\end{aligned}
\tag{9.2}
$$

where $x_1, x_2, x_3$ are the old co-ordinates of a point and $x_1', x_2', x_3'$ are the new co-ordinates of the same point. Equations (9.2) can be represented by the simple matrix equation

$$\mathbf{x}' = \boldsymbol{\alpha}\mathbf{x} \tag{9.3}$$

where $\boldsymbol{\alpha}$ is the matrix

$$\boldsymbol{\alpha} = \begin{pmatrix} \alpha_{11} & \alpha_{12} & \alpha_{13} \\ \alpha_{21} & \alpha_{22} & \alpha_{23} \\ \alpha_{31} & \alpha_{32} & \alpha_{33} \end{pmatrix}$$

and $\mathbf{x}$ and $\mathbf{x}'$ are the column vectors

$$\begin{pmatrix} x_1 \\ x_2 \\ x_3 \end{pmatrix} \quad \text{and} \quad \begin{pmatrix} x_1' \\ x_2' \\ x_3' \end{pmatrix}$$

respectively.

For an inversion, a rotation about an axis, or a reflection in a plane through the origin, the distance of a point from the origin remains constant.

$$r^2 = x_1^2 + x_2^2 + x_3^2 = x_1'^2 + x_2'^2 + x_3'^2. \tag{9.4}$$

The matrix $\boldsymbol{\alpha}$ is an orthogonal matrix and its determinant takes the value $\pm 1$, the negative sign belonging to an operation involving an odd number of reflections.

The effect of any operator on a function $f(\mathbf{x})$ is to transform it into another function. For example the differential operator $d/dx$ changes $x^n$ into $nx^{n-1}$. It is now necessary to define the effect of a symmetry operator $A$ on an arbitrary function. In the passive convention, the co-ordinate axes are rotated and the value of the function at each point in space remains unchanged. The transformed function is easily obtained by replacing the old co-ordinates by the new co-ordinates according to (9.3), i.e.

$$Af(\mathbf{x}) = f(\boldsymbol{\alpha}^{-1}\mathbf{x}') \tag{9.5}$$

where $\boldsymbol{\alpha}$ is the transformation matrix corresponding to the operator $A$. It is convenient to drop the dash and

$$Af(\mathbf{x}) = f(\boldsymbol{\alpha}^{-1}\mathbf{x}). \tag{9.6}$$

The operator is a substitutional one and is linear.

As an example consider the two functions

$$d_1(x, y) = 2xyf(r), \qquad d_2(x, y) = (x^2 - y^2)f(r). \tag{9.7}$$

The operator representing a rotation of the axes through $\pi/4$ radius in the anti-clockwise about the $z$-axis may be written $8_z$. The old co-ordinates are given in terms of the new by (9.1)

$$x = \frac{1}{\sqrt{2}}(x' - y'), \qquad y = \frac{1}{\sqrt{2}}(x' + y'), \qquad z = z'$$

i.e.

$$\alpha^{-1} = \frac{1}{\sqrt{2}} \begin{pmatrix} 1 & -1 & 0 \\ 1 & 1 & 0 \\ 0 & 0 & 1 \end{pmatrix}$$

$$\begin{aligned} \therefore \quad 8_z \, \mathrm{d}_1(x, y) &= 2 \times \tfrac{1}{2}(x' - y')(x' + y')f(r') \\ &= (x'^2 - y'^2)f(r') \\ &= \mathrm{d}_2(x', y') \end{aligned} \qquad (9.8)$$

The function $\mathrm{d}_1$ evaluated at the point $(x, y)$ is equal to the value of $\mathrm{d}_2$ at the same point with new co-ordinates $(x', y')$.

## 9.2 Groups

A group is a set of mathematical elements $\{A\}$ which obey certain specified relations. In this book the elements will be taken to be symmetry operators but the concept is much more general. The members of a group have the following properties.

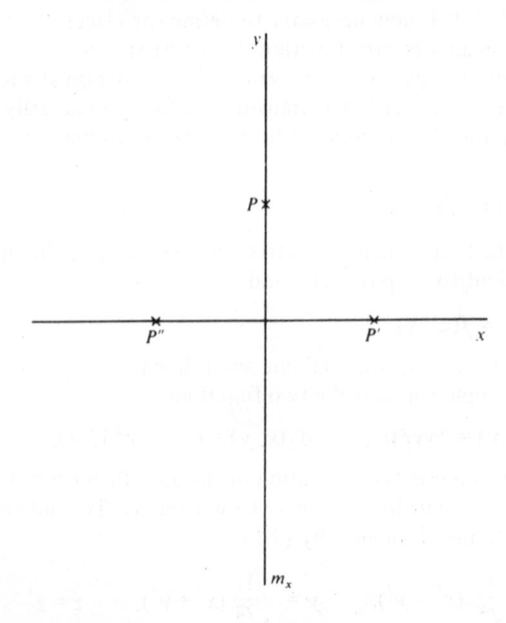


Figure 9.4

*1. Closure property*

It must be possible to combine the elements in pairs to form a new element of the group. That is, the group is closed with respect to some (binary) operation, e.g.

$$A = BC. \tag{9.9}$$

In this book the group elements are transformation operators and (9.9) means that if the operator $C$ is applied first, and $B$ second, the result is equivalent to applying the operator $A$ alone, where $A$ is another group element. In general the order of the operators is most important. Consider a point $P$ on the $y$-axis together with a plane of symmetry $m_x$ in the $yz$ plane and a four-fold rotation axis lying along the $z$-axis. If the operator $m_x$ is applied first, followed by an active-clockwise rotation $4z$, the point $P$ moves to $P'$. However if the order of the operations is reversed, $P$ moves to $P''$. The two operations do not commute.

However, any two rotations about the same axis do commute.

*2. Associative property*

The group operation is associative and the combination of three operators is uniquely defined, i.e.

$$A(BC) = (AB)C = ABC. \tag{9.10}$$

*3. Identity*

One of the group elements is the identity or unit element $E$ such that it commutes with all other members of the group, i.e.

$$EA = AE = A. \tag{9.11}$$

*4. Inverse*

Each element $A$ in a group has an inverse $A^{-1}$ so that the combination of an operator with its inverse, results in the unit operator, i.e.

$$AA^{-1} = A^{-1}A = E. \tag{9.12}$$

*The group 3m*

Consider a right pyramid with an equilateral triangular base and its central vertex above the centroid of the base. The pyramid is not a regular tetrahedron so that the three faces meeting at the central vertex are isosceles and not congruent to the base. This is the structure exhibited by the ammonia molecule $NH_3$ with the three hydrogen atoms situated at

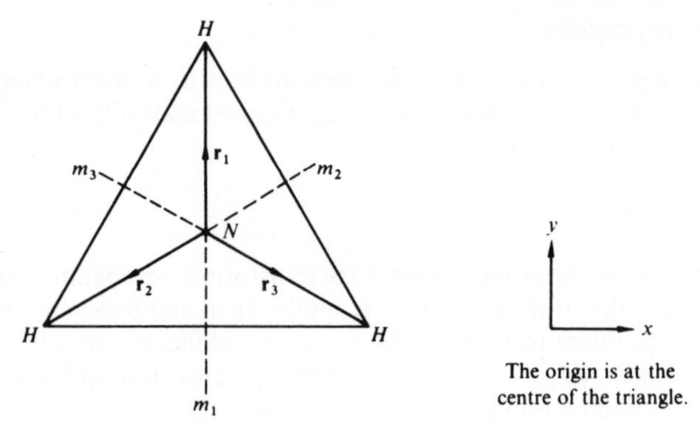


Figure 9.5 The ammonia molecule.

the vertices of the equilateral triangle and the nitrogen atom at the remaining vertex. The group $3m$ consists of those symmetry elements which leave this pyramid invariant. This is also the symmetry group of an equilateral triangle if the two plane sides are not identical.

There are six symmetry operations which leave this system invariant. They are,

(a) the identity operation $E$.
(b) the three mirror planes of the type $m_1$.
(c) the two rotations $3_z$, $3_z^3$ corresponding to an active clockwise rotation of 120° and 240° about the $z$-axis respectively.

These six operations may be shown to form a group, and Table 9.1 is the 'multiplication table' for the group. The fact that such a table can be constructed shows that the elements must form a group.

**Table 9.1** Multiplication table for the group $3m$.

| | | $E$ | $m_1$ | $m_2$ | $m_3$ | $3_z$ | $3_{z^2}$ | Applied first |
|---|---|---|---|---|---|---|---|---|
| | $E$ | $E$ | $m_1$ | $m_2$ | $m_3$ | $3_z$ | $3_{z^2}$ | |
| | $m_1$ | $m_1$ | $E$ | $3_{z^2}$ | $3_z$ | $m_3$ | $m_2$ | |
| | $m_2$ | $m_2$ | $3_z$ | $E$ | $3_{z^2}$ | $m_1$ | $m_3$ | |
| | $m_3$ | $m_3$ | $3_{z^2}$ | $3_z$ | $E$ | $m_2$ | $m_1$ | |
| | $3_z$ | $3_z$ | $m_2$ | $m_3$ | $m_1$ | $3_{z^2}$ | $E$ | |
| Applied second | $3_{z^2}$ | $3_{z^2}$ | $m_3$ | $m_1$ | $m_2$ | $E$ | $3_z$ | |

In Table 9.1 the elements in the top row are applied first and then the operation in the first column. For example, if the reflection $m_1$ is applied and then followed by the rotation $3_z$, the result is equivalent to the reflection $m_2$. However, if $3_z$ is applied first and then $m_1$, the result is $m_3$. This illustrates a point already mentioned that the order of the operations is important, as they do not necessarily commute.

*The group $\bar{6}m2$*

Suppose in the previous example the two faces of the triangle were identical. The system would then be invariant under twelve symmetry operations constituting the group $\bar{6}m2$. Apart from the six operations already discussed this group includes,

(d) the three rotations of the type $2_y$ through 180° about the vectors $\mathbf{r}_1$, $\mathbf{r}_2$ and $\mathbf{r}_3$ respectively.
(e) the mirror plane $m_z$ perpendicular to the $z$-axis and including the triangle.
(f) the two operations $\bar{6}_z$ involving a rotation through 60° and 300° respectively, about the $z$-axis followed by an inversion through the origin.

It is left to the reader to construct a multiplication table for this group.

This symmetry group is of importance in that it is related to that of an ion surrounded by three molecules of water in the hydrated crystal of a salt. The ozone molecule has almost this symmetry.

## 9.3 Matrix Representation

The reader was introduced to the concept of a function space in Chapter 3 (definition 6). A function space $\{\theta_i\}$ is said to be invariant under the set of operators $\{A\}$ if all transformed functions of the type $A\theta_i$ belong to the space. The Cartesian vector space, defined by the unit vectors $\{\hat{\mathbf{e}}_i\}$, is invariant under either of the two symmetry groups $3m$ or $\bar{6}m2$.

If the space $\{\theta_i\}$ is invariant under the group $\{A\}$ the transformed functions may be expressed in terms of the original set, i.e.

$$A\theta_i = \sum_m \alpha_{mi}\theta_m. \tag{9.13}$$

In matrix notation $A\boldsymbol{\theta} = \boldsymbol{\theta\alpha}$ where $\boldsymbol{\theta}$ is a row vector. The matrix $\boldsymbol{\alpha}$ represents the effect of the operator $A$ on the basis functions $\{\theta_i\}$ and each operator in the set has a matrix representation. If $B$ is another operator in the set then

$$B\boldsymbol{\theta} = \boldsymbol{\theta\beta}. \tag{9.14}$$

The effect of the multiple operator $AB$ is easily seen to be

$$AB\theta_i = \sum_j \sum_m \alpha_{jm}\beta_{mi}\theta_j \tag{9.15}$$

if the operators are linear, i.e.

$$AB\boldsymbol{\theta} = \boldsymbol{\theta}\boldsymbol{\alpha}\boldsymbol{\beta}. \tag{9.16}$$

The matrix corresponding to the product $AB$ is the product $\boldsymbol{\alpha\beta}$ of the separate matrices representing $A$ and $B$.

The matrices obey the same group multiplication table as the operators themselves and are said to form a representation of the group. The order of the matrices gives the dimensionality of the representation.

The set of functions $\{\theta_i'\}$ can be used to define the same vector space as $\{\theta_i\}$ if the number in each set is the same and if the new basis functions can be expressed in terms of the old by some linear transformation, i.e.

$$\theta_j' = \sum_i S_{ij}\theta_i. \tag{9.17}$$

In matrix notation,

$$\boldsymbol{\theta}' = \boldsymbol{\theta}\mathbf{S} \tag{9.18}$$

where $\boldsymbol{\theta}'$ and $\boldsymbol{\theta}$ are row vectors. Conversely

$$\boldsymbol{\theta} = \boldsymbol{\theta}'\mathbf{S}^{-1}. \tag{9.19}$$

The representation of the operator $A$ in the new basis is given by

$$\begin{aligned} A\boldsymbol{\theta}' &= A\boldsymbol{\theta}\mathbf{S} \\ &= \boldsymbol{\theta}\boldsymbol{\alpha}\mathbf{S} \\ &= \boldsymbol{\theta}'\mathbf{S}^{-1}\boldsymbol{\alpha}\mathbf{S}. \end{aligned}$$

That is, the new basis functions transform according to the representation

$$\mathbf{S}^{-1}\boldsymbol{\alpha}\mathbf{S}. \tag{9.20}$$

This new representation obtained by this similarity transformation is said to be equivalent to the original representation $\boldsymbol{\alpha}$. A change of basis functions, as described above, is equivalent to choosing a new set of vectors $\{\hat{e}_i'\}$ in Cartesian space related to the old set $\{\hat{e}_i\}$ by some rotation.

The matrices in equivalent representations corresponding to a given operator are not in general identical but their traces are equal (6.22). As the trace of a product of two matrices does not depend upon the order of the factors

$$\text{Trace}\,(\mathbf{S}^{-1}\boldsymbol{\alpha}\mathbf{S}) = \text{Trace}\,(\mathbf{S}\mathbf{S}^{-1}\boldsymbol{\alpha}) = \text{Trace}\,(\boldsymbol{\alpha}). \tag{9.21}$$

*Irreducible representations*

The function space $\{\theta_i\}$ is said to be irreducible under the set of operators $\{A\}$ if with any function $\chi_i$ in the space, the set of transformed functions $\{A\chi_i\}$ defines the same space where $A$ runs over all the members of $\{A\}$

in turn. The Cartesian space $\{\hat{e}_i\}$ is irreducible under the set of three operators consisting of a rotation of 90° about the $x$-axis, and 90° about $y$-axis and 90° about the $z$-axis.

On the other hand, this Cartesian space is not irreducible under the set of operators consisting of the identity and rotations of ±120° about the (111) axis. To see this, consider the unit vector along the (111) axis.

A set of matrices representing a group of transformation operators with respect to a set of basis functions $\{\theta_i\}$ is said to form an irreducible representation, if the function space defined by $\{\theta_i\}$ is irreducible under the group of operators. A reducible representation can always be brought into irreducible form by an equivalence transformation of the type (9.20).

The reducible matrices representing all the elements of the group are then reduced to block form, all with the same block structure. That is, each matrix is a direct sum of the square matrices $\mathbf{a}_1, \mathbf{a}_2, \ldots$ along the leading diagonal and the dimensions of corresponding submatrices are the same for every operator in the group, e.g.

$$\left(\begin{array}{c|c|c} \mathbf{a}_1 & 0 & 0 \\ \hline 0 & \mathbf{a}_2 & 0 \\ \hline 0 & 0 & \mathbf{a}_2 \end{array}\right).$$

A set of submatrices forms one of the constituent irreducible representations of the group.

The functions $xe^{-r}, ye^{-r}, ze^{-r}$ correspond to a reducible representation of the equilateral triangle group $3m$ (Table 9.1 and Fig. 9.5).

The matrices representing the different group operators are given in Table 9.2.

As an example consider the effect of applying the operator $m_1$ representing the mirror plane perpendicular to the $x$-axis,

$$m_1(xe^{-r}, ye^{-r}, ze^{-r}) = (xe^{-r}, ye^{-r}, ze^{-r})\begin{pmatrix} -1 & 0 & 0 \\ 0 & 1 & 0 \\ 0 & 0 & 1 \end{pmatrix}. \tag{9.22}$$

Inspection of the Table 9.2 shows that none of the six group elements mixes the functions $xe^{-r}$, $ye^{-r}$ with $ze^{-r}$ and the two sets $(xe^{-r}, ye^{-r})$ and $(ze^{-r})$ span separate spaces which are invariant under the group $3m$. There is no need to carry out a similarity transformation in this simple case. The two separate irreducible representations are listed under $A$ and $E$.

**Table 9.2** Matrix representation for the group $3m$.

| | $E$ | $m_1$ | $m_2$ | $m_3$ | $3_z$ | $3_{z^2}$ |
|---|---|---|---|---|---|---|
| | $\begin{pmatrix} 1 & 0 & 0 \\ 0 & 1 & 0 \\ 0 & 0 & 1 \end{pmatrix}$ | $\begin{pmatrix} -1 & 0 & 0 \\ 0 & 1 & 0 \\ 0 & 0 & 1 \end{pmatrix}$ | $\begin{pmatrix} \frac{1}{2} & \sqrt{3}/2 & 0 \\ \sqrt{3}/2 & -\frac{1}{2} & 0 \\ 0 & 0 & 1 \end{pmatrix}$ | $\begin{pmatrix} \frac{1}{2} & -\sqrt{3}/2 & 0 \\ -\sqrt{3}/2 & -\frac{1}{2} & 0 \\ 0 & 0 & 1 \end{pmatrix}$ | $\begin{pmatrix} -\frac{1}{2} & \sqrt{3}/2 & 0 \\ -\sqrt{3}/2 & -\frac{1}{2} & 0 \\ 0 & 0 & 1 \end{pmatrix}$ | $\begin{pmatrix} -\frac{1}{2} & -\sqrt{3}/2 & 0 \\ \sqrt{3}/2 & -\frac{1}{2} & 0 \\ 0 & 0 & 1 \end{pmatrix}$ |
| $ze^{-r}A$ | 1 | 1 | 1 | 1 | 1 | 1 |
| $\left.\begin{matrix} xe^{-r} \\ ye^{-r} \end{matrix}\right\} E$ | $\begin{pmatrix} 1 & 0 \\ 0 & 1 \end{pmatrix}$ | $\begin{pmatrix} -1 & 0 \\ 0 & 1 \end{pmatrix}$ | $\begin{pmatrix} \frac{1}{2} & \sqrt{3}/2 \\ \sqrt{3}/2 & -\frac{1}{2} \end{pmatrix}$ | $\begin{pmatrix} \frac{1}{2} & -\sqrt{3}/2 \\ -\sqrt{3}/2 & -\frac{1}{2} \end{pmatrix}$ | $\begin{pmatrix} -\frac{1}{2} & \sqrt{3}/2 \\ -\sqrt{3}/2 & -\frac{1}{2} \end{pmatrix}$ | $\begin{pmatrix} -\frac{1}{2} & -\sqrt{3}/2 \\ \sqrt{3}/2 & -\frac{1}{2} \end{pmatrix}$ |
| $\left.\begin{matrix} (x+iy)e^{-r} \\ (x-iy)e^{-r} \end{matrix}\right\} E$ | $\begin{pmatrix} 1 & 0 \\ 0 & 1 \end{pmatrix}$ | $\begin{pmatrix} 0 & -1 \\ -1 & 0 \end{pmatrix}$ | $\frac{1}{2}\begin{pmatrix} 0, & 1-i\sqrt{3} \\ 1+i\sqrt{3}, & 0 \end{pmatrix}$ | $\frac{1}{2}\begin{pmatrix} 0, & 1+i\sqrt{3} \\ 1-\sqrt{3}, & 0 \end{pmatrix}$ | $\frac{1}{2}\begin{pmatrix} -1+i\sqrt{3}, & 0 \\ 0, & -1-i\sqrt{3} \end{pmatrix}$ | $\frac{1}{2}\begin{pmatrix} -1-i\sqrt{3}, & 0 \\ 0, & -1+i\sqrt{3} \end{pmatrix}$ |

The letters most commonly used to represent a one-dimensional representation are $A$ and $B$, $E$ is used to represent a two-dimensional representation and $T$ to represent a three-dimensional representation.

Another basis for the two-dimensional irreducible representation of the group $3m$ is $(x+iy)e^{-r}$, $(x-iy)e^{-r}$. These functions are a linear combination of the previous basis functions and so produce an equivalent representations of the group. The transformation matrix representing the linear relationship between the two sets of basis functions is

$$S = \begin{pmatrix} 1 & 1 \\ i & -i \end{pmatrix} \quad \text{and } S^{-1} = \tfrac{1}{2}\begin{pmatrix} 1 & -i \\ 1 & i \end{pmatrix}.$$

The matrices of the new equivalent representation are displayed in the last row of Table 9.2, e.g. for $m_1$

$$\tfrac{1}{2}\begin{pmatrix} 1 & -i \\ 1 & i \end{pmatrix}\begin{pmatrix} -1 & 0 \\ 0 & 1 \end{pmatrix}\begin{pmatrix} 1 & 1 \\ i & -i \end{pmatrix} = \begin{pmatrix} 0 & -1 \\ -1 & 0 \end{pmatrix}.$$

Every group has an identity representation, such as $A$, in which all the 'matrices' are unity. This is not a 'faithful' representation since the 'matrices' obey more relations than the group elements themselves.

Apart from the two irreducible representations already mentioned the group $3m$ has another one-dimensional representation.

**Table 9.3** Character table for the group $3m$.

| $3m$ | $E$ | $m_1$ | $m_2$ | $m_3$ | $3_z$ | $3_{z^2}$ |
|---|---|---|---|---|---|---|
| $A_1$ | 1 | 1 | 1 | 1 | 1 | 1 |
| $A_2$ | 1 | −1 | −1 | −1 | 1 | 1 |
| $E$ | 2 | 0 | 0 | 0 | −1 | −1 |

The trace of a matrix representing an operator is called the 'character' of the operator in the representation and is independent of the choice of basis functions.

Table 9.3 is a character table for the group $3m$ and lists all three irreducible representations.

### *Classes*

The elements of a group may be partitioned into exclusive sets called 'classes'. Given any element $A$ of a group, the set of elements $B^{-1}AB$, where $B$ runs over all elements in the group, is called a class.

The repeated operation $B^{-1}AB$ is the result obtained by first rotating the system to some equivalent position by the rotation $B$, next carrying out the symmetry operation $A$, and then reversing the initial operation by $B^{-1}$. The members of a particular class must be operators of the same type, such as rotations through the same angle, but performed about different axes which are related to each other by members of the group.

As the characters of all the elements of a class in a given representation are identical (9.21), it is only necessary to specify the classes in a group character table.

The six elements of the group $3m$ (Table 9.3) are divided in three exclusive classes,

$$E$$

$$m_1, m_2, m_3$$

$$3_z, 3_{z^2}.$$

The identity operator always forms a class of its own.

It can be shown that the number of classes in a group is equal to the number of irreducible representations of the group. For the equilateral triangle group there are three classes and three irreducible representations.

There is a very useful relationship between the dimensionalities of the irreducible representations and the number $N$ of elements in the group. If the $i$th irreducible representation has dimension $n_i$ then

$$\sum_i n_i^2 = N, \tag{9.23}$$

where the sum is over all the possible irreducible representations. Consequently if the number of elements and classes in a group is known then (9.23) will give the dimensionality of the irreducible representations.

The group $3m$ has three classes and hence three irreducible representations. As the group has six elements then (9.23) becomes

$$1^2 + 1^2 + 2^2 = 6.$$

There are two one-dimensional and one two-dimensional irreducible representations in agreement with Table 9.3.

### *Axial rotation group*

The axial rotation group is the infinite group whose elements are all the possible rotations about some axis. Clearly a rotation through an angle $\phi_1$ followed by $\phi_2$, is exactly equivalent to $\phi_2$ followed by $\phi_1$. Such a group in which all the elements commute with one another is called an Abelian group and each element forms a class in itself. As the number of irreducible representations is equal to the number of elements, (9.23) implies that all these representations are one-dimensional.

If $\Gamma(\phi_1)$ is the number representing the rotation $\phi_1$

$$\Gamma(\phi_1)\Gamma(\phi_2) = \Gamma(\phi_1 + \phi_2).$$

This equation has solutions

$$\Gamma^{(m)}(\phi) = e^{im\phi}. \tag{9.24}$$

As a rotation through $2\pi$ is equivalent to zero rotation, if the function is single valued,

$$\Gamma^{(m)}(2\pi) = \Gamma^{(m)}(0) = 1$$

and $m$ is restricted to the values

$$m = 0, \pm 1, \pm 2, \ldots .$$

A suitable choice for basis functions is

$$\psi_m(r, \theta, \phi) = f(r, \theta)\, e^{im\phi}. \tag{9.25}$$

## 9.4 Symmetry and Quantum Mechanics

Many molecular systems and solids have symmetry properties which have important consequences for the solutions of Schrödinger's equation and are related to the occurrence of degeneracies. For the moment only one-electron systems will be considered and the Hamiltonian for a particle of mass $m$ is,

$$\mathscr{H}(x_1, x_2, x_3) \equiv -\frac{\hbar^2}{2m}\nabla^2 + V(x_1, x_2, x_3). \tag{9.26}$$

If the old co-ordinates $x_1, x_2, x_3$ are replaced by new co-ordinates $x_1', x_2', x_3'$ (9.2), by some linear transformation, it is well known that the Laplacian is invariant.

$$\frac{\partial^2}{\partial x_1^2} + \frac{\partial^2}{\partial x_2^2} + \frac{\partial^2}{\partial x_3^2} \equiv \frac{\partial^2}{\partial x_1'^2} + \frac{\partial_2}{\partial x_2'^2} + \frac{\partial^2}{\partial x_3'^2}. \tag{9.27}$$

Consider now a 'symmetry operation' that brings the system into self-coincidence. The potential function is invariant under the change in co-ordinates corresponding to this operation.

As a simple illustration consider the potential of an electron moving in the field of two protons. This arises in the hydrogen molecular ion.

A rotation through 180° about any axis perpendicular to and bisecting the line joining the two protons brings the system into self-coincidence. In particular the point $P$ is rotated into $P'$ and clearly the potential function has the same value at $P$ and $P'$, i.e.

$$V(x_1, x_2, x_3) = V(x_1', x_2', x_3') \tag{9.28}$$

where the co-ordinates $(x_1, x_2, x_3)$ of $P$ are related to those of $P'$ $(x_1', x_2', x_3'$ by a linear transformation of the form (9.2). When the transformation is interpreted passively as a rotation of co-ordinate axes instead of as a physical rotation, (9.28) requires that the potential function be invariant under the transformation.

The complete Hamiltonian is invariant under a linear transformation corresponding to a symmetry operation.

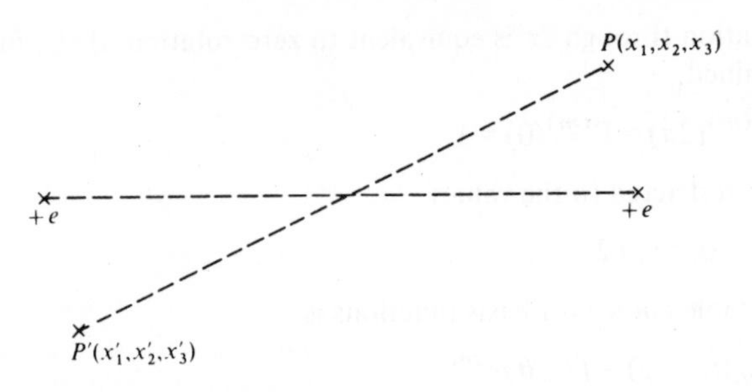


Figure 9.6 Hydrogen molecular ion.

Suppose the co-ordinates undergo an arbitrary linear transformation corresponding to the operator $A$. If $\psi(x_1, x_2, x_3)$ is any wave-function

$$\begin{aligned} A\mathscr{H}(x_1, x_2, x_3)\psi(x_1, x_2, x_3) &= \mathscr{H}'(x_1', x_2', x_3')\psi'(x_1', x_2', x_3') \\ &= \mathscr{H}'(x_1', x_2', x_3')A\psi(x_1, x_2, x_3) \end{aligned} \tag{9.29}$$

where $\mathscr{H}'$ and $\psi'$ are the transformed Hamiltonian and wave-function in terms of the new co-ordinates. If $A$ is a symmetry operator $\mathscr{H}'$ is the same operator as $\mathscr{H}$ and (9.29) implies that the Hamiltonian commutes with $A$.

The important relationships between commuting operators and degeneracy was mentioned in Chapter 5. If $\psi$ is an eigenfunction of $\mathscr{H}$

$$\mathscr{H}\psi = E\psi. \tag{9.30}$$

Transform this equation by operating on the left by the symmetry operator $A$.

$$A(\mathscr{H}\psi) = \mathscr{H}(A\psi) = E(A\psi). \tag{9.31}$$

In general the transformed wave-function

$$\psi'(x_1', x_2', x_3') = A\psi(x_1, x_2, x_3) \tag{9.32}$$

is not the same as $\psi(x_1, x_2, x_3)$ and they are two linearly independent wave-functions denoting degenerate states belonging to the same energy.

A simple illustration is obtained by considering the eigenfunctions for a particle in box. From equations (4.42) and (4.43) the three eigenfunctions

$$\begin{aligned} \psi(1,1,2) &= \sin\frac{\pi}{a}\left(x+\frac{a}{2}\right)\sin\frac{\pi}{a}\left(y+\frac{a}{2}\right)\sin\frac{2\pi}{a}\left(z+\frac{a}{2}\right) \\ \psi(1,2,1) &= \sin\frac{\pi}{a}\left(x+\frac{a}{2}\right)\sin\frac{2\pi}{a}\left(y+\frac{a}{2}\right)\sin\frac{\pi}{a}\left(z+\frac{a}{2}\right) \\ \psi(2,1,1) &= \sin\frac{2\pi}{a}\left(x+\frac{a}{2}\right)\sin\frac{\pi}{a}\left(y+\frac{a}{2}\right)\sin\frac{\pi}{a}\left(z+\frac{a}{2}\right) \end{aligned} \tag{9.33}$$

are degenerate and have the same energy,

$$E = 6h^2/8ma^2. \tag{9.34}$$

In (9.33) the origin is taken to be at the centre of the box. $\psi(1, 1, 2)$ can be transformed into $\psi(1, 2, 1)$ or $\psi(2, 1, 1)$ by considering the symmetry rotations of 90° about the $x$ and $y$ axes.

In general, each energy level possesses a set of degenerate eigenfunctions which span a space which is invariant under the action of any group of transformation operators which commute with the Hamiltonian. Consequently the eigenfunctions produce some representation of the group.

The important relationship between group theory and quantum mechanics is the postulate that the set of all degenerate wave-functions belonging to a given energy level form an irreducible representation of the group of all the symmetry operators of the Hamiltonian.

As a general rule, this statement is correct, but sometimes not all degenerate eigenfunctions are generated by applying the symmetry operators to a given eigenfunction. These extra degeneracies are called accidental. A well-known example that has already been referred to is the case of the hydrogen atom (Chapter 4).

The degeneracy between states with the same $n$ and $l$ quantum numbers is related to the spherical symmetry of the hydrogen atom and is discussed later in this chapter. However, the Coulomb degeneracy between the states with the same $n$ quantum number but with different values for $l$, does not arise in this way although Fock (1935) has related this degeneracy to the symmetry of the Schrödinger equation in momentum space. Often accidental degeneracies can be explained by deeper investigation.

## 9.5 Representative Operators and Symmetry Operators

A question that springs to mind is 'what relation is there between the operators representing the observables and the symmetry operators of a system?'. In Chapter 5 the reader was introduced to the concept of a complete set of commuting observables. When the eigenvalues of each is specified, the wave-function is known completely, apart from a phase factor. In the present chapter it has been stated that all degeneracy, apart from accidental degeneracy, can be explained by the symmetry operators that commute with the Hamiltonian. Consequently there must be a correspondence between representative operators and symmetry operators.

The simplest example is the relation between translational symmetry and linear momentum. The Hamiltonian for a single electron in a fieldless space is

$$\mathscr{H} \equiv -\frac{\hbar^2}{2m}\nabla^2$$

and the energy eigenfunctions are

$$\psi_{\mathbf{k}}(\mathbf{r}) = e^{i\mathbf{k}.\mathbf{r}}.$$

These are also eigenfunctions of the momentum operator

$$\frac{\hbar}{i}\nabla$$

with eigenvalues $\hbar\mathbf{k}$.

Consider the one-dimensional case. A translational operator $P(\delta x)$ may be defined by

$$P(\delta x)\psi(x) = \psi(x + \delta x). \tag{9.35}$$

An infinitesimal translation operator $I_x$ can be defined by

$$iI_x \equiv \lim_{\delta x \to 0} \frac{P(\delta x) - 1}{\delta x} \tag{9.36}$$

so that

$$iI_x\psi(x) = \lim_{\delta x \to 0} \frac{\psi(x + \delta x) - \psi(x)}{\delta x} = \frac{\mathrm{d}\psi}{\mathrm{d}x}. \tag{9.37}$$

From the last equation it is possible to deduce the equivalence

$$\frac{\hbar}{i}\frac{\mathrm{d}}{\mathrm{d}x} \equiv \hbar I_x$$

i.e.

$$\hat{p}_x \equiv \hbar I_x \tag{9.38}$$

where $\hat{p}_x$ is the linear momentum operator. This is a particular case of the relation

$$\hat{p} \equiv \hbar I_q \tag{9.39}$$

where $\hat{p}$ is the momentum operator conjugate to $q$. In general a representative operator that commutes with the Hamiltonian is related to a symmetry operator of the system.

## 9.6 The Full Rotation Group

The rotational symmetry of the hydrogen atom is intimately related to the angular momentum eigenfunctions discussed in Chapter 4.

Let $P(\phi, z)$ be an operator that rotates the co-ordinate axes by an angle $\phi$ about the $z$-axis in the positive sense. An infinitesimal rotation operator may be defined by

$$iI_z^{(r)} \equiv \lim_{\delta\phi \to 0} \frac{P(\delta\phi, z) - 1}{\delta\phi}. \tag{9.40}$$

Since, in spherical polar co-ordinates

$$P(\delta\phi, z)\psi(r, \theta, \phi) = \psi(r, \theta, \phi + \delta\phi)$$

then

$$iI_z^{(r)}\psi = \frac{\partial\psi}{\partial\phi}. \tag{9.41}$$

The $z$-component of angular momentum can be written

$$\mathscr{L}_z \equiv \frac{\hbar}{i}\frac{\partial}{\partial\phi}$$

and so

$$\mathscr{L}_z \equiv \hbar I_z^{(r)}. \tag{9.42}$$

This final equation is a special case of (9.39) and shows how axial rotational symmetry is related to angular momentum about the axis.

Consider a diatomic molecule and choose the $z$-axis to lie along the line joining the nuclei. The rotational symmetry of the molecule implies that the rotational operator $I_z^{(r)}$ and hence the angular momentum operator $\mathscr{L}_z$ commute with the Hamiltonian. The $z$-component of angular momentum is a constant of the motion and the eigenfunctions take the form (9.25) with eigenvalues $m\hbar$. The states with $|m| = 0, 1, 2, \ldots$ are denoted by $\Sigma, \Pi, \Delta, \ldots$.

Similarly to (9.42)

$$\mathscr{L}_x \equiv \hbar I_x^{(r)} \qquad \text{and} \qquad \mathscr{L}_y \equiv \hbar I_y^{(r)}. \tag{9.43}$$

Clearly, the infinitesimal rotation operators must obey the commutation relations similar to those for angular momentum operators, e.g.

$$I_x^{(r)}I_y^{(r)} - I_y^{(r)}I_x^{(r)} \equiv iI_z^{(r)}. \tag{9.44}$$

This result can be obtained without reference to the angular momentum operators.

Infinitesimal rotation operators (but not finite rotations) add like vectors and if $I_\alpha^{(r)}$ is an infinitesimal rotation about the $\alpha$-axis

$$I_\alpha^{(r)} \equiv lI_x^{(r)} + mI_y^{(r)} + nI_z^{(r)} \tag{9.45}$$

where $l$, $m$, $n$ are the direction cosines of the $\alpha$-axis. This is related to the vector properties of angular momentum.

From (9.40) it can be seen that a finite rotation through an angle $\delta\phi$ about the $z$-axis can be expressed by

$$P(\delta\phi, z) \equiv 1 + i\,\delta\phi I_z^{(r)} + \epsilon\delta\phi$$

where $\epsilon \to 0$ as $\delta\phi \to 0$.

The effect of the infinitesimal rotation operators on the co-ordinates of a point can be represented in matrix form. Consider a rotation of axes through $\delta\phi$ about the $z$-axis in the anti-clockwise sense. From (9.3) and (9.1)

$$\begin{pmatrix} x' \\ y' \\ z' \end{pmatrix} = P(\delta\phi, z)\begin{pmatrix} x \\ y \\ z \end{pmatrix} = \begin{pmatrix} \cos\delta\phi & \sin\delta\phi & 0 \\ -\sin\delta\phi & \cos\delta\phi & 0 \\ 0 & 0 & 1 \end{pmatrix}\begin{pmatrix} x \\ y \\ z \end{pmatrix}$$

The operator $P(\delta\phi, z)$ is represented by the matrix

$$\begin{pmatrix} \cos\delta\phi & \sin\delta\phi & 0 \\ -\sin\delta\phi & \cos\delta\phi & 0 \\ 0 & 0 & 1 \end{pmatrix} = \begin{pmatrix} 1 & 0 & 0 \\ 0 & 1 & 0 \\ 0 & 0 & 1 \end{pmatrix} + \begin{pmatrix} 0 & 1 & 0 \\ -1 & 0 & 0 \\ 0 & 0 & 0 \end{pmatrix}\delta\phi + \epsilon\delta\phi$$

where $\epsilon \to 0$ as $\delta\phi \to 0$ and so the infinitesimal rotation operator $I_z^{(r)}$ is represented by the matrix

$$i\begin{pmatrix} 0 & -1 & 0 \\ 1 & 0 & 0 \\ 0 & 0 & 0 \end{pmatrix}.$$

Similarly $I_x^{(r)}$ and $I_y^{(r)}$ are represented by

$$i\begin{pmatrix} 0 & 0 & 0 \\ 0 & 0 & -1 \\ 0 & 1 & 0 \end{pmatrix} \quad \text{and} \quad i\begin{pmatrix} 0 & 0 & 1 \\ 0 & 0 & 0 \\ -1 & 0 & 0 \end{pmatrix}$$

respectively. These three matrices do satisfy the commutation relations (9.44).

Any rotation about an axis can be expressed in terms of the corresponding infinitesimal rotation operator. Consider a rotation through the angle $\phi'$ about the $z$-axis. If the wave function is analytic then by Taylor's theorem

$$\begin{aligned} P(\phi', z)\psi(r, \theta, \phi) &= \psi(r, \theta, \phi + \phi') \\ &= \psi(r, \theta, \phi) + \sum_{n=1}^{\infty} \frac{\phi'^n}{n!} \frac{\partial^n \psi(r, \theta, \phi)}{\partial \phi^n}. \end{aligned}$$

Since

$$\frac{\partial}{\partial\phi} \equiv iI_z^{(r)}$$

then

$$P(\phi', z)\psi(r, \theta, \phi) = \left[1 + \sum_{n=1}^{\infty} \frac{1}{n!}(i\phi' I_z^{(r)})^n\right]\psi(r, \theta, \phi).$$

This can formally be written

$$P(\phi', z) \equiv \exp(i\phi' I_z^{(r)}) \tag{9.46a}$$

or in the general case for a rotation through $\theta$ about the $\alpha$-axis

$$P(\theta, \alpha) \equiv \exp(i\theta I_\alpha^{(r)}). \tag{9.46b}$$

This is an important result and together with (9.45) shows that any rotation can be expressed in terms of the three infinitesimal rotation operators $I_x^{(r)}$, $I_y^{(r)}$ and $I_z^{(r)}$.

The full rotation group is the infinite set of all proper rotations about all axes through a fixed point. If a function space is invariant under $I_x^{(r)}$, $I_y^{(r)}$, $I_z^{(r)}$ then it is also invariant under the full rotation group. The converse is also true.

*Angular momentum and irreducible representation*

In the next chapter spin momentum operators are defined through appropriate infinitesimal rotation operators and it follows that the spin operators satisfy the same commutation rules as those for angular momentum. It is worthwhile at this point to reconsider the derivation of the allowed eigenvalues for angular momentum by an argument based solely on the commutation relations. Essentially this is the problem of finding the irreducible representations for the rotation group. (See Heine, p. 55, for a direct method of finding the representations.)

Since the operators $\mathscr{L}^2$ and $\mathscr{L}_z$ commute, there exists a set of functions which are simultaneously eigenfunctions of both these operators and any other operators that are required to form a complete set. Let $\psi_m$ be the unique function that satisfies the eigenvalue equations

$$\mathscr{L}^2 \psi_m = a^2 \hbar^2 \psi_m$$

$$\mathscr{L}_z \psi_m = m\hbar \psi_m \tag{9.47}$$

and is also an eigenfunction with given eigenvalues of the remaining operators in the complete set. If the system possesses spherical symmetry these operators (apart from $\mathscr{L}_z$) will also commute with $\mathscr{L}_x$ and $\mathscr{L}_y$. This will be assumed and it is shown below how it is possible to generate from $\psi_m$ all the simultaneous eigenfunctions of the complete set that differ only in their eigenvalue of $\mathscr{L}_z$. Each of these functions is the basis for a one-dimensional irreducible representation of $I_z^{(r)}$, e.g.

$$I_z^{(r)} \psi_m = m \psi_m. \tag{9.48}$$

It is convenient to introduce two 'shift' (or 'ladder') operators $\mathscr{L}_\pm$ defined by

$$\mathscr{L}_+ \equiv \mathscr{L}_x + i\,\mathscr{L}_y \qquad \mathscr{L}_- \equiv \mathscr{L}_x - i\,\mathscr{L}_y. \tag{9.49}$$

These operators satisfy the commutation rules

$$\mathscr{L}_z \mathscr{L}_\pm - \mathscr{L}_\pm \mathscr{L}_z \equiv \hbar \mathscr{L}_\pm. \tag{9.50}$$

Consider the function $\mathscr{L}_+\psi_m$. Simple algebra gives

$$\mathscr{L}_z(\mathscr{L}_+\psi_m) = (\mathscr{L}_+\mathscr{L}_z + \hbar\mathscr{L}_+)\psi_m = (m+1)\hbar(\mathscr{L}_+\psi_m)$$

and

$$\mathscr{L}^2(\mathscr{L}_+\psi_m) = \mathscr{L}_+\mathscr{L}^2\psi_m = a^2\hbar^2(\mathscr{L}_+\psi_m).$$

All other operators in the complete set commute with $\mathscr{L}_+$ and so this shift operator generates from $\psi_m$ a sequence of simultaneous eigenfunctions that differ only in their $z$-component of angular momentum. The eigenvalues are

$$m\hbar, (m+1)\hbar, (m+2)\hbar, \ldots. \qquad (9.51)$$

Similarly, $\mathscr{L}_-$ generates from $\psi_m$ a sequence of simultaneous eigenfunction with decreasing $z$-component of angular momentum

$$(m-1)\hbar, (m-2)\hbar, \ldots. \qquad (9.52)$$

The number of eigenfunctions that can be generated in this way is finite. Any physical system will have a finite value for total angular momentum. That is, $a^2$ is finite. Also, since

$$\begin{aligned}(\mathscr{L}_x^2 + \mathscr{L}_y^2)\psi_m &= (\mathscr{L}^2 - \mathscr{L}_z^2)\psi_m \\ &= (a^2 - m^2)\hbar^2\psi_m\end{aligned} \qquad (9.53)$$

then

$$a^2 - m^2 > 0$$

as $\mathscr{L}_x^2 + \mathscr{L}_y^2$ is a positive definite operator and must have positive eigenvalues. The eigenvalues of $\mathscr{L}_z$ are bounded both above and below and let $l_+\hbar$ and $l_-\hbar$ be the largest and smallest eigenvalues with eigenfunctions $\psi_{l_+}$ and $\psi_{l_-}$ respectively.

Since $l_+\hbar$ is the largest eigenvalue

$$\mathscr{L}_+\psi_{l_+} = 0 \qquad (9.54)$$

and similarly

$$\mathscr{L}_-\psi_{l_-} = 0. \qquad (9.55)$$

From the commutation result (it is easy to verify it)

$$\mathscr{L}_\pm\mathscr{L}_\mp \equiv \mathscr{L}^2 - \mathscr{L}_z(\mathscr{L}_z \mp \hbar) \qquad (9.56)$$

then

$$\mathscr{L}_-\mathscr{L}_+\psi_{l_+} = 0 = [a^2 - l_+(l_+ + 1)]\hbar^2\psi_{l_+}$$

and

$$\mathscr{L}_+\mathscr{L}_-\psi_{l_-} = 0 = [a^2 - l_-(l_- - 1)]\hbar^2\psi_{l_-}$$

$$\therefore \quad a^2 = l_+(l_+ + 1) = l_-(l_- - 1)$$

i.e.

$$(l_+ + l_-)(l_+ - l_- + 1) = 0.$$

Clearly $(l_+ - l_- + 1) > 0$ as $l_+$ and $l_-$ are the largest and smallest quantum numbers and so

$$l_+ = -l_- = l \text{ (say)}$$

and

$$a^2 = l(l+1). \tag{9.57}$$

Since each step in the shift operation is by a unit of $\hbar$ then

$$l_+ - l_- = 2l \tag{9.58}$$

is either zero or a positive integer.

There are $(2l + 1)$ functions which are simultaneously eigenfunctions of the complete set of operators including $\mathscr{L}^2$ and $\mathscr{L}_z$ which have total angular momentum eigenvalue $\sqrt{l(l+1)}\hbar$. The functions have eigenvalues

$$l\hbar, (l-1)\hbar, \ldots, -l\hbar \tag{9.59}$$

for $\mathscr{L}_z$.

The above analysis shows that this set of $(2l + 1)$ functions span a space that is irreducible under the set of operators $\mathscr{L}_z$, $\mathscr{L}_+$ and $\mathscr{L}_-$ and hence under $\mathscr{L}_z$, $\mathscr{L}_x$, $\mathscr{L}_y$. From (9.42) and (9.43) this space is irreducible under the infinitesimal rotation operators and hence under the full rotation group. The irreducible representations may be denoted by $D^{(l)}$ where the dimensionality $(2l + 1)$ is a positive integer.

The argument has been based solely on the commutation relations satisfied by the angular momentum operators and is directly applicable to the spin and total angular momentum operators to be met in the succeeding chapters.

*The one-electron central field problem*

The central field Hamiltonian (5.23) is invariant under the full rotation group and so commutes with the angular momentum operators. Its eigenfunction must belong to one or other of the irreducible representations. $D^{(l)}$. These eigenfunctions are from (5.41)

$$\psi_{nlm_l}(r, \theta, \phi) = R_{nl}(r) P_l^{m_l}(\cos\theta) e^{im_l\phi}$$

and are clearly reduced with respect to rotations about the $z$-axis

$$l_z^{(r)} \psi_{nlm_l} = m_l \psi_{nlm_l}.$$

By using $\mathscr{L}_+$ and $\mathscr{L}_-$ it is possible to generate the degenerate set of eigenfunctions

$$\psi_{nlm_l}, m_l = +l, +(l-1), \ldots, -l. \tag{9.60}$$

This set forms the basis for the irreducible representation with dimension $(2l + 1)$.

The *s*-functions ($l = 0$) transform according to $D^{(0)}$, the *p*-functions transform like $D^{(1)}$ and the *d*-functions like $D^{(2)}$.

The degeneracy of the coulomb potential wave-functions with the same principle quantum number but different $l$ quantum numbers is not explained by the rotational symmetry of the atom in three-dimensional space.

In the above, $l$ is required to be an integer as the wave-functions must be single-valued and so the irreducible representations have odd dimension. However, representations of even dimension do exist. In particular, the representation $D^{(1/2)}$ of dimension two will be used in the discussion of electron spin in the next chapter.

## 9.7 Electron in a Periodic Field

The lattice periodicity of crystals allows the energy eigenstates to be classified by a real wave vector **k** and the eigenfunctions are called Bloch functions. The essentials of the problem are contained in the one-dimensional case considered below.

The Schrödinger equation for a line of identical atoms, distance '*a*' apart, is

$$-\frac{\hbar^2}{2m}\frac{d^2\psi}{dx^2} + V(x)\psi = E\psi \tag{9.61}$$

with

$$V(x - a) = V(x).$$

The substitutional operator representing the translation $na$ is written $\{\epsilon \mid na\}$ with

$$\{\epsilon \mid na\}\ V(x) = V(x - na). \tag{9.62}$$

If the line of atoms is infinitely long the number of lattice translations is infinite. To keep the number of distinct translations finite it is usual to introduce periodic boundary conditions. The finite line is divided up into identical 'micro-crystals' each containing $N$ atoms so that a translation through $Na$ produce no change at all, i.e.

$$\{\epsilon \mid 0\} \equiv \{\epsilon \mid Na\}. \tag{9.63}$$

The translation operators form a group of $N$ elements and as all translations commute with one another, the group is an Abelian group. As already mentioned previously in the discussion of the axial rotation group, each element in an Abelian group forms a class in itself and, the number of irreducible representations is equal to the number of elements $N$. All these representations are one-dimensional.

The operator $\nabla^2$ is invariant under a translation of the co-ordinate system and so the Hamiltonian is also invariant. Consequently the energy eigenfunctions must transform according to one of the irreducible representations of the group. An alternative view point is to note that the Hamiltonian and translation operators commute with each other so wave-functions may be chosen to be simultaneously eigenfunctions of $\mathscr{H}$ and all the operators $\{\epsilon|na\}$.

If $\psi$ is the basis for one of the irreducible representations

$$\{\epsilon|a\}\psi(x) = \lambda\psi(x) \qquad \text{and} \qquad \{\epsilon|na\}\psi(x) = \lambda^n\psi(x)$$

where $\lambda$ is a complex constant. The periodic boundary conditions require

$$\lambda^N = 1$$

and so $\lambda$ must be one of the $N$th roots of unity

$$\lambda = e^{-i2\pi r/N} \qquad r = 1, 2, \ldots, N.$$

If a real wave number $k$ is defined by

$$k = \frac{1}{N}\frac{2\pi}{a}r$$

then the irreducible representations and hence the energy eigenfunctions may be classified by the wave number and

$$\{\epsilon|na\}\psi_k(x) = e^{-ikna}\psi_k(x). \tag{9.64}$$

This is Bloch's theorem and a wave-function satisfying this condition is

$$\psi_k(x) = e^{ikx}u_k(x) \tag{9.65}$$

where $u_k(x)$ has the lattice periodicity.

## 9.8 Group Theory and Perturbation

As has been seen in Chapter 7, perturbation of a physical system often results in the removal of degeneracies. This is of course to be expected as the perturbed Hamiltonian has in general a lower symmetry than the unperturbed Hamiltonian. The previously degenerate wave-functions now form a basis for a reducible representation for the new smaller symmetry group and will split into two or more irreducible representations. A degenerate level will divide into two or more levels of diminished degeneracy.

In the discussion of the first-order Zeeman effect in Chapter 7 it was shown how the $(2l + 1)$ degenerate eigenfunctions of the hydrogen atom with given $n$ and $l$ quantum numbers split into $(2l + 1)$ separate levels when an axially symmetric magnetic field is applied.

### *The crystal field as a perturbation*

Important examples of the removal of degeneracies by perturbation occur in the splitting of the energy levels of paramagnetic ions of transition and rare earth metals in a crystal. The metallic ion and its surrounding neighbours (water molecules etc), should be solved as a many-body problem.

However, it is often a good approximation to consider the neighbours as producing an electrostatic crystal field which perturbs the degenerate ($d$ or $f$) eigenfunctions of the free metallic ion. Paramagnetic resonance is a tool that has proved useful in experimental measurements.

The Hamiltonian of an ion in the crystal is written

$$\mathscr{H} \equiv \mathscr{H}^0 + \mathscr{h}_c$$

where $\mathscr{H}^0$ is the spherically symmetrical free ion Hamiltonian and $\mathscr{h}_c$ is the electrostatic potential energy of the ion in the crystal field. In the intermediate crystal field case, $\mathscr{h}_c$ is greater than the spin-orbit coupling terms (spin is introduced in the next chapter) and the main effect is to split the central field orbital degeneracy. The full analysis involves the many-electron wave-function and will not be attempted here. A simplified discussion is given below.

Ions of the iron transition group with partially-filled $3d$-shells often belong to the intermediate field case. Often they occur in surroundings that are almost cubic, i.e. $\mathscr{h}_c$ has cubic symmetry.

The five degenerate $d$-wave-functions

$$2(x \pm iy)zf(r), \qquad (x \pm iy)^2 f(r), \qquad \sqrt{\tfrac{2}{3}}(3z^2 - r^2)f(r)$$

form a basis for the irreducible representation $D^{(2)}$ of the full rotation group. This 5-fold degenerate level splits up into a 2-fold and a 3-fold level under the perturbing action of the cubic crystal field. To see this it is only necessary to consider the 24 proper rotations of the group 432 that leaves a cube invariant. (The full cubic group $m3m$ contains 48 operations and can be expressed as the direct product $m3m = 432 \times \bar{I}$.) The 24 elements are divided into 5 classes as below

(a) The identity.
(b) The 4 3-fold diagonal axes making 8 elements denoted by 3.
(c) The 3 rotations through 180° parallel to the cube edges denoted by $2_z$.
(d) The 6 rotations through 180° parallel to the diagonals in the sides of the cube denoted by $2_d$.
(e) The 6 rotations through ±90° parallel to the cube edges denoted by $4_z$.

**Table 9.4** The character table for this group is given below.

| 432 | $E$ | 3 | $2_z$ | $2_d$ | $4_z$ |
|---|---|---|---|---|---|
| $A_1$ | 1 | 1 | 1 | 1 | 1 |
| $A_2$ | 1 | 1 | 1 | −1 | −1 |
| $E$ | 2 | −1 | 2 | 0 | 0 |
| $T_1$ | 3 | 0 | −1 | −1 | 1 |
| $T_2$ | 3 | 0 | −1 | 1 | −1 |
| $E + T_2$ | 5 | −1 | 1 | 1 | −1 |

The $5d$-wave functions transform according to a reducible representation of this group and the characters are given in the last row of Table 9.4. It is immediately clear from the table that this representation is composed of the two irreducible representations $E$ and $T_2$ and so the 5-fold degenerate level splits up into a 2-fold level and a 3-fold level.

In many cases the almost cubic crystal field may include smaller order terms with a further reduced symmetry and the degeneracy may be further reduced. The reader is referred to Heine, p. 148, and Schutte, p. 348, for further details.

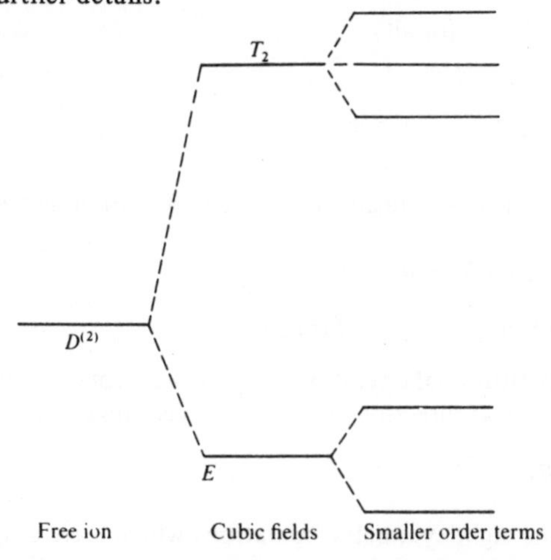


Figure 9.7 Splitting of 5-fold $D^{(2)}$-degeneracy by crystalline fields.

## 9.9 Selection Rules

Group theory is very useful in determining which transitions are allowed in the interaction of electromagnetic radiation and electrons in atoms and molecules. In Chapter 8 it was shown that the transition probabilities for a single electron are proportional to

$$\left|\int \chi_f^* \mathbf{r} \chi_i \, d\tau\right|^2 \tag{8.38}$$

where $\chi_i$ and $\chi_f$ are the initial and final state functions.
It can be shown that if the function $\phi$ is one of the basis functions for an irreducible representation of a symmetry group then the integral over all co-ordinates

$$\int \phi \, d\tau = 0$$

if the representation is not the unit representation. The three components of $\mathbf{r}$ transform like the three $p$ wave-functions, i.e., according to the irreducible representation $D^{(1)}$ of the full rotation group. This fact together with a knowledge of the symmetry properties of $\chi_i$ and $\chi_f$

make it possible to decide when the integral (8.38) is zero and hence which transitions are forbidden. This topic is dealt with in detail in Chapter 11 and will not be pursued here.

## PROBLEMS

1 Prove that the elements of the transformation matrix $\boldsymbol{\alpha}$ (see equation (9.3)) satisfy the relations

$$\sum_{j=1}^{3} \alpha_{ij}^2 = 1 \qquad \text{for all } i$$

$$\sum_{i=1}^{3} \alpha_{ij}\alpha_{im} = 0 \qquad j \neq m.$$

Prove also that the determinant of the transformation matrix is $\pm 1$.

2 Sketch the two functions

$$p_y = f(r) \sin \theta \qquad p_x = f(r) \cos \theta.$$

If $4_z$ is the substitutional operator representing a rotation of $\pi/2$ radians in the anti-clockwise direction about the $z$-axis, show that

$$4_z p_y = p_x.$$

3 Describe the eight symmetry operations which leave a square invariant. The two faces of the square are not identical. Construct a group multiplication table. (This is the group $4mm$.)

4 Repeat question three for a rectangle. (In this case the symmetry group is composed of four elements and the group symbol is $2mm$.)

5 Show that the four elements $1, i, -1, -i$ where $i = +\sqrt{-1}$ form a group. Draw up a group multiplication table.

6 Consider a system which has a single 4-fold rotation axis in the $z$-direction. Identify the four elements and draw up a group multiplication table. This is the group 4. Show that this table can be made to have the same form as that of question 5 by pairing off each element of the group 4 with a suitable element of the group of question 5. The two groups are said to be isomorphic.

7 A real operator $A$ is related to its Hermitian adjoint $A^+$ by

$$\int (A\phi^*)\psi \, d\tau = \int \phi^* A^+ \psi \, d\tau$$

where $\phi$, $\psi$ are arbitrary functions and the integration is over all space. If $A$ is a linear operator which possesses an inverse $A^{-1}$ it is said to be a unitary operator if $A^+ = A^{-1}$.

The integral over all space $\int \phi^*\psi \, d\tau$ is invariant under any reflection or rotation of axes. Explain how this implies that a symmetry operator is unitary.

8 Table 9.3 gives the group character table for the equilateral triangle group $3m$.

The character of the group element $A$ is the $\alpha$-irreducible representation is denoted by $\chi^{(\alpha)}(A)$. Verify the 'orthogonality relations',

$$\sum_A \chi^{(\alpha)*}(A)\chi^{(\beta)}(A) = 0 \qquad \alpha \neq \beta$$

$$\sum_A \chi^{(\alpha)*}(A)\chi^{(\alpha)}(A) = N.$$

The summation is over all the $N$ elements in the group.

9 Suppose $\chi(A)$ is the character of the group operator $A$ in the reducible representation $\Gamma$ and $\chi^{(i)}(A)$ is the character of $A$ in the irreducible representation $\Gamma^{(i)}$. Then

$$\chi(A) = \sum_i a_i \chi^{(i)}(A) \qquad \text{for all } A$$

where $a_i$ is the number of times $\Gamma^{(i)}$ occurs in $\Gamma$.

By multiplying both sides by $\chi^{(j)*}(A)$ and summing over all members of the group show that

$$a_j = \frac{1}{N} \sum_A \chi^{(j)*}(A)\chi(A) \tag{1}$$

where $N$ is the number of elements in the group. (Hint: Use the orthogonality relations given in question 8.)

Consider the reducible representations of the group $3m$ given by the first row of Table 9.2. Use equation (1) to show that this representation is composed of irreducible representations $A$, $E$.

10 Write down the 24 symmetry operations which leave a regular tetrahedron invariant. This is the group $\bar{4}3m$. Divide the group elements into the five classes and give the dimensions of all the irreducible representations.

11 In a free atom the atomic orbitals $(2s)$, $(2p_x)$, $(2p_y)$ and $(2p_z)$ are degenerate. In a tetrahedral molecule, such as methane the molecular field perturbs the free-atom carbon wave-functions and destroys this degeneracy. Choose one element from each of the five classes of the tetrahedral group $\bar{4}3m$ and construct the matrices of the reducible representations with basis $(2s)$, $(2p_x)$, $(2p_y)$ and $(2p_z)$. Show that this

representation is composed of a single one-dimensional and one three-dimensional irreducible representations. (Hint: Choose the co-ordinate axes so that a three-fold axis lies along the (111) direction.)

12 Suppose a uniform electric field is applied along one of the three-fold axes (C–H bond) of the methane molecule. The reduced symmetry group at the carbon atom is the group $3m$ composed of six elements and the group character table is given below.

| Number of elements in class | | 2 | 3 |
|---|---|---|---|
| Typical element in class | $E$ | $A$ | $B$ |
| $A_1$ | 1 | 1 | 1 |
| $A_2$ | 1 | 1 | −1 |
| $A_3$ | 2 | −1 | 0 |

Identify the elements $A$ and $B$.

Show that the three-dimensional irreducible representation of the group $\bar{4}3m$ with basis $(2p_x)$, $(2p_y)$ and $(2p_z)$ is composed of two irreducible representations of the reduced symmetry group $3m$.

This result shows that the three-fold degenerate level splits up into a non-degenerate and doubly-degenerate level when the electric field is applied.

13 Let $\alpha$ be an axis in the $yz$ plane through the origin at an angle of $\theta$ to the $z$-axis. A physical rotation through $\phi$ about the $\alpha$-axis is equivalent to a physical rotation $\theta$ about the $x$-axis so that the $\alpha$-axis coincides with the $z$-axis, followed by a physical rotation through $\phi$ about the $z$-axis followed finally by a physical rotation through $-\theta$ about the $x$-axis.

$$P(-\phi, \alpha) \equiv P(\theta, x)P(-\phi, z)P(-\theta, x)$$

where $P(\theta, x)$ denotes a passive rotation of the co-ordinate system through $\theta$ about the $x$-axis. Use (9.45) to express $I_\alpha^{(r)}$ in terms of $I_y^{(r)}$ and $I_z^{(r)}$ and with the help of (9.46a) show that

$$I_z^{(r)}I_x^{(r)} - I_x^{(r)}I_z^{(r)} \equiv iI_y^{(r)}$$

by equating coefficients of the product $\theta\phi$.

## References and Recommended Further Reading

Clark, H. *Solid State Physics–An introduction to its theory.* Macmillan, London (1968).

Fock, V. *Bull. de l'Academie des Sciences de L'URSS,* **179** (1935).

Hall, G. G. *Applied Group Theory.* Longmans, London (1967).
Heine, V. *Group Theory in Quantum Mechanics.* Pergamon Press, London (1960).
Johnston, D. F. *Group Theory in Solid State Physics.* Reports on Progress in Physics, XXIII, 66 (1960).
Landau, L. D. and Lifshitz, E. M. *Quantum Mechanics–Non-relativistic Theory.* Pergamon Press, London (1959).
Leech, J. W. and Newman, D. J. *How to use groups.* Methuen, London (1969).
Schutte, C. J. H. *The Wave Mechanics of Atoms, Molecules and Ions.* Edward Arnold, London (1968).
Tinkham, M. *Group Theory and Quantum Mechanics.* McGraw-Hill, New York (1964).

CHAPTER 10

# Electron Spin

## 10.1 Introduction

The quantum mechanical theory developed in the previous chapters is not sufficient to explain all the experimental results of atomic physics. It has already been shown that the magnitude of the angular momentum of a particle in a central field problem can take the values $\hbar\sqrt{l(l+1)}$ where $l$ is an integer. This follows from the differential form for the angular momentum operator together with the requirement that the wave-function be single-valued. However, half integral values of the angular momentum quantum numbers are necessary to explain the fine structure in the spectral series of some elements. The sodium D-line doublet is a well-known example.

Also the anomolous Zeeman effect cannot be explained by integral values of $l$ alone. Schrödinger theory indicates that the lowest state for the sodium atom is non-degenerate with zero angular momentum but the Zeeman effect in sodium can only be explained by assuming the ground state splits into two levels when a weak magnetic field is applied.

Apart from this spectroscopic evidence, Stern and Gerlach in 1921 demonstrated experimentally the existence of the magnetic moment of an electron. They did this by passing a narrow beam of silver atoms through an inhomogeneous magnetic field and observing that this beam split into two beams. (Note that a beam of electrons is not used since the Stern–Gerlach separation is then of the order of that imposed by the uncertainty principle.)

## 10.2 The Electron Spin

Uhlenbeck and Goudsmit in 1925 suggested that the electron has an intrinsic angular momentum and in 1927 Pauli introduced electron spin into quantum mechanics.

The experimental results can be understood if it is assumed that an electron has an intrinsic spin angular momentum of magnitude $\hbar\sqrt{s(s+1)}$ where $s = \frac{1}{2}$. Further, the component of the angular momentum can only be known along any one prescribed direction, say the $z$-direction and must take one of the values $m_s\hbar$ with

$$m_s = \pm\tfrac{1}{2}. \tag{10.1}$$

Also, it must be assumed that associated with the spin angular momentum the electron has a magnetic moment equal to $-e\hbar/m$ times the spin momentum where $m$ is the electron mass.

These assumptions arise naturally in the Dirac relativistic theory but must be postulated in the non-relativistic theory now under discussion.

In the Schrödinger theory, the three position co-ordinates of an electron form a complete set and when all three are known the electron state is specified completely. Now for a complete specification of a state the projection of the spin momentum along a prescribed direction must be known also.

*Postulate 7. Spin*

It is now postulated that a particle, in addition to its spatial co-ordinates, has an internal degree of freedom characterized by a spin co-ordinate $\sigma_z$. The particle state function is then a function of four co-ordinates $\psi(x, y, z, \sigma_z)$. The spin co-ordinate can only take a finite number of discrete values. For electrons, positrons, neutrons and protons, $\sigma_z$ can take either of the two values $\pm 1$. (This corresponds to the $z$-component of the spin momentum taking the values $\pm\hbar/2$.) The rest of this chapter is concerned only with such particles with spin 1/2.

The quantum mechanics based upon the seven postulates in this book is valid so long as the energies included are small compared with the energy associated with the particle rest mass (~0·5 MeV for electrons). At high energies a relativistic theory should be applied.

The operators representing the components of orbital angular momentum have been obtained from the classical expressions using Schrödinger's substitution for momentum. Because the spin co-ordinate can take only two values it follows (see below) that an electron spin momentum components can only take the values $\pm\hbar/2$ and hence in the classical limit $h \to 0$, spin vanishes. Spin does not occur in classical mechanics, and an alternative method must be used to define spin momentum.

In Chapter 9 it has been shown how $\hat{p}$, the momentum operator conjugate to $q$, is related to the infinitesimal translation (rotation) operator $I_q$ by

$$\hat{p} \equiv \hbar I_q. \tag{10.2}$$

The infinitesimal rotation operators discussed in section 9.7 transform the three space co-ordinates and satisfy the three commutation rules of the form (9.44). These rules are a basic property of rotation operators.

The spin co-ordinate will be transformed under a rotation of axes and the infinitesimal rotation operators $I_x^{(s)}$, $I_y^{(s)}$ and $I_z^{(s)}$ which carry out this transformation must satisfy the basic commutation rules of the type (9.44).

The fundamental result (10.2) can now be extended to define three spin momentum operators

$$s_x \equiv \hbar I_x^{(s)}, \qquad s_y \equiv \hbar I_y^{(s)}, \qquad s_z \equiv \hbar I_z^{(s)}. \tag{10.3}$$

The spin infinitesimal rotation operators (unlike the space co-ordinate infinitesimal rotation operators) do not have a differential operator form and so neither do the spin momentum operators. However, the spin momentum operators must satisfy the same commutation relations as $\mathscr{L}_x, \mathscr{L}_y$ and $\mathscr{L}_z$. That is

$$\begin{aligned} s_x s_y - s_y s_x &\equiv i\hbar s_z \\ s_y s_z - s_z s_y &\equiv i\hbar s_x \\ s_z s_x - s_x s_z &\equiv i\hbar s_y. \end{aligned} \tag{10.4}$$

As the spin component operators do not commute with each other only one of them, say $s_z$, can be known with certainty at any instant. The total spin momentum operator defined by

$$s_{op}^2 \equiv s_x^2 + s_y^2 + s_z^2 \tag{10.5}$$

commutes with all three components and so the total spin momentum can be measured simultaneously with any one of the three components.

## 10.3 Spin Wave-Functions and Representations

The wave-function for a single electron is now to be regarded as a function of four variables. The three space variables are continuous but the spin variable can only take discrete values $\pm 1$. So

$$\psi(x, y, z, \sigma_z) = \psi_+(x, y, z)\alpha(\sigma_z) + \psi_-(x, y, z)\beta(\sigma_z) \tag{10.6}$$

where $\alpha(\sigma_z)$ and $\beta(\sigma_z)$ are functions such that

$$\begin{aligned} \alpha(1) = 1 \quad &\text{and} \quad \alpha(-1) = 0 \\ \beta(1) = 0 \quad &\text{and} \quad \beta(-1) = 1. \end{aligned} \tag{10.7}$$

$\psi_+(x, y, z)$ is the 'component' of the wave-function with spin co-ordinate $\sigma_z = 1$ and $\psi_-(x, y, z)$ is the 'component' with spin co-ordinate $\sigma_z = -1$. If the electron is in a state defined by $\sigma_z = 1$ say, then

$$\psi(x, y, z, \sigma_z) = \psi_+(x, y, z)$$

In the general case $\int |\psi_+|^2 \, d\tau$ is the probability of observing the state $\sigma_z = 1$. The values of $\sigma_z$ correspond to the eigenvalues of $s_z$. This is explained below.

A rotation of axes corresponds to a linear transformation of $\alpha(\sigma_z)$ and $\beta(\sigma_z)$. From (10.6) all functions of $\sigma_z$ lie in the two-dimensional spin-space spanned by $\alpha(\sigma_z)$ and $\beta(\sigma_z)$ and so the space is invariant under the rotation group. For particles with spin $\frac{1}{2}$ this space must be irreducible under the group and $\alpha(\sigma_z), \beta(\sigma_z)$ form a basis for the representation $D^{(1/2)}$. Conventionally $\alpha(\sigma_z)$ and $\beta(\sigma_z)$ are taken to be the eigenfunctions of the rotation operator $I_z^{(r)}$ with eigenvalues $m_s = \frac{1}{2}, -\frac{1}{2}$ respectively (9.48), i.e.

$$I_z^{(s)}\alpha(\sigma_z) = \tfrac{1}{2}\alpha(\sigma_z) \qquad \text{with} \qquad m_s = \tfrac{1}{2}$$

or

$$s_z\alpha(\sigma_z) = \frac{\hbar}{2}\alpha(\sigma_3)$$

and

$$I_z^{(s)}\beta(\sigma_z) = -\tfrac{1}{2}\beta(\sigma_z) \qquad \text{with} \qquad m_s = -\tfrac{1}{2}$$

or

$$s_z\beta(\sigma_z) = -\frac{\hbar}{2}\beta(\sigma_z). \tag{10.8}$$

The spin operator $s_{op}^2 \equiv s_x^2 + s_y^2 + s_z^2$ has eigenvalues $\frac{1}{2}(\frac{1}{2} + 1)\hbar^2 = \frac{3}{4}\hbar^2$ in both states (see (9.59)). The results agree with the requirements mentioned in section 10.2.

If $P^s(\phi, z)$ is the operator that represents a rotation of $\phi$ about the $z$-axis and acts only on the spin co-ordinates, then from (9.46)

$$P^s(\phi, z)\alpha(\sigma_z) = \exp(i\phi/2)\alpha(\sigma_z) \tag{10.9}$$

and similarly for $\beta(\sigma_z)$

$$\therefore \quad P^s(2\pi, z)\alpha(\sigma_z) = -\alpha(\sigma_z) \tag{10.10}$$

and

$$P^s(4\pi, z)\alpha(\sigma_z) = +\alpha(\sigma_z).$$

Rotations of $4\pi$ must be considered before $\alpha(\sigma_z)$ and $\beta(\sigma_z)$ are transformed identically into themselves. These functions are called spinors.

It has been stated in postulate 1 that all wave-functions must be single-valued. This is not strictly true. The physical requirement is that integrals of the form

$$\int \psi_i^* A \psi_j \, d\tau$$

be single-valued. If $\psi_i$ and $\psi_j$ both change sign under a rotation of $2\pi$ applied to all co-ordinates then clearly the integral is invariant. However arguments have been given (Blatt and Weisskopf 1952, H. E. Rorschach 1962), to show that wave-functions depending on the three space co-ordinates $x, y, z$ must be single valued as already stated. This argument does not apply to wave-functions depending on spin.

*Matrix representation*

The spin momentum operators do not possess a differential operator representation. However, a set of representative matrices can be constructed. If the spin eigenfunctions are regarded as basis functions then an arbitrary wave-function $\psi(x, y, z, \sigma_z)$ is represented by the column vector (see (10.6)).

$$\begin{bmatrix} \psi_+(x, y, z) \\ \psi_-(x, y, z) \end{bmatrix}. \tag{10.11}$$

Normalization of the wave function $\psi(x, y, z, \sigma_z)$ requires

$$\sum_{\sigma_z} \int |\psi(x, y, z, \sigma_z)|^2 \, d\tau = \sum_{\sigma_z} \int [|\psi_+|^2\alpha^2(\sigma_z) + |\psi_-|^2\beta^2(\sigma_z)$$

$$+ (\psi_+^*\psi_- + \psi_+\psi_-^*)\alpha(\sigma_z)\beta(\sigma_z)] \, d\tau = 1$$

where the summation is over the two discrete values $\sigma_z = \pm 1$.

From (10.7)

$$\sum_{\sigma_z} \alpha(\sigma_z)\beta(\sigma_z) = 0$$

and

$$\sum_{\sigma_z} \alpha^2(\sigma_z) = \sum_{\sigma_z} \beta^2(\sigma_z) = 1$$

so

$$\int [|\psi_+(x, y, z)|^2 + |\psi_-(x, y, z)|^2] \, d\tau = 1. \tag{10.12}$$

The probability of the $z$-component of electron spin being measured as $+\hbar/2$ is

$$\int |\psi_+(x, y, z)|^2 \, d\tau$$

and similarly, the probability of spin $-\hbar/2$ is

$$\int |\psi_-(x, y, z)|^2 d\tau.$$

The column vectors representing $\alpha(\sigma_z)$ and $\beta(\sigma_z)$ are respectively

$$\boldsymbol{\alpha} = \begin{bmatrix} 1 \\ 0 \end{bmatrix} \quad \text{and} \quad \boldsymbol{\beta} = \begin{bmatrix} 0 \\ 1 \end{bmatrix}. \tag{10.13}$$

The matrices representing the spin momentum operators $s_x$, $s_y$ and $s_z$ must be 2 x 2 and the reader can easily verify that the choice

$$\mathbf{s}_x = \frac{\hbar}{2}\begin{pmatrix} 0 & 1 \\ 1 & 0 \end{pmatrix}, \qquad \mathbf{s}_y = \frac{\hbar}{2}\begin{pmatrix} 0 & -i \\ i & 0 \end{pmatrix}, \qquad \mathbf{s}_z = \frac{\hbar}{2}\begin{pmatrix} 1 & 0 \\ 0 & -1 \end{pmatrix} \tag{10.14}$$

satisfies the commutation relations (10.4). These can be written

$$\mathbf{s}_x = \frac{\hbar}{2}\boldsymbol{\sigma}_x, \qquad \mathbf{s}_y = \frac{\hbar}{2}\boldsymbol{\sigma}_y, \qquad \mathbf{s}_z = \frac{\hbar}{2}\boldsymbol{\sigma}_z \tag{10.15}$$

with

$$\boldsymbol{\sigma}_x = \begin{pmatrix} 0 & 1 \\ 1 & 0 \end{pmatrix}, \qquad \boldsymbol{\sigma}_y = \begin{pmatrix} 0 & -i \\ i & 0 \end{pmatrix}, \qquad \boldsymbol{\sigma}_z = \begin{pmatrix} 1 & 0 \\ 0 & -1 \end{pmatrix} \tag{10.16}$$

and $\boldsymbol{\sigma}_x$, $\boldsymbol{\sigma}_y$ and $\boldsymbol{\sigma}_z$ are known as the Pauli spin matrices. All these matrices are Hermitian and have real eigenvalues. The correctness of the choice (10.14) is confirmed by noting that the eigenvalue equations

$$\mathbf{s}_z\boldsymbol{\alpha} = \frac{\hbar}{2}\boldsymbol{\alpha} \qquad \text{and} \qquad \mathbf{s}_z\boldsymbol{\beta} = -\frac{\hbar}{2}\boldsymbol{\beta}$$

are represented by

$$\frac{\hbar}{2}\begin{pmatrix} 1 & 0 \\ 0 & -1 \end{pmatrix}\begin{pmatrix} 1 \\ 0 \end{pmatrix} = \frac{\hbar}{2}\begin{pmatrix} 1 \\ 0 \end{pmatrix} \qquad \text{and} \qquad \frac{\hbar}{2}\begin{pmatrix} 1 & 0 \\ 0 & -1 \end{pmatrix}\begin{pmatrix} 0 \\ 1 \end{pmatrix} = -\frac{\hbar}{2}\begin{pmatrix} 0 \\ 1 \end{pmatrix} \tag{10.17}$$

respectively.

Note also that from (10.13) and (10.14)

$$\mathbf{s}_x\boldsymbol{\alpha} = \frac{\hbar}{2}\boldsymbol{\beta}, \qquad \mathbf{s}_x\boldsymbol{\beta} = \frac{\hbar}{2}\boldsymbol{\alpha}, \qquad \mathbf{s}_y\boldsymbol{\alpha} = \frac{i\hbar}{2}\boldsymbol{\beta}, \qquad \mathbf{s}_y\boldsymbol{\beta} = -\frac{i\hbar}{2}\boldsymbol{\alpha}. \tag{10.18}$$

## 10.4 Total Angular Momentum

It is necessary to know how the one-electron operators which are functions of the space co-ordinates and momenta (and not of spin) are represented in the new two component theory. Clearly such operators must be represented by 2 x 2 matrices. The (1, 1) element of the matrix representing a Schrödinger operator $\gamma$ which is spin independent, is

$$\sum_{\sigma_z} \alpha(\sigma_z)\gamma\alpha(\sigma_z) = \gamma \sum_{\sigma_z} \alpha^2(\sigma_z) = \gamma.$$

Similarly the (2, 2) element is

$$\sum_{\sigma_z} \beta(\sigma_z)\gamma\beta(\sigma_z) = \gamma \sum_{\sigma_z} \beta^2(\sigma_z) = \gamma.$$

However the off-diagonal elements are both zero as

$$\sum_{\sigma_z} \alpha(\sigma_z)\gamma\beta(\sigma_z) = \gamma \sum_{\sigma_z} \alpha(\sigma_z)\beta(\sigma_z) = 0.$$

So, all spin independent operators, such as $\gamma$, are represented by a diagonal matrix

$$\begin{bmatrix} \gamma & 0 \\ 0 & \gamma \end{bmatrix} = \gamma \mathbf{I}.$$

All such diagonal matrices commute with the Pauli spin matrice.

*Total angular momentum*

The total angular momentum of an electron is defined to be the sum of the orbital angular momentum and the spin angular momentum. The 2 x 2 matrix operator representing the $z$-component is

$$\mathbf{J}_z \equiv \mathscr{L}_z \mathbf{I} + \mathbf{s}_z \equiv \begin{bmatrix} \frac{\hbar}{i}\frac{\partial}{\partial\phi} + \frac{\hbar}{2} & 0 \\ 0 & \frac{\hbar}{i}\frac{\partial}{\partial\phi} - \frac{\hbar}{2} \end{bmatrix}. \tag{10.19}$$

$\mathbf{J}_z$ has half-integer quantum numbers as $\mathscr{L}_z$ and $\mathbf{s}_z$ have integer and half-integer quantum numbers respectively. It is instructive to demonstrate this using the two component theory. The eigenvalues of $\mathbf{J}_z$ are denoted by $m_j\hbar$ and in two component form, the eigenvalue equation is (in spherical polars)

$$\left[\frac{\hbar}{i}\frac{\partial}{\partial\phi}\begin{pmatrix} 1 & 0 \\ 0 & 1 \end{pmatrix} + \frac{\hbar}{2}\begin{pmatrix} 1 & 0 \\ 0 & -1 \end{pmatrix}\right]\begin{pmatrix} \psi_+ \\ \psi_- \end{pmatrix} = m_j\hbar\begin{pmatrix} \psi_+ \\ \psi_- \end{pmatrix}.$$

This represents two equations,

$$\frac{1}{i}\left[\frac{\partial}{\partial\phi} + \frac{i}{2}\right]\psi_+ = m_j\psi_+$$

$$\frac{1}{i}\left[\frac{\partial}{\partial\phi} - \frac{i}{2}\right]\psi_- = m_j\psi_-.$$

The first equation may be written

$$\frac{\partial\psi_+}{\partial\phi} = i(m_j - \tfrac{1}{2})\psi_+$$

$$\therefore \quad \psi_+(r, \theta, \phi) = A_+(r, \theta) \exp\left[i(m_j - \tfrac{1}{2})\phi\right].$$

As $\psi_+$ must be invariant under a rotation of $2\pi$ about the $z$-axis

$$m_j - \tfrac{1}{2} = \text{integer}$$

$$\therefore \quad m_j = \text{integer} + \tfrac{1}{2}.$$

Similarly

$$\psi_-(r, \theta, \phi) = A_-(r, \theta) \exp [i(m_j + \tfrac{1}{2})\phi]$$

and

$$m_j = \text{integer} - \tfrac{1}{2}.$$

Therefore the eigenvalues of $\mathbf{J}_z$ are half-integral multiples of $\hbar$.

The matrix operators representing the $x$ and $y$ components of the total angular momentum are

$$\mathbf{J}_x \equiv \mathscr{L}_x\mathbf{I} + \mathbf{s}_x, \qquad \mathbf{J}_y \equiv \mathscr{L}_y\mathbf{I} + \mathbf{s}_y.$$

The matrix representing the squared total angular momentum is

$$\begin{aligned} \mathbf{J}^2 &\equiv \mathbf{J}_x^2 + \mathbf{J}_y^2 + \mathbf{J}_z^2 \\ &\equiv \mathscr{L}^2\mathbf{I} + \mathbf{s}^2 + 2[\mathscr{L}_x\mathbf{s}_x + \mathscr{L}_y\mathbf{s}_y + \mathscr{L}_z\mathbf{s}_z] \end{aligned} \tag{10.20}$$

as the matrices representing the orbital and spin components commute.

The matrices $\mathbf{J}_x$, $\mathbf{J}_y$ and $\mathbf{J}_z$ satisfy the same commutation relations as the orbital and spin angular momentum operators and a straight forward manipulation will show that each component commutes with $\mathbf{J}^2$. So, functions can be found which are simultaneously eigenfunctions of $\mathbf{J}_z$ and $\mathbf{J}^2$.

### *Rotation operators*

The total angular momentum can be associated with infinitesimal operators representing a simultaneous rotation of both space and spin co-ordinates. From the definition, the $z$-component of the total angular momentum is equivalent to the sum of infinitesimal rotation operators

$$\hbar [I_z^{(r)} + I_z^{(s)}]$$

and similarly for the $x$ and $y$ components.

Consider now an operator which produces a finite rotation of both space and spin co-ordinates through an angle $\theta$ about the $z$-axis. This operator can be expressed as a product of two operators one of which acts on the space co-ordinates and the other on the spin co-ordinate. If $I_z^{(t)}$ is the operator representing an infinitesimal rotation of all four co-ordinates, then from (9.46b)

$$\exp (i\theta I_z^{(t)}) = \exp (i\theta I_z^{(r)}) \exp (i\theta I_z^{(s)}).$$

This is valid for all $\theta$ and so by equating the coefficients of $\theta$

$$I_z^{(t)} \equiv I_z^{(r)} + I_z^{(s)}.$$

In conclusion, the $z$-component of the total angular momentum is equivalent to the 'total' infinitesimal rotation operator, $I_z^{(t)}$ multiplied by $\hbar$, i.e.

$$\hbar I_z^{(t)}$$

and similarly for the $x$ and $y$ components. These three new 'infinitesimal' rotation operators $I_x^{(t)}$, $I_y^{(t)}$ and $I_z^{(t)}$ clearly obey the commutation relations of the type (10.4).

*Eigenvalues*

The argument given in section 9.7, which depends only on the commutation relations, now indicates that $\mathbf{J}^2$ has a set of $(2j + 1)$ eigenfunctions, with $2j$ integer, with common eigenvalue $j(j + 1)\hbar^2$. Each of these eigenfunctions may be chosen to be simultaneously an eigenfunction of $\mathbf{J}_z$ with separate eigenvalues

$$m_j\hbar = j\hbar, (j - 1)\hbar, \ldots, -j\hbar. \tag{10.21}$$

For a single electron, $m_j$ and hence $j$ is half-integral.

*Addition of angular momenta*

Consider a one-electron system whose orbital angular momentum and spin momentum are conserved. It is shown below that for a given $l$ the $j$ quantum number can take only two values, namely $l \pm \frac{1}{2}$.

The four matrix operators $\mathbf{J}^2$, $\mathbf{J}_z$, $\mathscr{L}^2\mathbf{I}$ and $\mathbf{s}^2$ all commute with one another and the corresponding variables are simultaneously measurable. Alternatively, the four matrices $\mathscr{L}^2\mathbf{I}$, $\mathscr{L}_z\mathbf{I}$, $\mathbf{s}^2$, $\mathbf{s}_z$ also form a mutually commuting set and the observables are compatible.

Suppose these two sets of four operators are taken to be complete. Any other operators needed to make the sets complete will be assumed to have fixed eigenvalues. Normally the Hamiltonian is one of these and it must be spherically symmetrical and not include any spin-orbit coupling term. For a single electron $\mathbf{s}^2$ can only have the eigenvalue $3\hbar^2/4$ and any state can be specified by the three quantum numbers

(i) $j, m_j, l$ or by (ii) $l, m_l, m_s$

respectively.

In order to find the different possible eigenvalues of $\mathbf{J}^2$ associated with the first set of operators it is useful to consider the labelling of the different states arising from the second set.

Consider the second set. For a given $l$, $m_l$ can take $(2l + 1)$ values and $m_s$ can take two values and so there are $2(2l + 1)$ different degenerate states altogether. Linear combinations of these states can be constructed that are eigenfunctions of the first set of operators. Consequently, for a given $l$ there must be $2(2l + 1)$ different states.

From (10.19) $m_j = m_l + m_s$ and the third column of table 10.1 shows how the allowed values for $m_j$ in the second set are obtained while the fourth and fifth columns show how they arise in the first set.

**Table 10.1**

| Second set | | | First set | |
|---|---|---|---|---|
| | | | $j = l + \frac{1}{2}$ | $j = l - \frac{1}{2}$ |
| $m_l$ | $m_s$ | $m_j$ | $m_j$ | $m_j$ |
| $l$ | $+\frac{1}{2}$ | $l + \frac{1}{2}$ | $l + \frac{1}{2}$ | |
| | $-\frac{1}{2}$ | $l - \frac{1}{2}$ | | $l - \frac{1}{2}$ |
| $l - 1$ | $+\frac{1}{2}$ | $l - \frac{1}{2}$ | $l - \frac{1}{2}$ | |
| | $-\frac{1}{2}$ | $l - \frac{3}{2}$ | | $l - \frac{3}{2}$ |
| $l - 2$ | $+\frac{1}{2}$ | $l - \frac{3}{2}$ | $l - \frac{3}{2}$ | |
| | $-\frac{1}{2}$ | $l - \frac{5}{2}$ | | $l - \frac{5}{2}$ |
| . | . | . | . | . |
| . | . | . | . | . |
| . | . | . | . | . |
| $-l$ | $+\frac{1}{2}$ | $-l + \frac{1}{2}$ | | $-l + \frac{1}{2}$ |
| | $-\frac{1}{2}$ | $-l - \frac{1}{2}$ | $-l - \frac{1}{2}$ | |

It can be seen that all $2(2l + 1)$ states for a given $l$, that occur in the first set can be accounted for by letting $j$ take the values

$$j = l \pm \tfrac{1}{2} \qquad l > 0. \tag{10.22}$$

In the language of the 'vector model', the electron spin angular momentum is either parallel or antiparallel to the orbital angular momentum. (Note, when $l = 0, j$ must be equal to $\frac{1}{2}$).

## 10.5 Spin Magnetic Momentum

It is well known that a classical electron moving in a circular orbit with angular momentum $\mathbf{L}$ has a magnetic moment $-e\mathbf{L}/2m$ where $m$ is the electron mass. Perhaps not surprisingly there is also a spin magnetic moment associated with the electron spin momentum. Using relativistic arguments (F. Rohrlich, 1965), it can be shown that the relationship between the spin angular momentum of a point electron and its spin magnetic moment $\boldsymbol{\mu}$ is

$$\boldsymbol{\mu} = -g\, e\mathbf{s}/2m$$

where $g = 2$ to a first approximation.

(This value of $g$ is slightly in error due to the neglect of radiation reactions.) Observe that the ratio of spin magnetic moment to spin momentum is twice as large as the corresponding orbital ratio.

### *Spin spectroscopy*

Suppose an electron is in a uniform static field of induction **B**. This field interacts with the electron moment to give a contribution

$$-\boldsymbol{\mu}.\mathbf{B} \tag{10.23}$$

to the energy. If the spin-independent terms are omitted the Hamiltonian can be written

$$\frac{e}{m}\,[B_x s_x + B_y s_y + B_z s_z] \tag{10.24}$$

where $B_x, B_y$ and $B_z$ are the components of the field.

If the field is parallel to the $z$-axis, the time-independent Schrödinger equation becomes

$$\frac{e\hbar}{2m} B_z \begin{pmatrix} 1 & 0 \\ 0 & -1 \end{pmatrix} \begin{pmatrix} \psi_+ \\ \psi_- \end{pmatrix} = E \begin{pmatrix} \psi_+ \\ \psi_- \end{pmatrix}. \tag{10.25}$$

This two component equation has two eigensolutions. The upper level has energy

$$E_+ = \frac{e\hbar}{2m} B_z$$

with eigenvector $\boldsymbol{\alpha}$ and the lower level has energy

$$E_- = -\frac{e\hbar}{2m} B_z$$

with eigenvector $\boldsymbol{\beta}$. So the effect of an applied static magnetic field **B** is to split an electron energy level up into two levels with energies $\pm e\hbar B_z/2m$ corresponding to the two spin directions. This is essentially the explanation of the anomalous Zeeman effect in atoms with one valence electron.

One method of observing these levels is by electron spin resonance. The system in the static field **B** is subjected to an alternating electromagnetic field of angular frequency $\omega$. When the frequency satisfies

$$\hbar\omega = e\hbar B_z/m$$

i.e.

$$\omega = eB_z/m$$

transitions are induced and resonance occurs with energy absorption from the alternating field. Electron resonance experiments are generally carried out at fixed frequencies and the field $B_z$ varied. Such experiments can give useful information about the magnetic moments of more complicated systems and hence about their structures.

Atomic nuclei also have spin and associated magnetic moments. The splitting of the nuclear levels in a magnetic field can also be observed by resonance techniques. The resonance frequency for electrons is in the micro-wave region whereas the heavier mass of the nuclei result in a smaller (hyperfine) energy splitting and the corresponding resonance frequencies are in the radio frequencies (see problem 7).

### *Fine structure*

The energy expression (4.71) does not account completely for the observed spectrum of hydrogen. Many of the spectrum lines are composed of several components illustrating a fine structure. In 1916, Sommerfeld modified the Bohr atom theory by taking account of the relativistic mass change of the electron in an elliptic orbit. He found that this introduced a small energy correction, so that levels with the same principal quantum number but a different azimuthal quantum number have slightly different energies. The corresponding fine structure gave good agreement with the observed splitting for hydrogen.

Even before the development of quantum mechanics it was evident that this theory did not explain the spectra of the alkali atoms. An analysis of the penetration of the single valence electron into the ion core shows that the energy levels depend very much on the azimuthal quantum number and that Sommerfeld's relativistic mass effect is small in comparison.

The spectrum lines corresponding to a transition from the level $(n, l)$ to the level $(n', l')$ often show a fine structure which is of the order given by Sommerfeld's correction although his argument cannot apply as any splitting of the upper (lower) levels must produce two or more levels with the same azimuthal quantum number. The doublet splitting of the alkali atoms can be explained on the basis of electron spin and the theory is outlined below. The fine structure of hydrogen is really a combination of spin and relativistic mass change effects. Of course spin itself follows from a relativistic formulation of quantum mechanics.

The energy levels associated with the excitation of the single valence electron of an alkali atom can be evaluated using a single-electron model in which the combined effect of the core electrons is represented by a non-coulombic central field $V(r)$. The accidental degeneracy of the levels with the same principal quantum number but different $l$ is then removed.

Ignoring spin, the Hamiltonian

$$\mathscr{H}^0 \equiv -\frac{\hbar^2}{2m}\nabla^2 + V(r)$$

commutes with the orbital angular momentum operators and the energy levels have degeneracy $(2l + 1)$. Each eigenfunction may be specified by the three quantum numbers $n$, $l$. $m_l$. On the incorrect assumption that the Hamiltonian is unaffected by spin, each of the levels mentioned below is now composed of two levels corresponding to $m_s = \pm\frac{1}{2}$. The degeneracy is doubled to $2(2l + 1)$.

However, the Hamiltonian is really changed on the introduction of spin and the $2(2l + 1)$ levels are not in fact all degenerate. This effect is called 'spin-orbit' coupling.

In a frame of reference that is stationary relative to a nucleus an observer will only experience the electric field $\mathscr{E}$. However, in a frame of reference moving with an electron orbiting the nucleus, an observer would experience a magnetic induction,

$$\mathbf{B} = -\mathbf{V} \wedge \mathscr{E}/c^2 \tag{10.26}$$

where $\mathbf{V}$ is the orbiting electron velocity and $c$ is the speed of light. This is a relativistic effect (Panofsky and Phillips, p. 331).

This field interacts with the spin magnetic moment to give a contribution

$$-\tfrac{1}{2}\boldsymbol{\mu}.\mathbf{B} \tag{10.27}$$

to the total energy. The factor one-half is the Thomas factor (1926) and is due to the fact that the Hamiltonian is expressed in a field of reference in which the nucleus is at rest.

The electric field may be written

$$-\frac{1}{e}\frac{\mathrm{d}V}{\mathrm{d}r}\frac{\mathbf{r}}{r} \tag{10.28}$$

(take the nucleus at the origin) where $V(r)$ is the central field potential energy.

By combining (10.26), (10.27) and (10.28), it can be seen that the spin orbit interaction term to be added to the central field Hamiltonian $\mathscr{H}^0$ is

$$\mathscr{h} \equiv \frac{1}{2c^2m^2r}\frac{\mathrm{d}V}{\mathrm{d}r}\{\mathscr{L}_x\mathrm{s}_x + \mathscr{L}_y\mathrm{s}_y + \mathscr{L}_z\mathrm{s}_z\}. \tag{10.29}$$

This perturbation is a small effect but does give rise to fine structure of spectrum lines.

The 2 × 2 diagonal matrix $\mathscr{H}^0\mathbf{I}$ representing the original central field Hamiltonian commutes with the matrices representing the orbital angular momentum operators and the spin operators, and the five matrix operators

$$\mathscr{H}^0\mathbf{I}, \quad \mathscr{L}^2\mathbf{I}, \quad \mathscr{L}_z\mathbf{I}, \quad \mathbf{s}^2, \quad \mathbf{s}_z$$

form a complete set of commuting operators. For a single electron, $\mathbf{s}^2$ always has the eigenvalues $3\hbar^2/4$ and so a state is specified completely by the four quantum numbers $n$, $l$, $m_l$ and $m_s$. Alternatively the commuting set of matrix operators

$$\mathscr{H}^0\mathbf{I}, \quad \mathbf{J}^2, \quad \mathbf{J}_z, \quad \mathscr{L}^2\mathbf{I}, \quad \mathbf{s}^2$$

is complete and the quantum numbers $n$, $j$, $m_j$ and $l$ completely define a state.

The new Hamiltonian, including the spin-orbit coupling

$$\mathscr{H}^0\mathbf{I} + \mathscr{h}$$

no longer commutes with the orbital momentum operator $\mathscr{L}_z\mathbf{I}$ and the spin operator $\mathbf{s}_z$ and so $m_l$ and $m_s$ are no longer good quantum numbers. From a symmetry viewpoint, the Hamiltonian is no longer invariant under a rotation of space co-ordinates or of spin co-ordinates alone. It is however invariant under a simultaneous rotation of both the space and spin co-ordinates.

The new Hamiltonian forms a commuting set with the operators

$$\mathbf{J}^2, \quad \mathbf{J}_z, \quad \mathscr{L}^2\mathbf{I}, \quad \mathbf{s}^2.$$

However, whereas the unperturbed energies depend only upon the quantum numbers $n$, $l$ the perturbed energies depend also upon $j$. Under the action of the spin-orbit coupling (10.29), the $2(2l+1)$ degenerate levels with a given $n$, $l$ split into a number of distinct levels which differ only in their values of $j$.

The spin-orbit coupling term can be written (10.29)

$$\frac{1}{4c^2m^2r}\frac{\mathrm{d}V}{\mathrm{d}r}\{\mathbf{J}^2 - \mathscr{L}^2\mathbf{I} - \mathbf{s}^2\}. \tag{10.30}$$

This is strictly a problem involving perturbation of a degenerate level, but if the unperturbed wave-functions are specified by $n$, $j$, $m_j$, $l$ the off-diagonal terms of the perturbation determinant (Chapter 7) are already zero and to a first order, the change in energy of a level is the expectation value of (10.30), i.e.

$$\frac{1}{4c^2m^2}\left\langle\frac{1}{r}\frac{\mathrm{d}V}{\mathrm{d}r}\right\rangle_{n,l}[j(j+1) - l(l+1) - \tfrac{3}{4}]\hbar^2. \tag{10.31}$$

For a given $n$ and a given $l \neq 0$, the level splits into a doublet corresponding to $j = l \pm \frac{1}{2}$. The energy separation between the levels is

$$\delta E = \frac{\hbar^2}{4m^2c^2}(2l+1)\left\langle \frac{1}{r}\frac{\mathrm{d}V}{\mathrm{d}r} \right\rangle_{n,\,l}. \tag{10.32}$$

There is no splitting of an atomic level with $l = 0$ as in this case $j$ must equal $\frac{1}{2}$.

The sodium $D$-lines may be explained by this theory and (10.32) gives the correct order of magnitude although there is difficulty in specifying $V(r)$ exactly.

### *Landé g-factor*

An important quantity in the analysis of atomic spectra is the ratio of the magnetic moment of an atom to its angular momentum. This is the gyromagnetic ratio and it is usually written as

$$g\frac{e}{2m} \tag{10.33}$$

where $|g|$ is the Landé $g$-factor. The $g$-factor for spin is equal to 2.

In Chapter 7, in the discussion of the weak-field Zeeman effect for a central-field (spin-less) electron in a uniform external magnetic field $\mathbf{B}$ in the $z$-direction, it is shown that the extra terms to be added to the Hamiltonian due to $\mathbf{B}$ is

$$\frac{eB}{2m}\frac{\hbar}{i}\frac{\partial}{\partial\phi} \equiv \frac{e}{2m}B\mathscr{L}_z \tag{7.77}$$

where $m$ is the (reduced) electron mass. From (7.77) it is clear that the $g$-factor for orbital momentum is unity.

When spin is included the field interaction term to be added to the Hamiltonian is

$$\hbar \equiv \frac{eB}{2m}(\mathscr{L}_z\mathrm{I} + 2\mathrm{s}_z) \equiv \frac{eB}{2m}(\mathbf{J}_z + \mathbf{s}_z). \tag{10.34}$$

If the field is weak enough this term is small compared with the spin-orbit coupling and may be regarded as a perturbation to the Hamiltonian including spin-orbit coupling. The unperturbed eigenfunctions are specified by the quantum numbers $n$, $j$, $m_j$, $l$. The perturbed Hamiltonian does not commute with $\mathbf{J}^2$ and so $j$ is strictly no longer a good quantum number and off-diagonal terms do occur in the perturbation determinant. However it can be shown (Bethe, p. 114) that the non-zero terms occur between states of different $j$ and not between states of the same $j$ and different $m_j$. The off-diagonal terms can be neglected because of the relatively large energy separation between states of different $j$ due to spin-orbit coupling.

The perturbation produces a first-order shift in energy of the state $(n, j, m_j, l)$ equal to the expectation value of (10.34). It can be shown that (p. 61–64 in Condon and Shortley)

$$\langle s_z \rangle = c \langle J_z \rangle$$

where

$$c = \frac{j(j+1) - l(l+1) + s(s+1)}{2j(j+1)}. \tag{10.35}$$

Hence the change in energy of the state $(n, j, m_j, l)$ on introduction of the magnetic field is

$$(1 + c)\frac{e}{2m} B m_j \hbar. \tag{10.36}$$

The $(2j + 1)$ states with different values of $m_j$ now have different energies and the energy change is proportional to the component of total angular momentum parallel to the field.

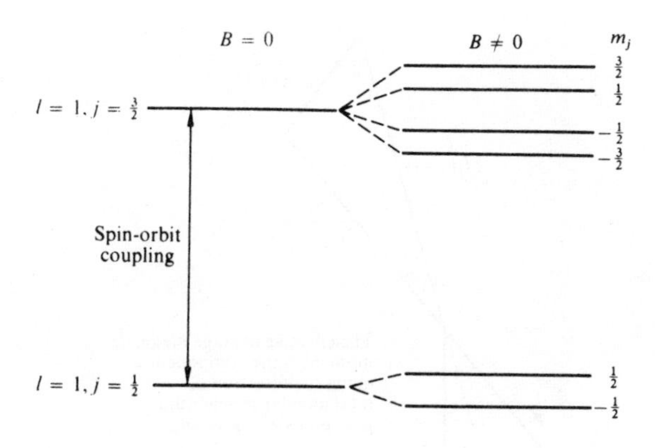


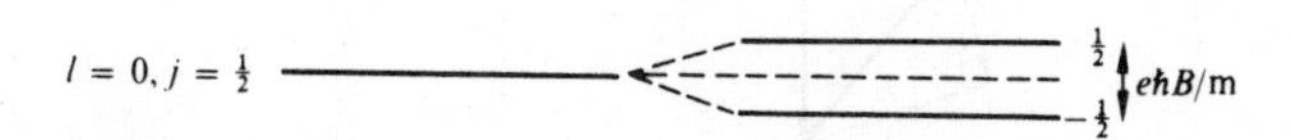


Figure 10.1 The Zeeman effect.

From (10.36) the gyromagnetic ratio is

$$(1+c)\frac{e}{2m}$$

and so the $g$-factor for the system is $(1+c)$, i.e.

$$g = 1 + \frac{j(j+1) - l(l+1) + s(s+1)}{2j(j+1)} \tag{10.37}$$

For a single electron $s = \frac{1}{2}$ and for

$$j = l + \tfrac{1}{2}, \qquad g = 1 + \frac{1}{2l+1}$$

and for

$$j = l - \tfrac{1}{2}, \qquad g = 1 - \frac{1}{2l+1}$$

For an $s$-state $l = 0$ and $g = 2$ and the two levels split by

$$e\hbar\frac{B}{m}$$

The anomalous Zeeman effect of the alkalis is explained in this way.

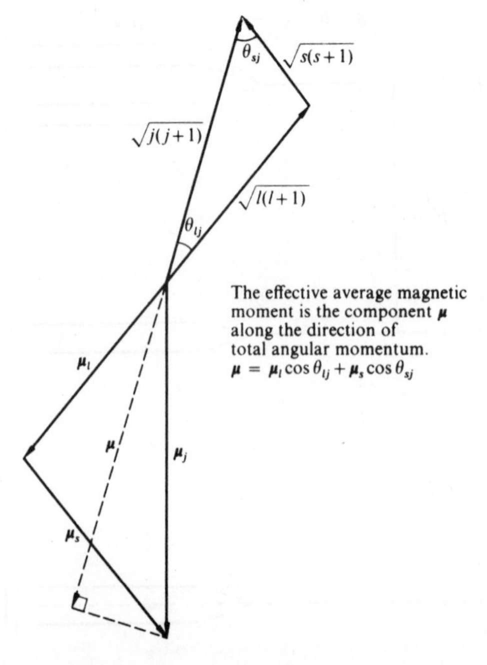


Figure 10.2 The vector model for the Lande-factor.

In a strong magnetic field the interaction term (10.34) may be large compared with the spin-orbit coupling. Then the spin-orbit interaction is regarded as the perturbation. This is the Paschen-Back effect.

## PROBLEMS

1 Verify that the matrix representation (10.14) satisfies the commutation relations (10.4).

Show that the matrices representing $s_x$ and $s_y$ have eigenvalues $\pm\hbar/2$. Show further that the matrix representing $s^2$ is

$$\frac{\hbar^2}{4}\begin{pmatrix}3 & 0\\ 0 & 3\end{pmatrix}$$

and deduce $s^2$ has only one eigenvalue $3\hbar^2/4$.

2 Define operators $\boldsymbol{\alpha}, \boldsymbol{\beta}, \boldsymbol{\gamma}$ by the relations

$$\alpha = 2s_x/i\hbar, \qquad \boldsymbol{\beta} = 2s_y/i\hbar \qquad \gamma = 2s_z/i\hbar$$

Using (10.14) show that

$$\boldsymbol{\alpha}^2 = \boldsymbol{\beta}^2 = \boldsymbol{\gamma}^2 = -\mathbf{I}$$

and further that

$$\boldsymbol{\alpha}\beta = -\boldsymbol{\beta}\boldsymbol{\alpha} = \boldsymbol{\gamma}$$

$$\boldsymbol{\beta}\boldsymbol{\gamma} = -\boldsymbol{\gamma}\boldsymbol{\beta} = \boldsymbol{\alpha}$$

$$\boldsymbol{\gamma}\boldsymbol{\alpha} = -\boldsymbol{\alpha}\boldsymbol{\gamma} = \boldsymbol{\beta}.$$

These results show the relation between the spin operators and quaternions.

3 Consider the Pauli spin matrices (10.16). Verify that

$$\boldsymbol{\sigma}_x^2 = \boldsymbol{\sigma}_y^2 = \boldsymbol{\sigma}_z^2 = +\mathbf{I}$$

and obtain the three commutation relations of the form

$$\boldsymbol{\sigma}_x\boldsymbol{\sigma}_y - \boldsymbol{\sigma}_y\boldsymbol{\sigma}_x = 2i\boldsymbol{\sigma}_z.$$

From these commutation relations show that

$$\boldsymbol{\sigma}_x\boldsymbol{\sigma}_y + \boldsymbol{\sigma}_y\boldsymbol{\sigma}_x = \mathbf{0}$$

and deduce

$$\boldsymbol{\sigma}_x\boldsymbol{\sigma}_y = i\boldsymbol{\sigma}_z.$$

Compare these results with the similar results obtained for the matrices $\boldsymbol{\alpha}, \boldsymbol{\beta}, \boldsymbol{\gamma}$ in question two.

4 From equations (10.3) and (10.14) deduce that the 2 x 2 matrices representing the infinitesimal rotation operators $I_x^{(s)}, I_y^{(s)}$ and $I_z^{(s)}$ in the irreducible representation $D^{(1/2)}$ are

$$\frac{1}{2}\begin{pmatrix} 0 & 1 \\ 1 & 0 \end{pmatrix} \quad \frac{1}{2}\begin{pmatrix} 0 & -i \\ i & 0 \end{pmatrix} \quad \frac{1}{2}\begin{pmatrix} 1 & 0 \\ 0 & -1 \end{pmatrix}$$

respectively. Verify that the commutation relations (10.4) are satisfied.

Use equation (9.46) to show that the corresponding matrix representing the finite rotation $P^s(\phi, z)$ is

$$\begin{pmatrix} e^{i\phi/2} & 0 \\ 0 & e^{-i\phi/2} \end{pmatrix}.$$

5 In the isotopic spin theory protons and neutrons are regarded as two distinct states of the 'nucleon'. The two states are defined by the two possible values of the isotopic spin co-ordinate $\tau_z$. $\tau_z = +1$ and $\tau_z = -1$ correspond to the proton and neutron respectively. A nucleon wave-function can formally be written

$$\psi(x, y, z, \tau_z) = \psi_+(x, y, z)\alpha(\tau_z) + \psi_-(x, y, z)\beta(\tau_z)$$

with

$$\alpha(+1) = 1 \quad \text{and} \quad \alpha(-1) = 0$$
$$\beta(+1) = 0 \quad \text{and} \quad \beta(-1) = 1.$$

In analogy to the electron spin matrix representation, confirm that the matrix representation

$$\boldsymbol{\alpha} = \begin{pmatrix} 1 \\ 0 \end{pmatrix} \quad \boldsymbol{\beta} = \begin{pmatrix} 0 \\ 1 \end{pmatrix} \quad \boldsymbol{\tau}_z = \begin{pmatrix} 1 & 0 \\ 0 & -1 \end{pmatrix}$$

is valid.

The electron charge of a nucleon is represented by the matrix

$$\mathbf{q} = \frac{e}{2}[\mathbf{I} + \boldsymbol{\tau}_z].$$

Confirm that the proton and neutron state vectors are eigenvectors of $\mathbf{q}$ with eigenvalues $+e$ and 0 respectively. (For further information the reader is referred to Heine's book.)

6 When a magnetic field of induction $\mathbf{B}$ is applied to an electron and if all other contributions to the energy are neglected, the Hamiltonian is

$$\frac{e}{m}[\mathbf{s}_x B_x + \mathbf{s}_y B_y + \mathbf{s}_z B_z].$$

Show that the eigenvalues of this Hamiltonian are

$$E = \pm \frac{e\hbar}{2m} B \qquad \text{where} \qquad B = (B_x^2 + B_y^2 + B_z^2)^{1/2}.$$

If **B** is at an angle $\theta$ to the $z$-axis then

$$B_z = B \cos\theta \qquad \text{and} \qquad B_x^2 + B_y^2 = B^2 \sin^2\theta.$$

Show that in this case if the normalized two component wave vector is

$$\begin{pmatrix} \psi_+ \\ \psi_- \end{pmatrix}$$

then

$$|\psi_+|^2 = \cos^2 \frac{\theta}{2} \qquad \text{and} \qquad |\psi_-|^2 = \sin^2 \frac{\theta}{2}.$$

7 Suppose, in question 6, the field is along the $z$-axis, then the Hamiltonian is

$$\frac{eB}{m} \mathbf{s}_z.$$

It has been shown in section 10.5 that the eigenvalues of this Hamiltonian are $E = \pm e\hbar B_z/2m$ corresponding to the $z$-component spin eigenvalues $\pm\hbar/2$.

However, if the state function is not an eigenfunction of $\mathbf{s}_z$ the equation that must be satisfied is the Schrödinger time-dependent equation.

$$\frac{e\hbar}{2m} B_z \begin{pmatrix} 1 & 0 \\ 0 & -1 \end{pmatrix} \begin{pmatrix} \psi_+ \\ \psi_- \end{pmatrix} = i\hbar \frac{\partial}{\partial t} \begin{pmatrix} \psi_+ \\ \psi_- \end{pmatrix}.$$

Verify that if initially at $t = 0$ the state function is an eigenvector of $\mathbf{s}_x$ with eigenvalue $+\hbar/2$, then at time $t$, the state function is

$$\begin{pmatrix} \psi_+ \\ \psi_- \end{pmatrix} = \frac{1}{\sqrt{2}} \begin{bmatrix} \exp\,[-i\omega t/2] \\ \exp\,[i\omega t/2] \end{bmatrix}$$

where

$$\omega = eB_z/m.$$

Confirm also, that at $t = \pi/2\omega$ the state function is an eigenvector of $\mathbf{s}_y$ with eigenvalue $\hbar/2$ and at $t = 2\pi/\omega$ the state function is once more an eigenvector of $\mathbf{s}_x$ with eigenvalue $\hbar/2$.

These results show how the spin magnetic moment vector processes about the applied magnetic field with angular frequency $eB_z/m$. (This is identical to the classical frequency.)

8 The relativistic change in mass of an electron may be taken into account by considering the perturbation

$$h \equiv -\hat{p}^4/8\mu^3c^2$$

where $\hat{p}$ is the operator representing linear momentum $\mu$ is the reduced mass and $c$ is the speed of light.

Explain why

$$\hat{p}^4\psi = 4\mu^2\left[E + \frac{Ze^2}{4\pi\epsilon_0 r}\right]^2\psi$$

where $\psi$ and $E$ are the eigenfunctions and eigenvalues for a hydrogen-type atom. Hence show that the first-order perturbation of the energy due to relativity, is for the state $(n, l, m_l)$

$$E^{(1)} = -\left[\frac{1}{(2l+1)n^3} - \frac{3}{8n^4}\right]\mu e^8 Z^4/(4\pi\epsilon_0)^4\hbar^4c^2.$$

This term should be added to the spin orbit coupling term in (10.31).

## References

Bethe, H. A. *Intermediate Quantum Mechanics.* Benjamin, New York (1964).

Blatt, J. M. and Weisskopf, V. F. *Theoretical Nuclear Physics,* p. 783. Wiley, New York (1952).

Condon, E. U. and Shortley, S. H. *The Theory of Atomic Spectra.* Cambridge (1959).

Heine, V. *Group Theory in Quantum Mechanics.* Pergamon, London (1960).

Panofsky, W. K. H. and Phillips, M. *Classical electricity and magnetism,* p. 331. Addison-Wesley, Cambridge, Mass. (2nd edition, 1962).

Rohrlich, F. *Classical charged particles.* Addison-Welsey, Reading, Mass. (1965).

Rorschach, H. E. *Bull. Am. Phys. Soc.* **7**, 121 (1962).

Thomas, L. H. *Nature* **117**, 514 (1926).

Uhlenbeck, G. E. and Goudsmith, S. *Nature* **117**, 264 (1926).

CHAPTER 11

# Systems of Particles

## 11.1 Introduction

The basic theory developed through the postulates in the earlier chapters, applies to many particle systems. In Chapter 4 it was shown that the Hamiltonian operator representing a system of two particles of masses $m_1$ and $m_2$ at positions $(x_1, y_1, z_1)$ and $(x_2, y_2, z_2)$ respectively is (4.45),

$$\mathscr{H} \equiv -\frac{\hbar^2}{2m_1}\left[\frac{\partial^2}{\partial x_1^2}+\frac{\partial^2}{\partial y_1^2}+\frac{\partial^2}{\partial z_1^2}\right]-\frac{\hbar^2}{2m_2}\left[\frac{\partial^2}{\partial x_2^2}+\frac{\partial^2}{\partial y_2^2}+\frac{\partial^2}{\partial z_2^2}\right]$$

$$+ V(x_1, y_1, z_1, x_2, y_2, z_2).$$

The state function $\Psi$ depends upon all six co-ordinates and the time and satisfies Schrödinger's time dependent equation (Postulate 3),

$$\mathscr{H}\Psi = -\frac{\hbar}{i}\frac{\partial \Psi}{\partial t}.$$

The normalization condition takes the form

$$\iint \Psi^*\Psi \, \mathrm{d}x_1 \, \mathrm{d}y_1 \, \mathrm{d}z_1 \, \mathrm{d}x_2 \, \mathrm{d}y_2 \, \mathrm{d}z_2 \equiv \iint \Psi^*\Psi \, \mathrm{d}\tau_1 \, \mathrm{d}\tau_2 = 1$$

where the field of integration is over all possible values of the six co-ordinates.

These results are easily extended to a system of $n$ particles. The classical Hamiltonian is

$$H = \sum_{i=1}^{n} \frac{p_i^2}{2m_i} + V(\mathbf{r}_1, \mathbf{r}_2, \ldots, \mathbf{r}_n)$$

where $\mathbf{r}_i$ is the position of the $i$th. The quantum mechanical Hamiltonian operator is

$$\mathscr{H} \equiv -\sum_{i=1}^{n} \frac{\hbar^2}{2m_i}\left(\frac{\partial^2}{\partial x_i^2} + \frac{\partial^2}{\partial y_i^2} + \frac{\partial^2}{\partial z_i^2}\right) + V(\mathbf{r}_1, \mathbf{r}_2, \ldots, \mathbf{r}_n). \quad (11.1)$$

The state function depends upon the values of all the co-ordinates and the time and again satisfies the time dependent Schrödinger equation.

An important Hamiltonian is that of a free atom with $n$ electrons. Ignoring the translational energy and the spin-orbit interaction terms, this operator is

$$\mathscr{H} \equiv -\frac{\hbar^2}{2m}\sum_{i=1}^{n}\nabla_i^2 - \sum_{i=1}^{n}\frac{Ze^2}{4\pi\epsilon_0 r_i} + \frac{1}{2}\sum_{i=1}^{n}\sum_{\substack{j=1 \\ (i \neq j)}}^{n}\frac{e^2}{4\pi\epsilon_0 |\mathbf{r}_i - \mathbf{r}_j|} \quad (11.2)$$

where $m$ is the electron mass and $Z$ is the atomic number. The operator $\nabla_i^2$ operates only on the co-ordinates of the $i$th electron and $|\mathbf{r}_i - \mathbf{r}_j|$ is the separation between the $i$th and $j$th electrons.

## 11.2 Angular Momentum of Systems

*Orbital angular momentum*

Classically, the orbital angular momentum, of a system of $n$ particles, about the origin is

$$\mathbf{L} = \sum_{i=1}^{n} \mathbf{r}_i \wedge \mathbf{p}_i$$

where $\mathbf{p}_i$ is the linear momentum of the $i$th particle at position $\mathbf{r}_i$. The quantum mechanical operator representing the $z$-component of the angular momentum of the system is

$$\mathscr{L}_z \equiv \frac{\hbar}{i}\sum_{i=1}^{n}\left(x_i\frac{\partial}{\partial y_i} - y_i\frac{\partial}{\partial x_i}\right) \equiv \sum_{i=1}^{n}\mathscr{L}_{iz} \quad (11.3)$$

where

$$\mathscr{L}_{iz} \equiv \frac{\hbar}{i}\left(x_i\frac{\partial}{\partial y_i} - y_i\frac{\partial}{\partial x_i}\right)$$

is the operator representing the $z$-component of the $i$th particle. Similarly it is possible to define the operators $\mathscr{L}_x$ and $\mathscr{L}_y$ representing the $x$ and

$y$ components of the orbital angular momentum of the system and the square of the angular momentum of the system is

$$\mathscr{L}^2 \equiv \mathscr{L}_x^2 + \mathscr{L}_y^2 + \mathscr{L}_z^2.$$

It is easy to deduce that the angular momentum operators of the system satisfy the same commutation rules as those for a single particle, thus

$$\mathscr{L}_x \mathscr{L}_y - \mathscr{L}_y \mathscr{L}_x \equiv i\hbar \mathscr{L}_z \qquad \text{etc.}$$

(Note that $\mathscr{L}_{ix}$, $\mathscr{L}_{jy}$, $i \neq j$, etc. commute as they refer to different co-ordinates.) Also $\mathscr{L}^2$ commutes with $\mathscr{L}_x$, $\mathscr{L}_y$ and $\mathscr{L}_z$.

The angular momentum operators can be related to infinitesimal rotation operators. From (11.3)

$$\mathscr{L}_z \equiv \hbar \sum_{i=1}^{n} I_{iz}^{(r)} \tag{11.4}$$

where $I_{iz}^{(r)}$ is the infinitesimal operator that operates on the space co-ordinate of the $i$th particle only. By a similar argument to that given in section 10.4 concerning rotation operators and total angular momentum, it can be shown that the summation on the right hand side of (11.4) can be replaced by a single infinitesimal operator producing a simultaneous rotation of the space co-ordinates of all the particles. Consequently, orbital angular momentum is conserved if the system is invariant under a simultaneous rotation of all the space co-ordinates. This is the case for the approximate Hamiltonian (11.2) for a free atom.

The argument given in section 9.6 implies that $\mathscr{L}^2$ has a discrete set of $(2L + 1)$ eigenfunctions, $2L$ an integer, with common eigenvalues $L(L + 1)\hbar^2$. Each of these functions may be chosen to be an eigen-function of $\mathscr{L}_z$ with separate eigenvalues

$$M_L \hbar = L\hbar, (L - 1)\hbar, \ldots, -L\hbar.$$

Single-valuedness of the space wave-function once again requires $L$ to be an integer. (Note that $L$, $M_L$ are used as the quantum numbers for a system of particles and $l$, $m_l$ are used for a single particle.)

The approximate free atom Hamiltonian (11.2) is invariant under a simultaneous rotation of all the co-ordinates and so commutes with the system angular momentum operators. The quantum number $L$ is a good quantum number and may be used in the classification of states. The states with $L = 0, 1, 2, \ldots$ are labelled by the letters S, P. D, . . ..

*Product space*

In a two-particle system where the two particles do not interact, the Schrödinger equation is separable and the two-particle wave-function can be expressed as a product of one-particle wave-functions

$$\psi(\mathbf{r}_1, \mathbf{r}_2) = \psi_1(\mathbf{r}_1)\psi_2(\mathbf{r}_2).$$

Products of this type can be used to define a product space.

The term 'function space' was defined in Chapter 3. Suppose the linearly independent set of functions $\{\theta_i\}, i = 1, \ldots, n$ forms the basis for an $n$-dimensional function space and similarly $\{\phi_j\}, j = 1, \ldots, m$ forms the basis for an $m$-dimensional space. The $n \times m$ linearly independent products $\theta_i\phi_j$ are the basis functions for an $n \times m$-dimensional space which is called the product of the original spaces. Any function in this product space can be expressed

$$\sum_i^n \sum_j^m A_{ij}\theta_i\phi_j$$

where the $n \times m$ coefficients $A_{ij}$ are constants.

*Addition of angular momenta*

Consider a system of two particles and suppose that the operators representing the angular momentum components of each particle commute with the system Hamiltonian so that the angular momentum of each particle is conserved. This implies that the Hamiltonian is invariant under a rotation of the co-ordinates of each particle separately. This would happen in the case of a two-electron atom where the electron–electron interaction term (the last term in (11.2)) is ignored or approximated by a smoothed out, separable, central-field potential.

The operators $\mathscr{L}^2$, $\mathscr{L}_z$, $\mathscr{L}_1^2$, $\mathscr{L}_2^2$ where $\mathscr{L}^2$, $\mathscr{L}_z$ refer to the angular momentum of the system and $\mathscr{L}_1^2$, $\mathscr{L}_2^2$ refer to the separate particles, commute with one another and with the Hamiltonian and may be measured simultaneously. Alternatively the operators $\mathscr{L}_1^2$, $\mathscr{L}_{1z}$, $\mathscr{L}_2^2$, $\mathscr{L}_{2z}$ where $\mathscr{L}_{1z}$, $\mathscr{L}_{2z}$ refer to the separate particles, commute with one another and with the Hamiltonian and so their eigenvalues may be known simultaneously. Either of these four sets of four angular momentum operators, when combined with certain other operators will form a complete set. Assuming the other operators have fixed eigenvalues the state of the system is specified by either the set of quantum numbers (i) $L$, $M_L$, $l_1, l_2$ or by (ii) $l_1, m_1, l_2, m_2$. The quantum numbers $L, M_L$ refer to the system while $l_1$, $m_1$ and $l_2$, $m_2$ refer to the first and second particles respectively.

To find the different possible values of $\mathscr{L}^2$ associated with the first set of operators it is useful to consider the labelling of the different states arising from the second set.

Consider the second set $\mathscr{L}_1^2$, $\mathscr{L}_{1z}$, $\mathscr{L}_2^2$, $\mathscr{L}_{2z}$. For a given $l_1$, the quantum number $m_1$, can take $(2l_1 + 1)$ values and similarly for given $l_2, m_2$ can take $(2l_2 + 1)$ values. So for fixed $l_1$ and $l_2$ there are $(2l_1 + 1)(2l_2 + 1)$ different degenerate states that span a sub-space of the function space of the system. The eigenfunctions may be written $\psi(l_1, l_2, m_1, m_2)$ and the largest allowed values for $m_1, m_2$ are $l_1, l_2$ respectively. The space of the two-particle system can be expressed as the product of two spaces each describing a single particle, i.e.

$$\psi(l_1, l_2, m_1, m_2) = \phi(l_1, m_1)\phi(l_2, m_2)$$

where $\phi(l_1, m_1)$ and $\phi(l_2, m_2)$ describe the first and second particles respectively.

Consider now the first set of operators $\mathscr{L}^2$, $\mathscr{L}_z$, $\mathscr{L}_1^2$, $\mathscr{L}_2^2$. For a fixed $l_1$ and $l_2$ there must be $(2l_1 + 1)(2l_2 + 1)$ different states spanning the same sub-space of the system space. These eigenfunctions may be written $\psi(L, M_L, l_1, l_2)$ and are obtained by taking appropriate linear combinations of the eigenstates of the second set of operators. The coefficients are the Clebsch–Gordon coefficients, i.e.

$$\psi(L, M_L, l_1, l_2) = \sum_{m_1} \sum_{m_2} c(l_1, l_2, m_1, m_2 | L, M_L)\psi(l_1, l_2, m_1, m_2).$$

Since

$$\mathscr{L}_z \equiv \mathscr{L}_{1z} + \mathscr{L}_{2z}$$

the largest allowed value for $M_L$ is $(l_1 + l_2)$ obtained by combining $m_1 = l_1$ and $m_2 = l_2$. It follows that the largest $L$ is also $(l_1 + l_2)$. This system wave-function is the product $\phi(l_1, l_1)\phi(l_2, l_2)$.

There are two states in the second set with the next value for $M_L$.

$$M_L = (l_1 + l_2) - 1.$$

These occur with $m_1 = l_1, m_2 = l_2 - 1$ and $m_1 = l_1 - 1, m_2 = l_2$. By suitably combining these there must be two states in the first set with this value for $M_L$. These are

$$L = (l_1 + l_2),\quad M_L = L - 1 \qquad \text{and} \qquad L = (l_1 + l_2 - 1),\quad M_L = L.$$

There are three states in the second set with

$$M_L = (l_1 + l_2) - 2$$

and hence three such states in the first set with $L = (l_1 + l_2), (l_1 + l_2 - 1)$ and $(l_1 + l_2 - 2)$ respectively. For the next value

$$M_L = (l_1 + l_2) - 3$$

there are four states and so on.

The number of states with a given $M_L$ does not increase indefinitely and in fact the maximum number occurs for

$$M_L = |l_1 - l_2|.$$

This implies that the minimum value for $L$ is $|l_1 - l_2|$.

Finally, for a given $l_1, l_2$ the system quantum number $L$ can take the positive values

$$L = (l_1 + l_2), (l_1 + l_2 - 1), \ldots, |l_1 - l_2|$$

i.e.

$$(l_1 + l_2) \geqslant L \geqslant |l_1 - l_2|. \tag{11.5}$$

This is the triangle inequality. If $l_1 \geqslant l_2$ there are $(2l_2 + 1)$ distinct values for $L$. For each value of $L$ there are $(2L + 1)$ values of $M_L$ and the total number of states is

$$\sum_{(l_1 - l_2)}^{(l_1 + l_2)} (2L + 1) = (2l_1 + 1)(2l_2 + 1)$$

as required.

The vector coupling rule (11.5) can be used to add three or more angular momenta of particles together by successive applications. The reader is advised to compare the work above with the discussion of total angular momentum in section 10.4.

There is an equivalent group theoretic way of discussing the addition of angular momentum.

*Product representation of a group*

Let the two function spaces $\{\theta_i\}$ and $\{\phi_j\}$ each be invariant under the group of operators $\{A\}$. The product space is also invariant under the same group of operators and the set of functions $\{\theta_i \phi_j\}$ forms an $n \times m$-dimensional representation of the group. In general the product space will be reducible into a sum of sub-spaces each irreducible under the group.

As an illustration consider the two sets of functions $\{\psi_{3m_l}\} l = 3$, $m_l = 3, 2, 1, 0, -1, -2, -3$ and $\{\psi_{1m_l}\} l = 1, m_l = 1, 0, -1$ which transform according to the irreducible representations $D^{(3)}$ and $D^{(1)}$ respectively of the full rotation group. (Note that $\psi_{3m_l} = P_3^{m_l}(\cos\theta) e^{im_l\phi}$ etc.) The 21-dimensional product space is invariant under the rotation group and transforms according to some reducible representation which is denoted by $D^{(3)} \times D^{(1)}$. The 21 functions in the product space can be split up into three sets, one containing nine functions forming the basis for the irreducible representation $D^{(4)}$ another containing seven functions and transforming according to $D^{(3)}$ and finally a set of five functions transforming according to $D^{(2)}$. The result can be written

$$D^{(3)} \times D^{(1)} = D^{(4)} + D^{(3)} + D^{(2)}.$$

For a more complete discussion the reader is referred to *Group Theory in Quantum Mechanics,* by V. Heine, or *Group Theory and Quantum Mechanics*, by M. Tinkham.

More generally, if two sets of functions form the basis for irreducible representations $D^{(j_1)}$ and $D^{(j_2)}$, the product space transforms according to the reducible representation

$$D^{(j_1)} \times D^{(j_2)} = D^{(j_1+j_2)} + D^{(j_1+j_2-1)} + \ldots + D^{|j_1-j_2|}. \qquad (11.6)$$

This is of course another statement of the vector coupling rule (11.5).

### *Parity*

Consider the inversion operator $I$ which is defined by

$$I\psi(\mathbf{r}_1, \mathbf{r}_2, \ldots, \mathbf{r}_n) = \psi(-\mathbf{r}_1, -\mathbf{r}_2, \ldots, -\mathbf{r}_n).$$

That is, it changes the sign of each of the position variables of all the particles in the system. The eigenvalue equation for the inversion operator is

$$I\psi = p\psi$$

where $p$ is an eigenvalue and $\psi$ is an eigenfunction. If the inversion operator is applied twice the function must be converted back into itself. The operator $I^2$ is equivalent to the identity $E$ and

$$I^2\psi = p^2\psi = \psi$$

$$\therefore \quad p^2 = 1 \qquad \text{and} \qquad p = \pm 1.$$

The inversion operator has two eigenvalues $p = \pm 1$ and the eigenfunctions are either

(a) unchanged in sign when acted upon by the inversion operator (even functions) or

(b) changed in sign when acted upon by $I$ (odd functions).

The state is said to have even or odd parity respectively.

For some systems the parity eigenvalues are good quantum numbers. Consider a closed system of particles with no external forces acting. Then the potential energy depends only upon the particle separations and the quantum mechanical Hamiltonian is

$$\mathcal{H} \equiv -\frac{\hbar^2}{2}\sum_i \frac{1}{m_i}\nabla_i^2 + \frac{1}{2}\sum_{i \neq j}\sum V(|\mathbf{r}_i - \mathbf{r}_j|). \qquad (11.7)$$

Clearly $\mathcal{H}$ commutes with the inversion operator, i.e.

$$\mathcal{H}I - I\mathcal{H} \equiv 0.$$

Functions can be chosen which are simultaneously eigenfunctions of both $\mathcal{H}$ and $I$.

In a central field there is a term

$$\sum_i V(|\mathbf{r}_i'|)$$

to be added to the Hamiltonian above (see (11.2)). This Hamiltonian also commutes with $I$ and the parity eigenvalues remain good quantum numbers.

The inversion operator also commutes with the angular momentum operators of the system. In particular

$$\mathcal{L}_z I - I\mathcal{L}_z \equiv 0$$

and

$$\mathcal{L}^2 I - I\mathcal{L}^2 \equiv 0.$$

For a central field problem (or for the Hamiltonian (11.7) the inversion operator, the square of the angular momentum, the $z$-component of the angular momentum and the Hamiltonian form a set of commuting operators and can be simultaneously measured.

Consider a single-particle, hydrogen-type wave-function. In spherical polar co-ordinates

$$\psi_{nlm_l} = R_{nl}(r) P_l^{m_l}(\cos\theta) e^{im_l\phi}.$$

The inversion operator corresponds to the transformation

$$r \to r, \qquad \theta \to \pi - \theta, \qquad \phi \to \pi + \phi.$$

As

$$e^{im_l(\pi+\phi)} = (-1)^{m_l} e^{im_l\phi}$$

and

$$P_l^{m_l}(\cos(\pi-\theta)) = P_l^{m_l}(-\cos\theta) = (-1)^{l-m_l} P_l^{m_l}(\cos\theta),$$

the wave-function is multiplied by $(-1)^l$ when operated on by the inversion operator, i.e.

$$I\psi_{nlm_l} = (-1)^l \psi_{nlm_l}.$$

The parity of the state is independent of the quantum number $m_l$ and depends only upon $l$. If $l$ is an even integer, the state has even parity whereas if $l$ is an odd integer the state has odd parity.

The parity of an $n$-particle wave-function which can be expressed as a product of $n$ single particle central field wave-functions is

$$(-1)^{l_1+l_2+\ldots+l_n}$$

where $l_i$ is the angular momentum quantum number for the $i$th particle.

The parity concept is useful in considering transition probabilities and is particularly useful in classifying nuclear energy states (*Theoretical Nuclear Physics*, by Blatt and Weisskopf).

The conservation of parity in quantum mechanics has no equivalent in classical mechanics as the invariance of the classical Hamiltonian under inversion does not lead to any interesting results.

*Spin angular momentum*

The operator representing the spin momentum of an electron is defined in terms of the infinitesimal rotation operators, which transform the spin-co-ordinates, by equation (10.3). The operator representing the $z$-component of the spin momentum of the system of $n$ particles is

$$S_z \equiv \sum_{i=1}^{n} s_{iz} \tag{11.8}$$

where

$$s_{iz} \equiv \hbar I_{iz}^{(s)}$$

is the operator representing the $z$-component of the $i$th particle. Similarly it is possible to define the operators $S_x$ and $S_y$ representing the $x$ and $y$ components of the spin momentum of the system and the square of the angular momentum of the system is

$$S_{\text{op}}^2 \equiv S_x^2 + S_y^2 + S_z^2 .$$

The spin momentum operators of the system satisfy the same commutation rules as those for a single particle.

From (11.8)

$$S_z \equiv \hbar \sum_{i=1}^{n} I_{iz}^{(s)} \tag{11.9}$$

where $I_{iz}^{(s)}$ is the infinitesimal operator that transforms the spin co-ordinate of the $i$th particle only. The right hand side of (11.9) can be replaced by a single infinitesimal operator producing a simultaneous rotation of the spin co-ordinates of all the particles. So, spin momentum is conserved if the system is invariant under a simultaneous rotation of all the spin co-ordinates. This is the case for the approximate free atom Hamiltonian (11.2). In fact this Hamiltonian is invariant under a rotation of the $i$th co-ordinate alone and so the spin momentum of each separate particle is conserved.

The argument given in section 9.6 implies that $S_{\text{op}}^2$ has a discrete set of $(2S + 1)$ eigenfunctions, $2S$ an integer, with common eigenvalue

$S(S+1)\hbar^2$. Each of these functions may be chosen to be an eigenfunction of $S_z$ with eigenvalues

$$M_s\hbar = S\hbar,\ (S-1)\hbar, \ldots, -S\hbar. \tag{11.10}$$

(The reader is warned not to confuse the operator $S^2_{\text{op}}$ with the quantum number $S$.) For a single electron $S \equiv s = \frac{1}{2}$.

Consider a system composed of two electrons. If the spin interaction terms are ignored, the system Hamiltonian is invariant under a separate rotation of each spin co-ordinate and the spin momentum of each electron is conserved. For each separate electron $s = \frac{1}{2}$ and by following an argument similar to that leading to equation (11.5), the system spin quantum number $S$ can take either of the two values

$$S = 1 \quad \text{with} \quad M_s = 1, 0, -1 \quad \text{(triplet)}$$

$$S = 0 \quad \text{with} \quad M_s = 0 \quad \text{(singlet)}.$$

Alternatively, from (11.6)

$$D^{(1/2)} \times D^{(1/2)} = D^{(1)} + D^{(0)}.$$

The four products

$$\alpha(\sigma_{1z})\alpha(\sigma_{2z}), \quad \beta(\sigma_{1z})\beta(\sigma_{2z}), \quad \alpha(\sigma_{1z})\beta(\sigma_{2z}), \quad \beta(\sigma_{1z})\alpha(\sigma_{2z})$$

where $\sigma_{1z}$, $\sigma_{2z}$ are the spin co-ordinates of the two electrons, form the basis for the reducible representation $D^{(1/2)} \times D^{(1/2)}$. From these it is possible to construct the three functions

| | spin quantum number $S$ | $M_s$ |
|---|---|---|
| $\alpha(\sigma_{1z})\alpha(\sigma_{2z})$ | 1 | 1 |
| $\beta(\sigma_{1z})\beta(\sigma_{2z})$ | 1 | −1 |
| $\frac{1}{\sqrt{2}}[\alpha(\sigma_{1z})\beta(\sigma_{2z}) + \beta(\sigma_{1z})\alpha(\sigma_{2z})]$ | 1 | 0 |

which transform according to $D^{(1)}$ and the single function

| | $S$ | $M_s$ |
|---|---|---|
| $\frac{1}{\sqrt{2}}[\alpha(\sigma_{1z})\beta(\sigma_{2z}) - \beta(\sigma_{1z})\alpha(\sigma_{2z})]$ | 0 | 0 |

which transforms according to $D^{(0)}$. All of these functions are orthonormal. From the spin operator viewpoint, each of the three functions that transforms according to $D^{(1)}$ is simultaneously an eigenfunction of $S^2_{\text{op}}$ with eigenvalue $2\hbar^2$ and also an eigenfunction of $S_z$ with eigenvalues $\hbar, 0, -\hbar$ respectively. The single function that transforms according to $D^{(0)}$ is an eigenfunction of $S^2_{\text{op}}$ and $S_z$ with common eigenvalue zero.

The reader is advised to verify the results using equations (10.17), (10.18) and noting that (from, (11.8))

$$S_{\text{op}}^2 \equiv (s_{1x}^2 + s_{1y}^2 + s_{1z}^2) + (s_{2x}^2 + s_{2y}^2 + s_{2z}^2)$$
$$+ 2(s_{1x}s_{2x} + s_{1y}s_{2y} + s_{1z}s_{2z}).$$

*Total angular momentum*

The total angular momentum of the system is defined to be the sum of the orbital angular momentum and the spin angular momentum of the system. In particular

$$J_z \equiv \mathscr{L}_z + S_z \tag{11.11}$$

and similarly for $J_x$ and $J_y$. The square of the total angular momentum is

$$J^2 \equiv J_x^2 + J_y^2 + J_z^2 \tag{11.12}$$

and these operators satisfy the usual commutation rules.

From (11.4), (11.8) and (11.11)

$$J_z \equiv \hbar \sum_{i=1}^{n} (I_{iz}^{(r)} + I_{iz}^{(s)}) \equiv \hbar \sum_{i=1}^{n} I_{iz}^{(t)} \tag{11.13}$$

where $I_{iz}^{(t)}$ is an infinitesimal operator producing a simultaneous rotation of both the space and spin co-ordinates of the $i$th particle. The right hand side of (11.13) can be replaced by a single infinitesimal operator producing a simultaneous rotation of all the co-ordinates of all the particles. Consequently, total spin momentum is conserved if the system is invariant under a simultaneous rotation of all the co-ordinates, both space and spin. This is the case for the general free atom Hamiltonian even when spin-orbit coupling is included.

The argument of section 9.6 implies that $J^2$ has a discrete set of $(2J + 1)$ eigenfunctions, $2J$ an integer, with common eigenvalue $J(J + 1)\hbar^2$. Each of these may be chosen to be an eigenfunction of $J_z$ with separate eigenvalues

$$M_J\hbar = J\hbar, (J - 1)\hbar, \ldots, -J\hbar. \tag{11.14}$$

The addition of the total angular momenta of the particles in a system follows along the same lines as that leading to equations (11.5) and (11.6).

## 11.3 Identical Particles

In quantum mechanics the uncertainty principle makes it impossible to define precisely a particle trajectory. Because of this a given particle, in a collection of identical particles, cannot be labelled. This is not the case in classical mechanics. A classical particle has a definite trajectory and it is possible to distinguish individual particles by following their paths.

So, in quantum mechanics, unlike classical mechanics, there is, in principle, no physical way of distinguishing between identical particles. This can be formalized by introducing permutation operators. A permutation operator $P_{ij}$ interchanges the labels of the $i$ and $j$ particles. In particular

$$P_{12}\psi(\mathbf{r}_1, \sigma_{1z}; \mathbf{r}_2, \sigma_{2z}) = \psi(\mathbf{r}_2, \sigma_{2z}; \mathbf{r}_1, \sigma_{1z}) \tag{11.15}$$

where $\mathbf{r}_i$, $\sigma_{iz}$ are the space and spin co-ordinates of the $i$th particle. As identical particles are indistinguishable it is not possible to observe any change in a system if two such particles are interchanged. This means that if particles are relabelled, a normalized function must only change by a phase factor, i.e.

$$P_{12}\psi(\mathbf{r}_1, \sigma_{1z}; \mathbf{r}_2, \sigma_{2z}) = e^{i\alpha}\psi(\mathbf{r}_1, \sigma_{1z}; \mathbf{r}_2, \sigma_{2z}). \tag{11.16}$$

If the interchange is repeated the original wave-function must be obtained and so $P_{12}$ has two eigenvalues given by (compare the inversion operator),

$$e^{i2\alpha} = +1$$

i.e.

$$e^{i\alpha} = \pm 1. \tag{11.17}$$

The quantum mechanical principle of indistinguishability requires that a many-particle wave-function behaves in one of two ways when two particle are interchanged (i) either the wave-function is unchanged, i.e. symmetric, or (ii) the wave-function changes sign, i.e. antisymmetric. For consistency, it is clear that a wave-function describing a system with three or more identical particles cannot be symmetric with respect to some interchanges and antisymmetric with respect to others. The wave-function is either symmetric or antisymmetric with respect to all interchanges.

The Hamiltonian describing a system of identical particles must be invariant when the particle co-ordinates including spin are relabelled. The permutation operators must commute with the Hamiltonian and the permutation quantum numbers (11.17) are good quantum numbers. So the wave-function is either permanently symmetric or it is permanently antisymmetric, which of these being determined by the nature of the particles rather than the state of the system.

Particles described by symmetric wave-functions are said to satisfy Bose–Einstein statistics whereas particles described by antisymmetric wave-functions satisfy Fermi–Dirac statistics. It can be shown from relativistic quantum theory that particles with integral spin are Bose–Einstein particles (Bosons), and that particles with half-integer spin (e.g. electrons, protons, neutrons) obey the Fermi–Dirac statistics.

It is easily shown that two Fermi particles cannot occupy the same position in space and also have the same spin. Consider a wave-function describing a system of Fermi particles. If two of the particles are interchanged the wave-function must change sign. However if these two particles have the same set of four co-ordinates (space and spin) then the wave-function is clearly invariant.

This implies that the wave-function vanishes when two particles have the same set of co-ordinates indicating that this state does not exist. This is one form of the Pauli exclusion principle.

If the relativistic spin dependent terms in the Hamiltonian are ignored the wave-function can be written as a product of a function depending on the space co-ordinates multiplied by a function of the spin co-ordinates. Consider the case of two Fermi particles. The product wave-function is

$$\psi(\mathbf{r}_1, \sigma_{1z}; \mathbf{r}_2, \sigma_{2z}) = \phi(\mathbf{r}_1, \mathbf{r}_2)\chi(\sigma_{1z}, \sigma_{2z}). \tag{11.18}$$

As the complete wave-function must be antisymmetric in the co-ordinates (including spin) then $\phi(\mathbf{r}_1, \mathbf{r}_2)$ must be either symmetric in the space co-ordinates, in which case $\chi(\sigma_{1z}, \sigma_{2z})$ is antisymmetric in the spin co-ordinates or vice-versa. Then the wave-function (11.18) is a simultaneous eigenfunction of three commuting operators, the Hamiltonian, the particle interchange operator and the operator that interchanges the spin co-ordinates only. The system spin operators $S_{\text{op}}^2$ and $S_z$ commute with the three mentioned above and so the wave-function can also be chosen to be a simultaneous eigenfunction of all five operators. These wave-functions are the correct zero order functions for a perturbation treatment regarding the spin terms as a perturbation.

*Non-interacting particles*

Consider a system composed of $n$ identical particles whose mutual interaction can be ignored. The Hamiltonian $\mathscr{H}$ of the system is the sum of $n$ equal Hamiltonians for the individual particles, i.e.

$$\mathscr{H} \equiv \sum_{i=1}^{n} \mathscr{H}_i$$

with

$$\mathscr{H}_i \equiv -\frac{\hbar^2}{2m} \nabla_i^2 + V(\mathbf{r}_i, \sigma_{iz})$$

where $\nabla_i^2$ operates on the position co-ordinates of the $i$th particle at $\mathbf{r}_i$. The potential $V$ is the same for each particle.

As the individual particle Hamiltonians $\mathscr{H}_i$ all commute with each other and with $\mathscr{H}$, wave-functions can be found that are simultaneously eigenfunctions of $\mathscr{H}$ and all the $\mathscr{H}_i$. These are the variable separable eigenfunctions

$$\psi(\mathbf{r}_1, \sigma_{1z}; \mathbf{r}_2, \sigma_{2z}; \ldots; \mathbf{r}_n, \sigma_{nz}) = \phi_{Q_1}(\mathbf{r}_1, \sigma_{1z}) \ldots \phi_{Q_n}(\mathbf{r}_n, \sigma_{nz}) \tag{11.19}$$

where

$$\mathscr{H}_1 \phi_{Q_i}(\mathbf{r}_1, \sigma_{1z}) = E_{Q_i} \phi_{Q_i}(\mathbf{r}_1, \sigma_{1z})$$

and

$$\mathscr{H}\psi = E\psi \qquad \text{with} \qquad E = \sum_{i=1}^{n} E_{Q_i}.$$

The subscripts $Q_1, \ldots, Q_n$ indicate the completely defined (including spin orientation) single particle energy states occupied. If the co-ordinates of any two particles are interchanged the wave-function (11.19) remains an eigenfunction of the system Hamiltonian with the same energy. As there are $n!$ permutations of $n$ objects then $n!$ such eigenfunctions can be generated. Note that none of these eigenfunctions is in general symmetric or antisymmetric with respect to particle interchanges. However the sum of all $n!$ functions is clearly symmetric. If the particles are electrons (i.e. Fermi particles) the many electron wave-function must be antisymmetric. Such a wave-function is obtained by forming the sum of all wave-functions obtained from (11.19) by an even number of particle interchanges and substracting the sum of all wave-functions obtained by an odd number of interchanges. This antisymmetric wave-function can be expressed as the determinant

$$\psi(\mathbf{r}_1, \sigma_{1z}; \ldots; \mathbf{r}_n, \sigma_{nz}) = \frac{1}{\sqrt{n!}} \begin{vmatrix} \phi_{Q_1}(\mathbf{r}_1, \sigma_{1z}) & \phi_{Q_1}(\mathbf{r}_2, \sigma_{2z}) \ldots \phi_{Q_1}(\mathbf{r}_n, \sigma_{nz}) \\ \phi_{Q_2}(\mathbf{r}_1, \sigma_{1z}) & \phi_{Q_2}(\mathbf{r}_2, \sigma_{2z}) \ldots \phi_{Q_2}(\mathbf{r}_n, \sigma_{nz}) \\ \vdots & \\ \phi_{Q_n}(\mathbf{r}_1, \sigma_{1z}) \ldots & \phi_{Q_n}(\mathbf{r}_n, \sigma_{nz}) \end{vmatrix}. \tag{11.20}$$

If the single particle wave-functions are normalized (and orthogonal) then so is $\psi$. Clearly if the co-ordinates of any two particles are interchanged the determinantal wave-function (11.20) changes spin. (Note a determinant changes sign when two rows or two columns are interchanged.)

Note that if two of the occupied one-electron states are the same the determinant (11.20) has two rows equal and so vanishes. The system Hamiltonian has no solution for which the completely defined one electron states are occupied by more than one electron. This is of course the Pauli exclusion principle for non-interacting particles and helps to explain the periodic system of the elements. At most there can be two electrons in each hydrogen type atomic orbit and they must have opposite spins.

*The Helium atom*

In section 7.2 the ground state of the Helium atom was discussed using the variational method. The ground state and lowest excited states will now be investigated using first-order perturbation theory. The Hamiltonian operator for the Helium atom, ignoring spin dependent terms and the translational energy, is

$$\mathscr{H} \equiv -\frac{\hbar^2}{2m}(\nabla_1^2 + \nabla_2^2) - \frac{2e^2}{4\pi\epsilon_0 r_1} - \frac{2e^2}{4\pi\epsilon_0 r_2} + \frac{e^2}{4\pi\epsilon_0 |\mathbf{r}_1 - \mathbf{r}_2|} \tag{7.17}$$

where the two electrons are at $\mathbf{r}_1$ and $\mathbf{r}_2$ respectively. The spin momenta operators of each particle and of the system all commute with $\mathscr{H}$ and the total spin quantum number can only take the values $S = 0$ or $S = 1$ corresponding to antiparallel and parallel spins respectively (section 11.2). It will be shown below that even when spin dependent terms in the Hamiltonian are ignored, the allowed energies of the system do depend upon the total spin quantum number. This 'exchange' effect arises because of the use of a determinantal wave-function.

If the electron–electron term in (7.17) is ignored the exact ground state occurs when the two electrons with antiparallel spins occupy the hydrogen type $(1s)$ state (Pauli principle). This is the atomic configuration $(1s)^2$ and the wave-function is

$$\begin{aligned}\psi(\mathbf{r}_1, \sigma_{1z}; \mathbf{r}_2, \sigma_{2z}) &= \frac{1}{\sqrt{2!}}\begin{vmatrix} \phi_{1s}(\mathbf{r}_1)\alpha(\sigma_{1z}) & \phi_{1s}(\mathbf{r}_2)\alpha(\sigma_{2z}) \\ \phi_{1s}(\mathbf{r}_1)\beta(\sigma_{1z}) & \phi_{1s}(\mathbf{r}_2)\beta(\sigma_{2z}) \end{vmatrix} \\ &= \phi_{1s}(\mathbf{r}_1)\phi_{1s}(\mathbf{r}_2)\frac{1}{\sqrt{2}}[\alpha(\sigma_{1z})\beta(\sigma_{2z}) \\ &\quad - \beta(\sigma_{1z})\alpha(\sigma_{2z})] \end{aligned} \tag{11.21}$$

where $\phi_{1s}(\mathbf{r})$ is the hydrogen type wave-function

$$\phi_{1s}(\mathbf{r}) = \frac{1}{\sqrt{\pi}}\left(\frac{2}{a_0}\right)^{3/2} e^{-2r/a_0}.$$

The spatial part of the determinantal wave-function is symmetric whereas the spin part is antisymmetric. The operators $S_{\text{op}}^2$ and $S_z$ have zero eigenvalues as the electrons have antiparallel spins and the orbital angular momentum number is $L = 0$. (Note, the operator representing the orbital angular momentum of the system commutes with the Hamiltonian (7.17).) All states with a given $L$ and $S$ form a 'term'. The wave-function (11.21) represents the 'singlet' term $^1S$. The $S$ being the spectroscopic notation for a state with $L = 0$ and the superscript represents the spin degeneracy $(2S + 1)$ of the state. (It is hoped the reader will not confuse the two distinct meanings given to $S$.)

The nondegenerate determinantal wave-function (11.21) can be used to give the energy of the ground state of the Hamiltonian (7.17) correct to first order, with the electron–electron interaction term regarded as a perturbation, i.e.

$$E = 2E(1s) + \iint \phi_{1s}^*(\mathbf{r}_1)\phi_{1s}^*(\mathbf{r}_2)\frac{e^2}{4\pi\epsilon_0|\mathbf{r}_1 - \mathbf{r}_2|}\phi_{1s}(\mathbf{r}_1)\phi_{1s}(\mathbf{r}_2)\,\mathrm{d}\tau_1\,\mathrm{d}\tau_2$$

where

$$E(1s) = \int \phi_{1s}^*(\mathbf{r}_1)\left[-\frac{\hbar^2}{2m}\nabla_1^2 - \frac{2e^2}{4\pi\epsilon_0 r_1}\right]\phi_{1s}(\mathbf{r}_1)\,\mathrm{d}\tau_1$$

is the energy of a single electron in the $(1s)$ state assuming no other electrons present. The first terms represents the kinetic energy of the two electrons together with their potential energy in the field of the nucleus. The second term represents the electrostatic interaction between the electrons. The value of the ground state energy obtained in this way is $E = -2{\cdot}75e^2/4\pi\epsilon_0 a_0$ and is not as good as that obtained in section 7.2 using the variational method.

The more important excited configurations occur when one of the electrons is excited into a hydrogen type level with $n = 2$ to produce the configurations $(1s)(2s)$ or $(1s)(2p)$. It is not difficult to show that matrix elements between these two types of excited configurations vanish so that the $(1s)(2p)$ states can be dealt with separately. Consider the $(1s)(2s)$ configuration. There are four possible determinantal wave-functions depending upon the spin orientations of the two electrons. By suitably combining two of these determinantal functions, the four possible antisymmetric wave-functions can each be expressed as a product of a spatial function multiplied by a spin function and can be divided into two classes depending upon whether the spin part is symmetric or antisymmetric. These are the correct zeroth-order wave-functions for a perturbation treatment of the spin-orbit coupling.

One of the wave-functions has an antisymmetric spin part (and hence symmetric spatial part) and corresponds to the singlet term $^1S$ with $L = 0$ and $S = 0$.

$$\psi_1 = \frac{1}{\sqrt{2}}[\phi_{1s}(\mathbf{r}_1)\phi_{2s}(\mathbf{r}_2) + \phi_{2s}(\mathbf{r}_1)\phi_{1s}(\mathbf{r}_2)] \times \frac{1}{\sqrt{2}}[\alpha(\sigma_{1z})\beta(\sigma_{2z}) - \beta(\sigma_{1z})\alpha(\sigma_{2z})].$$

The remaining three wave-functions have a symmetric spin part and correspond to the triplet term $^3S$ with $L = 0$ and $S = 1$.

$$\psi_2 = \frac{1}{\sqrt{2}} [\phi_{1s}(\mathbf{r}_1)\phi_{2s}(\mathbf{r}_2) - \phi_{2s}(\mathbf{r}_1)\phi_{1s}(\mathbf{r}_2)]$$

$$\times \frac{1}{\sqrt{2}} [\alpha(\sigma_{1z})\beta(\sigma_{2z}) + \beta(\sigma_{1z})\alpha(\sigma_{2z})]$$

$$\psi_3 = \frac{1}{\sqrt{2}} [\phi_{1s}(\mathbf{r}_1)\phi_{2s}(\mathbf{r}_2) - \phi_{2s}(\mathbf{r}_1)\phi_{1s}(\mathbf{r}_2)]\,\alpha(\sigma_{1z})\alpha(\sigma_{2z})$$

$$\psi_4 = \frac{1}{\sqrt{2}} [\phi_{1s}(\mathbf{r}_1)\phi_{2s}(\mathbf{r}_2) - \phi_{2s}(\mathbf{r}_1)\phi_{1s}(\mathbf{r}_2)]\,\beta(\sigma_{1z})\beta(\sigma_{2z}).$$

The last three wave-functions all have the same antisymmetric spatial part, and vanish if the electrons occupy the same point in space. The electrons tend to keep apart in this case and the positive contribution to the energy arising from the repulsive electron–electron interaction term will be smaller than that arising from the symmetric spatial part of the term $^1S$. So, it is to be expected that the perturbed $^3S$ triplet term will lie deeper than the $^1S$ term. This will be shown to be the case.

When the electron–electron interaction term is ignored all four states (singlet and triplet) are degenerate. Although the unperturbed states are degenerate, the perturbation matrix is already diagonal with the above choice for the excited wave-functions. To the first order, the perturbed energy of the excited singlet state $^1S$ is the expectation value of (7.17) using $\psi_1$ and is

$$E(^1S) = E(1s) + E(2s) + J + K \tag{11.22}$$

where $E(1s)$ and $E(2s)$ are the energies of a single electron in the $(1s)$ and $(2s)$ states respectively and

$$J = \iint \phi_{1s}^*(\mathbf{r}_1)\phi_{2s}^*(\mathbf{r}_2) \frac{e^2}{4\pi\epsilon_0 |\mathbf{r}_i - \mathbf{r}_j|} \phi_{1s}(\mathbf{r}_1)\phi_{2s}(\mathbf{r}_2)\, d\tau_1\, d\tau_2$$

$$K = \iint \phi_{1s}^*(\mathbf{r}_1)\phi_{2s}^*(\mathbf{r}_2) \frac{e^2}{4\pi\epsilon_0 |\mathbf{r}_i - \mathbf{r}_j|} \phi_{1s}(\mathbf{r}_2)\phi_{2s}(\mathbf{r}_1)\, d\tau_1\, d\tau_2.$$

$J$ is called the Coulomb integral and represents the average Coulomb energy of the two electrons in the $(1s)$ and $(2s)$ states respectively. $K$ is called the exchange integral and is a direct consequence of using a determinantal wave-function instead of a simple product. Unlike $J$ it does not have an electrostatic interaction explanation and arises as identical particles cannot be distinguished.

The perturbed energy of the excited triplet term $^3S$ is

$$E(^3S) = E(1s) + E(2s) + J - K. \tag{11.23}$$

Note that the triplet degeneracy of this term is not removed by the electron–electron interaction. The difference between the $^1S$ and $^3S$ terms is $2K$ and as it can be shown that $K > 0$, the triplet terms lies deeper than the singlet term as expected.

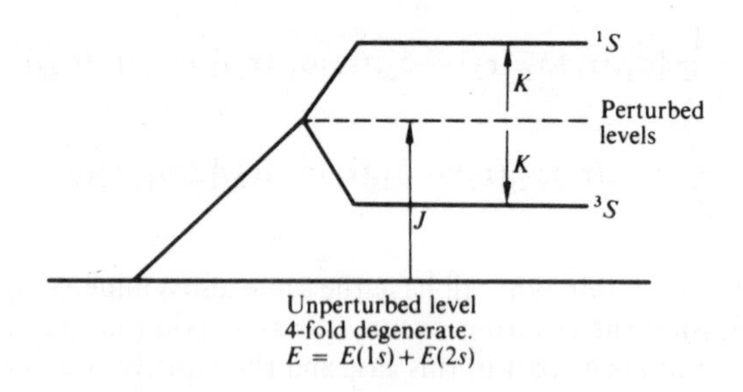


Figure 11.1 Displacement and splitting of the configuration $(1s)(2s)$ by the electron–electron interaction.

Actually this first-order perturbation treatment is not very good as the electron–electron interaction term is too large. However, from symmetry considerations, the quantum numbers $L$ and $S$ are still good in an exact treatment and the $^3S$ term is still triply degenerate. When the spin-orbit coupling term is introduced, the Hamiltonian is no longer invariant under a rotation of the space or spin co-ordinates separately and $L$ and $S$ are no longer good quantum numbers. However, it is invariant under a simultaneous rotation of all co-ordinates and the states can be classified by the total angular momentum quantum number $J$.

*The hydrogen molecule*

Even though this is the simplest molecule involving more than one electron no exact solution exists. The Hamiltonian for the molecule, ignoring spin dependent terms and taking the nuclei to be fixed, is

$$\mathscr{H} \equiv -\frac{\hbar^2}{2m}(\nabla_1^2 + \nabla_2^2) - \frac{1}{4\pi\epsilon_0}\left(\frac{e^2}{r_{1a}} + \frac{e^2}{r_{1b}} + \frac{e^2}{r_{2a}} + \frac{e^2}{r_{2b}}\right)$$

$$+ \frac{e^2}{4\pi\epsilon_0 r_{12}} + \frac{e^2}{4\pi\epsilon_0 R}$$

where $r_{1a}, r_{1b}$ are the distances of the first electron from the protons which are at $a$ and $b$ and similarly $r_{2a}, r_{2b}$ are the distances of the second electron from the protons. The electron separation is $r_{12}$ and the protons are a distance $R$ apart.

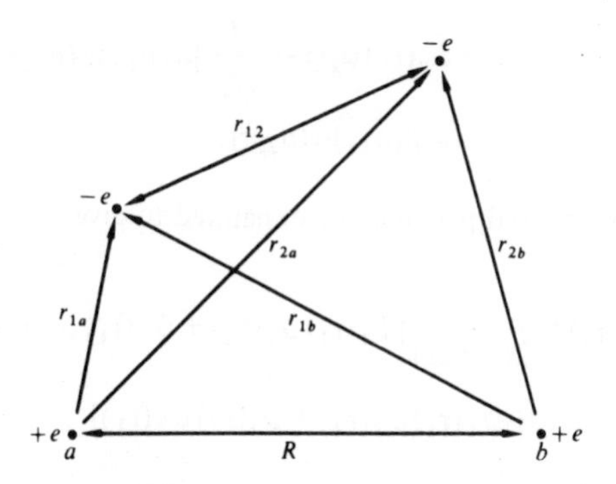


Figure 11.2 The hydrogen molecule.

In Chapter 7 the molecular orbital method was applied to the hydrogen molecular ion and it was shown that the ground state orbital is the gerade, g state

$$\psi_g = \frac{(\psi_1 + \psi_2)}{\sqrt{2(1 + \Delta)}}$$

where $\psi_1, \psi_2$ are the normalized (1*s*) wave-functions centred on the two protons and $\Delta$ is the overlap integral. This represents a bonding state in which electron charge is transferred to the bond. This molecular orbital corresponds to a binding energy for $H_2^+$ of ~1·8 eV compared with the observed binding energy of 2·8 eV. (Binding energy of a molecule = ground state energy of isolated atoms − ground state energy of molecule.)

Clearly this simple model is not very good although it does illustrate the bond concept. Electron charge is transferred to the space between the atoms and holds them together. Confining an electron in a bond will increase its kinetic energy and binding will only result if the decrease in potential energy is greater than the increase in kinetic energy.

The hydrogen ion bond orbital can take up another electron with opposite spin and the energy of the electron pair in the bonding orbital will be lower than that of the two isolated hydrogen atoms and binding energy will result. In this crude model, the two-electron wave-function for the hydrogen molecule is (c.f. (11.21))

$$\psi_g(\mathbf{r}_1, \sigma_{1z}; \mathbf{r}_2, \sigma_{2z}) = \frac{1}{\sqrt{2}} \begin{vmatrix} \psi_g(\mathbf{r}_1)\alpha(\sigma_{1z}) & \psi_g(\mathbf{r}_2)\alpha(\sigma_{2z}) \\ \psi_g(\mathbf{r}_1)\beta(\sigma_{1z}) & \psi_g(\mathbf{r}_2)\beta(\sigma_{2z}) \end{vmatrix}$$

$$= \psi_g(\mathbf{r}_1)\psi_g(\mathbf{r}_2) \frac{1}{\sqrt{2}} [\alpha(\sigma_{1z})\beta(\sigma_{2z}) - \beta(\sigma_{1z})\alpha(\sigma_{2z})].$$

The symmetric spatial part may be expanded to give

$$\psi_g(\mathbf{r}_1)\psi_g(\mathbf{r}_2) = \frac{1}{2(1+\Delta)} [\psi_1(\mathbf{r}_1)\psi_1(\mathbf{r}_2) + \psi_2(\mathbf{r}_1)\psi_2(\mathbf{r}_2) + \psi_1(\mathbf{r}_1)\psi_2(\mathbf{r}_2) + \psi_1(\mathbf{r}_2)\psi_2(\mathbf{r}_1)].$$

This wave-function gives equal weight to the ionic states in which the two electrons are on the same atom and the remaining states in which each atom has one electron. Electrons will repel each other and this suggests that the ionic states should not be as important as this. This correlation may be taken into account by configuration interaction in which the two-electron wave-function is expressed as a linear combination of determinants.

The lowest determinant that will mix in is that which is obtained by exciting both electrons to the antibonding orbital $\psi_u(\mathbf{r})$. The spin part of the two-electron wave-function $\psi_u(\mathbf{r}_1, \sigma_{1z}; \mathbf{r}_2, \sigma_{2z})$ is antisymmetric as above and the spatial part is

$$\psi_u(\mathbf{r}_1)\psi_u(\mathbf{r}_2) = \frac{1}{2(1-\Delta)} [\psi_1(\mathbf{r}_1)\psi_1(\mathbf{r}_2) + \psi_2(\mathbf{r}_1)\psi_2(\mathbf{r}_2) - \psi_1(\mathbf{r}_1)\psi_2(\mathbf{r}_2) - \psi_1(\mathbf{r}_2)\psi_2(\mathbf{r}_1)].$$

By taking a suitable combination of the two determinantal wave-functions the ionic terms may be reduced or eliminated. The energy may be is minimized in this way and the resulting combination is such that ionic terms are relatively unimportant.

It would appear that a better first wave-function would be to take the electrons on different atoms to begin with. This is the basis of the Heitler-London approach. There are two possible functions to consider.

(i) There is the singlet state with symmetric spatial part ($S = 0$)

$$\frac{1}{\sqrt{2(1+\Delta^2)}} [\psi_1(\mathbf{r}_1)\psi_2(\mathbf{r}_2) + \psi_2(\mathbf{r}_1)\psi_1(\mathbf{r}_2)] \qquad {}^1\Sigma$$

(ii) and also the triplet state with antisymmetric spatial part ($S = 1$)

$$\frac{1}{\sqrt{2(1-\Delta^2)}}[\psi_1(\mathbf{r}_1)\psi_2(\mathbf{r}_2) - \psi_2(\mathbf{r}_1)\psi_1(\mathbf{r}_2)]. \qquad {}^3\Sigma$$

(The $\Sigma$ denotes zero orbital angular momentum about the molecular axis.)

Detailed analysis indicates (problem 8) that the singlet state lies deepest and so represents the ground state. This is the opposite situation to that occurring in helium. However it is to be expected that the strongest binding for hydrogen will occur when the electrons tend to accumulate between the protons. This will occur when the electrons can occupy the same space state and hence have antiparallel spins.

Some electron correlation is already built into this method and the energy can be minimized by varying parameters in $\psi_1$ and $\psi_2$. This method is obviously correct in the limit of infinite separation of the atoms.

The Heitler–London wave-function and the best two-determinant molecular wave-functions are of equivalent accuracy and both involve an error of about 0·5 eV in the calculated binding energy. The observed binding energy of the hydrogen molecule is 4·7 eV. For more details the reader is referred to the article by Coulson and Lewis in *Quantum Theory*, Vol. II, edited by Bates.

### *Rotational spectra of homonuclear molecules*

The previous discussions have concerned the identity of electrons. A homonuclear diatomic molecule (e.g. $H_2$) contains another pair of identical particles these being the two nuclei. The complete wave-function describing the system of electrons and nuclei must be either symmetric or antisymmetric under an interchange of nuclei depending on whether they have integral nuclear spin or half-integral nuclear spin respectively. The proton is a Fermi particle and so the hydrogen molecular wave-function, including nuclear terms, must be antisymmetric under an interchange of the nuclei. The nucleus in $({}^{16}_{8}O)_2$ has zero spin and so the complete wave-function must be symmetric under an interchange of nuclei.

Using the Born–Oppenheimer approximation the molecular wave-function may be written

$$\psi = \psi(\text{el})\psi(\text{nuc})$$

where $\psi(\text{el})$ describes the electron motion relative to fixed nuclei (this is the molecular wave-function discussed in the previous section), and $\psi(\text{nuc})$ describes the nuclear motion.

Further, the nuclear wave-function may be written (ignoring nuclear-spin-molecular coupling)

$$\psi(\text{nuc}) = \psi(\text{vib})\psi(\text{rot})\psi(\text{sp})$$

where $\psi$(vib) describes the vibrational motion of the nuclei, $\psi$(rot) describes the rotational motion and $\psi$(sp) depends on the nuclear spin. The vibrational function depends only upon the separation of the nuclei (4.30) and is unchanged by interchanging the nuclei. The rotational function can be written (Chapter 4) (c.f. hydrogen atom functions)

$$\psi(\text{rot}) = P_K^m(\cos\theta)e^{im\phi} \qquad K = 0, 1, 2, \ldots.$$

Interchanging the nuclei corresponds to

$$\theta \to \pi - \theta, \qquad \phi \to \pi + \phi$$

and from the discussion of parity earlier in this chapter it is clear that $\psi$(rot) is symmetric (anti-symmetric) if $K$ is even (odd).

Consider the hydrogen molecule. The protons are Fermi particles with spin $\frac{1}{2}$ and the complete wave-function must be antisymmetric with respect to interchange of the protons. The analysis given earlier for the electron spin functions of helium applies and there is one anti-symmetric nuclear spin function ($S = 0$) and three symmetric spin functions ($S = 1$). The normal ground state of $\psi$(el) is symmetric with respect to proton interchange and because of the nuclear spin degeneracy there are three times as many odd rotational states as even. At moderate and high temperatures in thermal equilibrium three times as many molecules will be in odd as in even rotational states.

If coupling terms between the nuclear spin and the molecule are ignored then nuclear spin is a constant of the motion. In fact these coupling terms are extremely weak and so a molecule with anti-symmetric spin is very unlikely to undergo a transition to a state with symmetric spin. The hydrogen molecule may be regarded as composed of two independent parts, one with zero nuclear spin (para-hydrogen) and the other with spin one (ortho-hydrogen). At normal temperatures hydrogen is one-quarter para and three-quarters ortho-hydrogen. Hence in a transition between two electron states the rotational fine structure of hydrogen shows a 3 : 1 alternation of intensities. The selection rule $SK = \pm 1$ (applicable for $\Sigma$ electron states) prevents any transitions between electronic states with the same symmetry and explains the general lack of infra-red absorption spectra for homonuclear diatomic molecules.

In general, for a homonuclear diatomic molecule, where nuclei have spin $I$, there are $(I + 1)(2I + 1)$ symmetric spin states and $I(2I + 1)$ anti-symmetric spin states. The intensities of alternate lines in the rotational fine structure will then vary as $(I + 1)$: $I$. Measurements of rotational intensities have been used to determine nuclear spins.

For a molecule with zero nuclear spin, the nuclei are bosons and the total wave-function must be symmetric under an interchange of nuclei. The nuclear spin part of the wave-function is a constant and so if $\psi$(el) is symmetric (antisymmetric) the only allowed states are those with $K$ even (odd). In transitions between two electronic states of different symmetry, alternate lines in the rotational fine structure will be missing because of the selection rule for $K$.

If the two nuclei composing the molecule are different isotopes of the same element then they are no longer identical and the above discussion does not apply.

## 11.4 Atomic Energy Levels

It is very difficult to solve problems involving more than one electron and complicated numerical procedures are necessary even in the simplest case. The difficulties arise because the electron–electron interaction contains terms of the form $e^2/4\pi\epsilon_0|\mathbf{r}_i - \mathbf{r}_j|$ involving the spatial co-ordinates of two electrons. However there are approximate methods for reducing the many electron problem to a one-electron self-consistent field problem.

### *Hartree's self-consistent field*

Ignoring the translational energy and the spin-orbit interaction terms, the Hamiltonian operator for a free atom of atomic number $Z$ with $n$ electron is

$$\mathscr{H} \equiv \sum_{i=1}^{n}\left[-\frac{\hbar^2}{2m}\nabla_i^2 - \frac{Ze^2}{4\pi\epsilon_0 r_i}\right] + \frac{1}{2}\sum_{\substack{i=1 \\ (i \neq j)}}^{n}\sum_{j=1}^{n}\frac{e^2}{4\pi\epsilon_0|\mathbf{r}_i - \mathbf{r}_j|}. \tag{11.2}$$

The first term represents the kinetic energy of the electrons and the second term represents the attraction between the nucleus and the electrons. The third represents the electron–electron interaction, the summation being over all pairs of electrons.

The Hartree equations may be obtained by assuming a product wave-function of the form

$$\psi(\mathbf{r}_1, \mathbf{r}_2, \ldots, \mathbf{r}_n) = \phi_{Q_1}(\mathbf{r}_1)\phi_{Q_2}(\mathbf{r}_2)\ldots\phi_{Q_n}(\mathbf{r}_n). \tag{11.24}$$

It is essential that the one-electron functions be normalizable. Of course (11.24) is not antisymmetric with respect to interchange of any two electrons but the single electron states are chosen in agreement with the Pauli exclusion principle in that each completely defined occupied state occurs once only.

The best approximation to the ground state by a solution of the form (11.24) is obtained by minimizing the expectation value of the Hamiltonian. So, the best Hartree product solution is that which minimizes the integral

$$\int \ldots \int \psi^*(\mathbf{r}_1, \mathbf{r}_2, \ldots, \mathbf{r}_n)\mathscr{H}\psi(\mathbf{r}_1, \mathbf{r}_2, \ldots, \mathbf{r}_n)\, \mathrm{d}\tau_1 \mathrm{d}\tau_2 \ldots \mathrm{d}\tau_n \tag{11.25}$$

subject to the restrictive conditions

$$\int \phi^*_{Q_i}(\mathbf{r}_1)\phi_{Q_i}(\mathbf{r}_1)\, \mathrm{d}\tau_1 = 1 \qquad i = 1, 2, \ldots, n. \tag{11.26}$$

The method of Lagrange multipliers may be used. Consider the function

$$I = \int \ldots \int \phi^*_{Q_1}(\mathbf{r}_1) \ldots \phi^*_{Q_n}(\mathbf{r}_n)$$

$$\times \left[ \sum_{i=1}^{n} h_i + \frac{1}{2} \sum^{n}\sum^{n}_{i \neq j} \frac{e^2}{4\pi\epsilon_0 |\mathbf{r}_i - \mathbf{r}_j|} \right] \phi_{Q_1}(\mathbf{r}_1) \ldots \phi_{Q_n}(\mathbf{r}_n)\, \mathrm{d}\tau_1 \ldots \mathrm{d}\tau_n$$

$$+ \sum_{j=1}^{n} \lambda_j \left[ \int \phi^*_{Q_j}(\mathbf{r}_1)\phi_{Q_j}(\mathbf{r}_1)\, \mathrm{d}\tau_1 - 1 \right]$$

where

$$h_i \equiv -\frac{\hbar^2}{2m}\nabla_i^2 - \frac{Ze^2}{4\pi\epsilon_0 r_i},$$

and $\lambda_j, j = 1, \ldots, n$ are the Lagrange multipliers.

Suppose each of the one-electron functions $\phi_{Q_i}(\mathbf{r}_i)$ undergoes a small change $\delta\phi_{Q_i}(\mathbf{r}_i)$. For a stationary value of the integral (11.25) subject to the conditions (11.26) it is necessary that

$$\delta I = 0$$

i.e. $$\sum_{k=1}^{n} \int \ldots \int \phi^*_{Q_1}(\mathbf{r}_1) \ldots \phi^*_{Q_k}(\mathbf{r}_k) \ldots \phi^*_{Q_n}(\mathbf{r}_n)$$

$$\times \left[ \sum_{i=1}^{n} h_i + \frac{1}{2} \sum^{n}\sum^{n}_{i \neq j} \frac{e^2}{4\pi\epsilon_0 |\mathbf{r}_i - \mathbf{r}_j|} \right] \phi_{Q_1}(\mathbf{r}_1) \ldots \delta\phi_{Q_k}(\mathbf{r}_k) \ldots$$

$$\phi_{Q_n}(\mathbf{r}_n)\, \mathrm{d}\tau_1 \ldots \mathrm{d}\tau_n + \sum_{k=1}^{n} \int \ldots \int \phi^*_{Q_i}(\mathbf{r}_1) \ldots \delta\phi^*_{Q_k}(\mathbf{r}_k) \ldots$$

$$\phi^*_{Q_n}(\mathbf{r}_n) \left[ \sum_{i=1}^{n} h_i + \frac{1}{2} \sum^{n}\sum^{n}_{i \neq j} \frac{e^2}{4\pi\epsilon_0 |\mathbf{r}_i - \mathbf{r}_j|} \right] \phi_{Q_1}(\mathbf{r}_1) \ldots$$

$$\phi_{Q_k}(\mathbf{r}_k)\ldots\phi_{Q_n}(\mathbf{r}_n)\,\mathrm{d}\tau_1\ldots\mathrm{d}\tau_n+\sum_{k=1}^{n}\lambda_j\left[\int\phi^*_{Q_k}(\mathbf{r}_k)\delta\phi_{Q_k}(\mathbf{r}_k)\,\mathrm{d}\tau_k\right.$$
$$\left.+\int\delta\phi^*_{Q_k}(\mathbf{r}_k)\phi_{Q_k}(\mathbf{r}_k)\,\mathrm{d}\tau_k\right]=0$$

i.e. $$\sum_{k=1}^{n}\int\delta\phi^*_{Q_k}(\mathbf{r}_k)\left[\sum_{i\neq k}\int\phi^*_{Q_i}(\mathbf{r}_i)\,h_i\phi_{Q_i}(\mathbf{r}_i)\,\mathrm{d}\tau_i\right.$$
$$+\frac{1}{2}\sum_{\substack{i\neq j\\ i,j\neq k}}^{n}\sum^{n}\frac{e^2}{4\pi\epsilon_0}\iint\frac{|\phi_{Q_i}(\mathbf{r}_i)|^2\times|\phi_{Q_j}(\mathbf{r}_j)|^2}{|\mathbf{r}_i-\mathbf{r}_j|}\,\mathrm{d}\tau_i\,\mathrm{d}\tau_j+h_k$$
$$\left.+\sum_{j\neq k}\frac{e^2}{4\pi\epsilon_0}\int\frac{|\phi_{Q_j}(\mathbf{r}_j)|^2}{|\mathbf{r}_k-\mathbf{r}_j|}\,\mathrm{d}\tau_j+\lambda_k\right]\phi_{Q_k}(\mathbf{r}_k)\,\mathrm{d}\tau_k$$
$$+\text{(a corresponding expression in }\delta\phi_{Q_k}(\mathbf{r}_k))=0. \tag{11.27}$$

$\delta\phi^*_{Q_k}$ and $\delta\phi_{Q_k}$ may be regarded as independent variations and the coefficients of these in (11.27) must be zero. The coefficient of $\delta\phi^*_{Q_k}$ is

$$h_k\phi_{Q_k}(\mathbf{r}_k)+\left[\sum_{j\neq k}\frac{e^2}{4\pi\epsilon_0}\int\frac{|\phi_{Q_j}(\mathbf{r}_j)|^2}{|\mathbf{r}_k-\mathbf{r}_j|}\,\mathrm{d}\tau_j\right]\phi_{Q_k}(\mathbf{r}_k)-\epsilon_k\phi_{Q_k}(\mathbf{r}_k)$$
$$=0\qquad k=1,2,\ldots,n \tag{11.28}$$

where

$$\epsilon_k=-\sum_{i\neq k}\int\phi^*_{Q_i}(\mathbf{r}_i)\,h_i\phi_{Q_i}(\mathbf{r}_i)\,\mathrm{d}\tau_i$$
$$-\frac{1}{2}\sum_{\substack{i\neq j\\ i,j\neq k}}\sum\frac{e^2}{4\pi\epsilon_0}\int\frac{|\phi_{Q_i}(\mathbf{r}_i)|^2|\phi_{Q_j}(\mathbf{r}_j)|^2}{|\mathbf{r}_i-\mathbf{r}_j|}\,\mathrm{d}\tau_i\,\mathrm{d}\tau_j-\lambda_k. \tag{11.29}$$

Note that $\epsilon_k$ is real and that the coefficient of $\delta\phi_{Q_k}$ is simply the complex conjugate of (11.28). (Note that the operators $h_i$ are Hermitian.)

In conclusion, the best Hartree product, of the type (11.24), is composed of normalized one-electron functions that satisfy the $n$ Hartree equations (11.28) with $k=1,2,\ldots,n$. The one-electron function $\phi_{Q_k}(\mathbf{r}_1)$ is an eigenfunction of the Hamiltonian

$$\mathscr{H}_k\equiv-\frac{\hbar^2}{2m}\nabla_1^2-\frac{Ze^2}{4\pi\epsilon_0 r_1}+\sum_{j\neq k}\frac{e^2}{4\pi\epsilon_0}\int\frac{|\phi_{Q_j}(\mathbf{r}_2)|^2}{|\mathbf{r}_1-\mathbf{r}_2|}\,\mathrm{d}\tau_2 \tag{11.30}$$

with eigenvalue $\epsilon_k$. Each of these terms may be given a physical significance. The first term represents the kinetic energy of the $k$th state electron, the second terms represents the interaction between the electron and the nucleus and the final term represents the Coulomb energy of the electron in the average field produced by all the other electrons. The Hamiltonian is not the same for all electrons and the eigenfunctions are not orthogonal.

The set of $n$ non-linear integro-differential equations (11.28) is too difficult to solve directly. Hartree (1928) suggested a self-consistent approximation. An initial set of one-electron functions is assumed and used to obtain the Hamiltonians (11.30). The Hartree equations (11.28) are then solved numerically and a new set of one-electron functions obtained. This process is continued until the functions are self-consistent to the required order of accuracy. In practice the third term in the Hamiltonian (11.30) is approximated by a spherically symmetric term by averaging over all directions and the one-electron functions can then be characterized by $l$ and $m_l$ quantum numbers.

### *The Hartree–Fock self-consistent field*

The variational method used above was suggested independently by Fock and Slater in 1930 after Hartree obtained the equation (11.28) directly by physical arguments.

The simple product wave-function (11.24) is not antisymmetric with respect to interchange of electrons. The Hartree–Fock method starts by assuming the wave-function to be the antisymmetric determinant

$$\psi(\mathbf{r}_1, \sigma_{1z}; \ldots; \mathbf{r}_{nz}\sigma_{nz}) = \frac{1}{\sqrt{n!}} \begin{vmatrix} \phi_{Q_1}(\mathbf{r}_1, \sigma_{1z}) \ldots \phi_{Q_1}(\mathbf{r}_n, \sigma_{nz}) \\ \phi_{Q_2}(\mathbf{r}_1, \sigma_{1s}) \\ \vdots \\ \phi_{Q_n}(\mathbf{r}_1, \sigma_{1z}) \qquad \phi_{Q_n}(\mathbf{r}_n, \sigma_{nz}) \end{vmatrix}. \tag{11.31}$$

The expectation value of the Hamiltonian (11.2) obtained using the wave-function (11.31), is minimized subject to the condition that each one-electron function is normalized. These one-electron functions must be linearly independent so that it is possible to form an orthogonal set from them (Adams, 1961).

The set of equations which the spatial parts of the one-electron functions satisfy can be written in the 'standard form'

$$\mathscr{H}_k \phi_{Q_k}(\mathbf{r}_1) = \epsilon_k \phi_{Q_k}(\mathbf{r}_1) \qquad k = 1, 2, \ldots, n \tag{11.32a}$$

where the one-electron Hamiltonians are given by

$$\mathscr{H}_k \equiv -\frac{\hbar^2}{2m}\nabla_1^2 - \frac{Ze^2}{4\pi\epsilon_0 r_1} + \frac{e^2}{4\pi\epsilon_0}\sum_{j=1}^{n}\int \frac{|\phi_{Qj}(\mathbf{r}_2)|^2}{|\mathbf{r}_1 - \mathbf{r}_2|}\,d\tau_2$$

$$-\frac{e^2}{4\pi\epsilon_0}\sum_{j}^{\parallel}\int \frac{\phi_{Qj}^*(\mathbf{r}_2)\phi_{Qk}^*(\mathbf{r}_1)\phi_{Qk}(\mathbf{r}_2)\phi_{Qj}(\mathbf{r}_1)}{|\mathbf{r}_1 - \mathbf{r}_2|\,\phi_{Qk}^*(\mathbf{r}_1)\phi_{Qk}(\mathbf{r}_1)}\,d\tau_2. \tag{11.32b}$$

The first two terms are just the first two terms of Hartree's-Hamiltonian (11.30). The third term is summed over all the ground state wave-functions including the $k$th and represents the Coulomb energy of the $k$th state electron in the average field of all the electrons including itself. This Coulomb interaction of an electron with itself is clearly not acceptable and must be compensated somehow by the fourth term. This new 'exchange' term is summed over all wave-functions with spin parallel to the $k$th. In fact if only the element in this summation with $j = k$ is used, the Hartree-Fock and Hartree–Hamiltonians are identical.

Of course, the equations (11.32a) have an infinite number of solutions. There are the $n$ one-electron wave-functions which are occupied in the ground state and also an infinite number of solutions corresponding to unoccupied excited states. All the Hartree-Fock solutions are orthogonal to each other. This is not the case for the Hartree solutions.

The total energy of the system is the expectation value of the Hamiltonian (11.2) obtained using the determinantal wave-function (11.31) and is less than the sum of single particle energies $\Sigma\,\epsilon_k$ as this sum counts each electron–electron interaction twice. However the magnitude of the single particle energies are approximately equal to the ionization energies when only one particle is excited.

It is to be expected that as an electron moves it will repel other electrons and a 'coulomb hole' will be formed about it. There is correlation between the motion of the electrons. The Hartree-Fock wave-function does take some account of correlation between electrons with parallel spins but it ignores correlation between electrons of unlike spin. Of course correlation reduces the energy of the system by keeping electrons apart and the Hartree-Fock method is a better approximation than the Hartree method which completely ignores correlation. The difference between the energy of the system obtained using the Hartree and the Hartree-Fock methods is called the 'exchange energy'. The reader is warned not to give too much physical significance to this as it is simply the difference between two mathematical approximations. The Hartree-Fock energy $E_{\text{H.F.}}$ is an upper bound to the exact ground state energy $E$. The energy difference $(E - E_{\text{H.F.}})$ is called the 'correlation energy' and is a significant measure of the accuracy of the Hartree–Fock method. The correlation energies for atoms are of the order of a few electron volts.

In practice the Hartree–Fock equations are solved numerically by iteration until self-consistent solutions are obtained.

The infinite set of solutions of the Hartree–Fock equations forms a complete set of one-electron functions. The infinite set of $n \times n$ determinantal wave-functions constructed by taking different combinations of $n$ Hartree–Fock solutions form a complete set of $n$-electron antisymmetric wave-functions. The exact solutions of Schrödinger's many-electron equation may be expressed as linear combinations of these determinantal wave-functions. However in many cases a single determinant is fairly reasonable.

*The periodic table and the shell model*

Usually the coulomb and exchange terms in (11.32) are averaged over all directions to make them spherically symmetric and the one-electron functions are then characterized by $l$ and $m_l$ angular momentum quantum numbers as well as by a principle quantum number $n$. (The number of nodes, excluding the origin, of the radial part of the one-electron function is equal to $n - l - 1$.)

The one-electron states are denoted by

(1$s$) when $n = 1, \quad l = 0$

(2$s$) when $n = 2, \quad l = 0$

(2$p$) when $n = 2, \quad l = 1$

⋮

with the same notation used for hydrogen type states. In this general (not coulomb) central field approximation the energy of the $i$th electron depends upon both $n_i$ and $l_i$.

Suppose the Hartree–Fock equations (11.32) are equivalent to a variable separable, central field Hamiltonian for the system with the form

$$\mathscr{H}_0 \equiv \sum_{i=1}^{n} \left[ -\frac{\hbar^2}{2m} \nabla_i^2 + u_i(r_i) \right] \tag{11.33}$$

where $u_i(r_i)$ are suitable one-electron potentials. (By further approximating the exchange terms in (11.32) Slater (1951) has been able to show this.) Then the **determinantal** wave-function formed by solutions of the Hartree–Fock equations are eigenfunctions of the central field

Hamiltonian $\mathscr{H}_0$. This Hamiltonian is invariant under separate rotations of the co-ordinates of each particle and so commutes with the separate particle angular momentum operators. Its eigenvalues depend only upon the set of quantum numbers $(n_i, l_i), i = 1, 2, \ldots, n$ associated with the one-electron functions in the determinantal eigenfunctions. The energy is independent of the quantum numbers $m_{li}$ and $m_{si}$ as there are no spin-orbit interaction terms.

This central field 'atomic' energy level is called a configuration (sometimes 'configuration' refers to the set of $n$ associated one-electron functions) and its degeneracy is less than or equal to

$$2^n \prod_{i=1}^{n} (2l_i + 1) \tag{11.34}$$

as each one-electron energy has degeneracy $2(2l_i + 1)$. In general the Pauli principle prevents the degeneracy (11.34) from being attained and also limits the configurations allowed.

A central field single electron state is completely defined by the set of four quantum numbers $(n, l, m_l, m_s)$. Electrons are Fermi particles and the antisymmetry of the many-particle wave-function demands that only one electron can occupy a completely defined state. This is the Pauli exclusion principle. The combination of the Pauli principle with the Hartree–Fock method gives an explanation of the periodic system of elements.

In the ground state the one-electron atomic levels of lowest energy are filled up first, the Pauli principle preventing all electrons from occupying the lowest $(1s)$ state. The chemical properties are mainly determined by the 'valence' electrons in the shell (defined by $n, l$) of greatest energy. If the outermost shell is completely filled and there is a considerable energy gap to the next empty shell then it is to be expected that the element will be inert. The rare gases are typical examples.

Atomic hydrogen has the configuration $(1s)$ in the ground state. The next atom, helium, has both electrons in this shell and the configuration is denoted by $(1s)^2$. This is a closed shell and helium is inert. The next element, lithium, has the configuration $(1s)^2(2s)$ and has one electron outside a closed shell. Such elements with one electron in an $s$-shell (apart from hydrogen) are chemically similar and these are the alkalis. Atoms with a full $p$-shell (or $(1s)$-shell) are inert and are the rare gases.

The configuration of atoms with more and more electrons can be obtained in this way and imply periodicities in the chemical properties of the elements. The number of electrons in a completely filled shell depends only upon the $l$-value and is given by $2(2l + 1)$. For the $s, p, d$ states the number of electrons is 2, 6, 10, respectively.

Complications do arise because the one-electron energies depend upon both $n$ and $l$. For example, the shells $(4s)$ and $(3d)$ have similar energy and they may not be filled up in sequence. Another transition period occurs because of the similarity in energy of the shells $(5s)$ and $(4d)$.

The nuclear shell model is an attempt to explain nuclear properties in a similar manner. An atomic nucleus is composed of protons and neutrons both of which are Fermi particles with spin $\frac{1}{2}$ and with similar masses. These nucleons interact through a short range ($\sim 10^{-15}$ m) attractive nuclear force. (The protons of course also repel each other by the long range coulomb force.) The shell model assumes that each nucleon moves in an averaged field due to the other nucleons. For many nuclei, this field can be taken to be spherically symmetric and this means that each nucleon state can be classified by $(n, l)$ quantum numbers. However spin-orbit coupling is relatively important and the energy levels also depend to a large extent upon $j$. States of similar energy may be grouped together and called nucleon shells but the numbers of allowed nucleons in the shells are different from the electrons because of the strong spin-orbit coupling. The neutrons and the protons form their own shells but the energy difference is very small for the lighter nuclei. The three lowest shells can contain up to 2, 6 or 12 protons or neutrons respectively. A nucleus containing 2, 8 or 20 . . . protons (or neutrons) will have completed proton (neutron) shells. These are the 'magic numbers'. A nucleus with completed proton and neutron shells may be expected to be stable. The isotope ${}^4_2\mathrm{He}$ is a typical example. The nuclear shell model has proved useful in considering nuclear properties.

*Electrostatic interaction and terms*

The atomic configuration degeneracy described above does not really occur as the averaged central field potential used in the Hartree–Fock equations (and hence in $\mathscr{H}_0$) is only an approximation to the actual electrostatic interaction between the electrons. The difference between the two Hamiltonians is (from (11.2) and (11.33))

$$\left[-\sum_{i=1}^{n} \frac{Ze^2}{4\pi\epsilon r_i} + \frac{1}{2}\sum_{\substack{i=1\\(i\neq j)}}^{n}\sum_{j=1}^{n} \frac{e^2}{4\pi\epsilon_0|\mathbf{r}_i - \mathbf{r}_j|}\right] - \sum_{i=1}^{n} u_i(r_i) \quad (11.35)$$

and may be regarded as a perturbation. This is a degenerate perturbation problem and the correct zeroth order functions must be used. The functions required are characterized by $L$, $S$ quantum numbers. This is explained below.

The Hamiltonian (11.2) unlike $\mathscr{H}_0$ is not invariant under a rotation of the spatial co-ordinates of any one particle taken alone and so does not commute with the individual electron angular momentum operators. The $l_i$, $i = 1, 2, \ldots, n$ are not good quantum numbers. However this Hamiltonian is invariant under a simultaneous rotation of all the spatial (or spin) co-ordinates and so commutes with the system angular (and spin) momentum operators. The energy of the system depends upon the system momentum quantum numbers $L$ and $S$. The Hamiltonian (11.2) is spin-independent and it may appear that the energy should only depend upon $L$. In fact the energy does also depend upon $S$, and this is an exchange effect. (See discussion of the Helium atom (11.3).)

The determinantal wave-functions belonging to a given configuration form the basis for a reducible representation of the symmetry group of the Hamiltonian (11.2).

Repeated use of the vector coupling rule (11.5) or (11.6) may be used to give the possible irreducible representations and hence the allowed values of $L$ and $S$ that are included. Once more, application of the Pauli principle may exclude some of these values of $L$ and $S$.

By taking suitable combinations of the determinantal wave-functions belonging to a configuration it is possible to construct sets of

$$(2L+1)(2S+1) \tag{11.36}$$

functions for each allowed pair of values $(L, S)$ which transform according to $D^{(L)}$ and $D^{(S)}$ under spatial and spin rotations respectively. Each set of wave-functions belongs to the same energy level, called a spectral term, of the Hamiltonian (11.2) and the degeneracy of the term defined by $(L, S)$ is given by (11.36).

The wave-functions for a given term $(L, S)$ do not consist entirely of linear combinations of the wave-functions from the corresponding configuration. They include small contributions from terms, with the same values of $L$ and $S$, belonging to different configurations. This effect is called configuration interaction. The energy difference between terms associated with a given configuration is small compared with the energy difference between configurations.

As an illustration consider the following examples. The helium atom ground state has the configuration with both electrons in the $(1s)$ state.

$$(1s)^2. \tag{11.37}$$

These electrons have the same values of $n$ and $l$ and are said to be equivalent (11.34) gives this configuration a maximum possible degeneracy of $2^2 = 4$. However this is clearly incorrect as the Pauli principle requires that the two electrons have opposite spins and this configuration is not degenerate. The allowed values of $L$ are given by the coupling rule

$$D^{(0)} \times D^{(0)} = D^{(0)}$$

i.e.

$$L = 0. \tag{11.38}$$

The allowed values of $S$ are given by

$$D^{(1/2)} \times D^{(1/2)} = D^{(1)} + D^{(0)}$$

i.e.

$$S = 0, 1. \tag{11.39}$$

So it would appear that the eigenfunction (11.37) includes the terms

$$^1S, \quad ^3S.$$

This is incorrect since the Pauli principle does not permit the term $^3S$ and the configuration corresponds to the single term $^1S$.

The first excited state of helium has the configuration

$$(1s)(2s)$$

(11.34) does give the degeneracy correctly as $2^2 = 4$ in this case. The allowed values of $L$ and $S$ are given by (11.38) and (11.39) and both the terms $^1S$ and $^3S$ are included in the configuration.

The next excited state of helium has the configuration

$$(1s)(2p)$$

and (11.34) gives the correct degeneracy $2^2 \times 1 \times 3 = 12$. The allowed values of $L$ are given by

$$D^{(0)} \times D^{(1)} = D^{(1)}$$

i.e.

$$L = 1$$

and $S = 0, 1$ as before. The allowed terms in the configuration are $^1P$ and $^3P$. These terms have degeneracy three and nine respectively making twelve wave-functions in all as required.

These last two configurations each have different terms with the same value of $L$ but different values of $S$. Even though the Hamiltonian (11.2) does not depend upon the spin co-ordinates these terms do have different energies. This is an exchange effect and arises because determinantal wave-functions require antisymmetry in all the co-ordinates including spin.

In the previous discussion of the helium atom it was explained that the term, arising from the excited configuration $(1s)(2s)$, in which the electrons have parallel spins ($S = 1$) has a vanishing probability for the electrons to be in the same point in space at the same instant. This reduces the positive contribution to the energy arising from the electron-electron interaction. This result can be generalized to give Hund's rule. The term with the lowest energy obtained from a given configuration is that term with the largest value of $S$ and the largest value of $L$ associated with that $S$.

*Fine structure*

The Hamiltonian (11.2) neglects relativistic effects and excludes all terms that depend upon the spin co-ordinates. The coupling between the spin and the orbital motion of each electron adds a contribution,

$$\frac{1}{2c^2m^2}\sum_{i=1}^{n}\frac{1}{r_i}\frac{\mathrm{d}V_i}{\mathrm{d}r_i}\{\mathscr{L}_{ix}s_{ix}+\mathscr{L}_{iy}s_{iy}+\mathscr{L}_{iz}s_{iz}\} \tag{11.40}$$

to the Hamiltonian where $\mathscr{L}_{ix}$ and $s_{ix}$ are the $x$-components of the momenta of the $i$th-particle. In addition there are interactions between the spin of one electron and the orbital motion of the others and also interactions between different spins.

These terms are of the order of $(v/c)^2$ and produce the fine structure of term energy levels. All spin interaction terms (e.g. (11.40)) involve first or second powers of inner products of the vector operators $(\mathscr{L}_{ix}, \mathscr{L}_{iy}, \mathscr{L}_{iz})$ and $(s_{ix}, s_{iy}, s_{iz})$. These inner products are not invariant under a separate rotation of the spatial or spin co-ordinates of all the electrons but they, and hence the complete Hamiltonian are invariant under a simultaneous rotation of all the co-ordinates both spatial and spin. That is, the complete Hamiltonian does not commute with the operators representing the system angular momentum or the spin angular momentum but does commute with the total angular momentum operators. $L$ and $S$ are no longer good quantum numbers but $J$ and $M_J$ are good quantum numbers. The energy depends only upon $J$ and so each energy level has degeneracy $(2J + 1)$.

For the lighter atoms the spin terms are smaller than the electrostatic terms and the spin terms can be regarded as a perturbation of the spectral term energy level. The $(2L + 1)(2S + 1)$ degenerate wave-functions belonging to a given spectral term $(L, S)$ span a function space that is reducible under the new (reduced) symmetry group of the Hamiltonian and so this degeneracy will be split by the spin perturbation to give the spectral term 'fine structure'. This function space can be separated into sub-spaces each of which transform irreducibly under the simultaneous orbital-spin rotation group according to $D^{(J)}$ where $J$ is given by the vector coupling rule (11.6).

$$J = (L + S), (L + S - 1), \ldots, |L - S|. \tag{11.41}$$

Each of these values of $J$ corresponds to a separate energy level. The values of the fine-structure splittings can be obtained from (11.40) by calculating the appropriate matrix elements.

This approximate method for constructing the wave-functions for an atomic energy level from a given term is based on the assumption that the 'fine structure' energy differences are small compared with the differences between terms. This is often the case and is referred to as Russell–Saunders or $LS$ coupling.

The exact wave-functions for an atomic level with angular momentum quantum numbers $(J, M_J)$ will also contain small contributions from other terms.

For heavier atoms the relativistic spin terms are not small compared with the non-central field electrostatic terms and cannot be treated as perturbation in this way. A better approximation to the atomic wave-functions is obtained by characterizing each electron by its total angular momentum $j$. The non-central field electrostatic terms are then used as a perturbation to construct wave-functions with total angular momentum $J$. This scheme is called $jj$-coupling. Actually this coupling method is never completely valid although it may apply for highly excited states. It is the opposed limiting case to the $LS$ scheme and the true coupling scheme for heavy atoms may lie somewhere in between.

In addition to the fine-structure splitting due to the electron spin there is also a very small splitting of energy levels arising from the weak interaction between the spin of the nucleus and the electrons. The nuclear spin and the total angular momentum of the electrons can be coupled together. This produces the hyperfine structure of the levels and the splitting is about $10^{-3}$ of the fine structure splitting.

*Selection rules*

It can be shown that when the radiation wave-length is large compared with atomic dimensions and the electron-dipole approximation is valid, the spontaneous and induced transition probabilities are proportional to the square of the modulus of the dipole integral.

$$\left| \int \psi_f^* \left[ \sum_{k=1}^{n} \mathbf{r}_k \right] \psi_i \, \mathrm{d}\tau \right|^2 \tag{11.42}$$

$\psi_i$ and $\psi_f$ are the initial and final many-electron states, $\mathbf{r}_k$ is the position of the $k$th electron and the integration is over all electron co-ordinates. (Compare (8.38) for a single electron.)

$\Sigma\, \mathbf{r}_k$ changes sign under an inversion and since the atomic wave-functions defined by $(J, M_J)$ have either positive or negative parity the dipole integral (11.42) will vanish unless the transition involves a change in parity of the atomic wave-functions. This selection rule is called Laporte's rule.

There are other selection rules and group theory will now be used to determine the non-zero transition probabilities.

Suppose the function $\phi$ is one of the basis functions for an irreducible representation of a symmetry group. It can be shown that the integral over all co-ordinates

$$\int \phi \, \mathrm{d}\tau = 0 \tag{11.43}$$

if the representation is not the unit representation (see Landau and Lifshitz, p. 343). As examples consider the hydrogen type one-electron functions. The functions which transform according to the irreducible representation $D^{(1)}$ of the rotation group are $\psi_{Px}$, $\psi_{Py}$ and $\psi_{Pz}$ and clearly

$$\int \psi_{P_x}\,d\tau = \int \psi_{P_y}\,d\tau = \int \psi_{P_z}\,d\tau = 0.$$

However the function which transforms according to the unit representation $D^{(0)}$ is $\psi_s$ and

$$\int \psi_s\,d\tau \neq 0.$$

Now the three components of $\Sigma_{k=1}^{n}\,\mathbf{r}_k$ transform amongst themselves as a vector under simultaneous rotation of all the electron co-ordinates. That is, they transform according to the three-dimensional irreducible representation $D^{(1)}$.

The initial state wave-function for the atom $\psi_i$ must transform according to some irreducible representation $D^{(J_i)}$ and so the set of all functions (remember $\psi_i$ is $2J_i + 1$ fold degenerate)

$$\sum_{k=1}^{n} \mathbf{r}_k \psi_i \tag{11.44}$$

transforms according to the sum of irreducible representations

$$\begin{aligned} & = D^{(J_i+1)} + D^{(J_i)} + D^{(J_i-1)} && J_i \geqslant 1 \\ D^{(1)} \times D^{(J_i)} & = D^{(3/2)} + D^{(1/2)} && J_i = \tfrac{1}{2} \\ & = D^{(1)} && J_i = 0. \end{aligned}$$

Of course the final state of the atom must also transform according to some irreducible representation $D^{(J_f)}$ and so the integrand of (11.42) transforms according to

$$D^{(J_f)} \times D^{(1)} \times D^{(J_i)}. \tag{11.45}$$

This triple product will contain the unit representation only if $D^{(J_f)}$ is contained in the product $D^{(1)} \times D^{(J_i)}$. That is, if

$$\begin{aligned} J_f & = J_i + 1,\, J_i,\, J_i - 1 && J_i \geqslant 1 \\ & = \tfrac{3}{2}, \tfrac{1}{2} && J_i = \tfrac{1}{2} \\ & = 1 && J_i = 0. \end{aligned}$$

The dipole integral (11.42) may be different from zero only for pairs of states such that

$$\begin{aligned} & \delta J = +1, -1 \\ & \delta J = 0 \end{aligned} \tag{11.46}$$

except the transition $J_i = 0$ to $J_f = 0$. These are the selection rules for allowed transitions. It can be shown that $\delta M_J = +1, 0, -1$.

For lighter atoms in the Russell–Saunders approximation, the spatial part of an atomic level wave-function transforms according to $D^{(L)}$ under space co-ordinate rotations and the spin part transforms according to $D^{(s)}$ under spin rotations. Since $\sum_{k=1}^{n} \mathbf{r}_k$ transforms according to $D^{(1)}$ and $D^{(0)}$ under space and spin rotations respectively, then it can easily be shown that the selection rules for transitions between Russell–Saunders states are

$$\delta L = +1, -1$$

$$\delta L = 0$$

except the transition

$$L_i = 0 \text{ to } L_f = 0$$

and

$$\delta S = 0. \qquad (11.47)$$

Observe that the total spin is conserved.

*The sodium D-lines*

In the ground state the sodium atom in the self-consistent field approximation has eleven electrons in the configuration

$$(1s)^2(2s)^2(2p)^6(3s).$$

The Pauli principle drastically reduces the possible 'degeneracy' of this configuration. The inner closed core of ten electrons transforms according to $D^{(0)}$ in both orbital and spin space and the only degeneracy allowed occurs from the possible values for the spin quantum number $s$ for the outer $(3s)$ electron. The ground state configuration is doubly degenerate and is represented by the doublet term $L = 0, S = \frac{1}{2}$, i.e. $^2S$. The spin-orbit coupling does not reduce this degeneracy and the atomic level may be written

$$^2S_{1/2} \qquad \text{i.e.} \qquad L = 0, \quad S = \tfrac{1}{2}, \quad J = \tfrac{1}{2}.$$

The subscript gives the $J$-value.

The sodium atom may be excited into a higher energy perhaps by applying a suitable field or by heating it in an electric arc. The lowest excited configuration is

$$(1s)^2(2s)^2(2p)^6(3p).$$

There are six such 'degenerate' states corresponding to the three different values of $l$ and the two different values of $s$ taken by the outer

($3p$) electron. For the complete configuration $L = 1, S = \frac{1}{2}$ and so it contains the single term $^2P$. The allowed values of the total angular momentum of the system are

$$J = (1 + \tfrac{1}{2}), (1 - \tfrac{1}{2})$$

i.e.

$$J = \tfrac{3}{2}, \tfrac{1}{2}.$$

Consequently, the term $^2P$ splits up into the separate energy levels

$$^2P_{3/2}, \ ^2P_{1/2}$$

under the spin-orbit interaction.

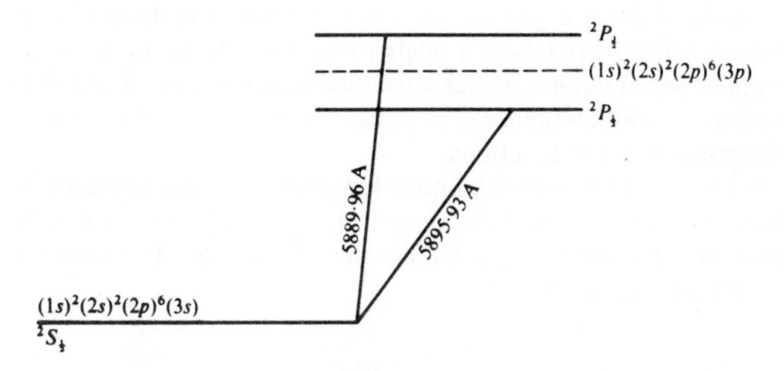


Figure 11.3 The sodium D-line transition.

The downward-transition from either of the levels $^2P_{3/2}$ or $^2P_{1/2}$ to the level $^2S_{1/2}$ is possible as

$$\delta L = -1, \qquad \delta S = 0.$$

These two transitions are the cause of the D-line doublet in the sodium spectrum and correspond to wave-lengths 5.889·96 Å and 5895·93 Å respectively.

## 11.5 Second Quantization

The method of second quantization is very useful for studying systems of identical particles. It is necessary in the quantization of fields where the number of particles may actually change. For example, photons may be created or annihilated in a radiation field. However this work is outside the scope of an introductory text and the second quantization scheme will only be applied to electron systems below.

Let $\{\phi_{Q_i}(\mathbf{x})\}$ be a complete set (infinite) of orthonormal one-particle functions. $\mathbf{x}$ denotes the set of four co-ordinates including the spin-orientation of the particle. This set may be combined to form anti-symmetric (or symmetric) $n$-particle wave-functions. For electrons (i.e. Fermi particles) the $n$-particle wave-function is anti-symmetry. Let $\Phi^{(n)}(\mathbf{x}_1, \mathbf{x}_2, \ldots, \mathbf{x}_n)$ be such an anti-symmetric $n$-particle function constructed from the first $n$ functions in the complete set. Then

$$\Phi^{(n)}(\mathbf{x}_1, \mathbf{x}_2, \ldots, \mathbf{x}_n) = \frac{1}{\sqrt{n!}} \begin{vmatrix} \phi_{Q_1}(\mathbf{x}_1) \ldots \phi_{Q_1}(\mathbf{x}_n) \\ \cdot \\ \vdots \\ \phi_{Q_n}(\mathbf{x}_1) \ldots \phi_{Q_n}(\mathbf{x}_n) \end{vmatrix} . \tag{11.48}$$

Similarly, other $n$-particle determinantal wave-functions can be constructed using other sets of $n$ single-particle functions. In this way a complete set of determinantal wave-functions is obtained and an arbitrary $n$-particle wave-functions can be expanded in terms of these $n \times n$ determinantal wave-functions.

Each of the determinantal wave-functions is completely specified when the number of particles in each single-particle state is specified. Let $n_i$ be the number of particles in the state $\phi_{Q_i}(\mathbf{x})$. The wave-function (11.48) corresponds to

$$n_1 = n_2 = \ldots = n_n = 1, \qquad n_{n+1} = n_{n+2} = \ldots = 0. \tag{11.49}$$

The 'occupation numbers' $n_i$ can only take the values 0 or 1 for Fermi particles (Pauli principle). This restriction does not apply to particles obeying Bose–Einstein statistics.

The $n$-particle determinantal wave-functions will now be regarded as functions of the occupation numbers of the states $\phi_{Q_i}(\mathbf{x})$ rather than as functions of the particle co-ordinates. With this in mind the wave-functions will be written

$$\Phi(n_1, n_2, n_3, \ldots) \tag{11.50}$$

where

$$\sum_{i=1}^{\infty} n_i = n \tag{11.51}$$

and for Fermi particles $n_i = 0, 1$.

The wave-function (11.48) is written

$$\Phi(\underbrace{1, 1, 1, \ldots, 1}_{n}, 0, 0, \ldots).$$

In this new scheme the operators representing dynamical variables must act on the occupation numbers rather than on the particle coordinates. At this point it is necessary to introduce creation and annihilation operators. The creation operator $a_i^+$ operates on a $n$-electron wave-function such as (11.50) to add a new particle in the state $\phi_{Q_i}(\mathbf{x})$, i.e.

$$a_i^+\Phi(n_1, n_2, \ldots, n_i, \ldots) = (-1)^{s_i}(1 - n_i)\Phi(n_1, n_2, \ldots, n_i + 1, \ldots) \tag{11.52}$$

where

$$s_i = \sum_{k=1}^{i-1} n_k, \tag{11.53}$$

$s_i$ is equal to the number of occupied states up to but excluding $\phi_{Q_i}(\mathbf{x})$. $(-1)^{s_i}$ is a phase factor, the significance of which will become apparent later. Observe that if $n_i = 1$, the right hand side of (11.52) is zero in agreement with the Pauli principle. A determinantal wave-function with two particles in the same state is identically zero.

An annihilation operator $a_i$ is defined to remove a particle from the state $\phi_{Q_i}(\mathbf{x})$

$$a_i\Phi(n_1, n_2, \ldots, n_i, \ldots) = (-1)^{s_i} n_i \Phi(n_1, n_2, \ldots, n_i - 1, \ldots) \tag{11.54}$$

($a_i^+$ is the Hermitian adjoint of $a_i$). Similar but different expressions define the action of the creation and annihilation operators if the particles are bosons.

As $n_i$ can only equal one or zero for Fermi particles then from (11.52)

$$a_i^+ a_i^+ \equiv 0 \tag{11.55}$$

and from (11.54)

$$a_i a_i \equiv 0. \tag{11.56}$$

The vacuum state is defined to be that state containing no particles. Hence

$$a_1^+\Phi_{\text{vac}} = \Phi(n_1 = 1, n_2 = 0, \ldots) = \phi_{Q_1}(\mathbf{x})$$

represents the creation of a single particle in the state $\phi_{Q_1}(\mathbf{x})$. A general determinantal wave-function can be constructed from the vacuum state.

$$\Phi(n_1, n_2, \ldots, n_i, \ldots) = (a_1^+)^{n_1}(a_2^+)^{n_2} \ldots (a_i^+)^{n_i} \ldots \Phi_{\text{vac}}. \tag{11.57}$$

Changing the order of the creation operators in a product corresponds to interchanging the particles. When two Fermi particles are interchanged the wave-function changes sign. Hence it is necessary that

$$(a_i^+ a_j^+)\Phi = -(a_j^+ a_i^+)\Phi \tag{11.58}$$

for arbitrary $\Phi$, i.e.

$$[a_i^+, a_j^+]_+ \equiv 0 \tag{11.59}$$

where $[\alpha, \beta]_+ \equiv \alpha\beta + \beta\alpha$ is the anticommunicator of $\alpha$ and $\beta$. It is to satisfy this requirement that the phases are chosen as in (11.52) and (11.54). To see this consider

$$\begin{aligned} &a_i^+ a_j^+ \Phi(n_1, \ldots, n_i = 0, \ldots, n_j = 0, \ldots) \\ &\quad = (-1)^{s_j} a_i^+ \Phi(n_1, \ldots, n_i = 0, \ldots, n_j = 1, \ldots) \\ &\quad = (-1)^{s_j}(-1)^{s_i} \Phi(n_1, \ldots, n_i = 1, \ldots, n_j = 1, \ldots) \qquad j > i. \end{aligned} \tag{11.60}$$

Also

$$\begin{aligned} &a_j^+ a_i^+ \Phi(n_1, \ldots, n_i = 0, \ldots, n_j = 0, \ldots) \\ &\quad = (-1)^{s_i} a_j^+ \Phi(n_1, \ldots, n_i = 1, \ldots, n_j = 0) \\ &\quad = (-1)^{s_i}(-1)^{s_j'} \Phi(n_1, \ldots, n_i = 1, \ldots, n_j = 1, \ldots) \qquad j > i. \end{aligned} \tag{11.61}$$

Since $s_j' = s_j + 1$ then (11.58) is indeed satisfied.

The annihilation operators satisfy a relationship similar to (11.59)

$$[a_i, a_j]_+ \equiv 0. \tag{11.62}$$

There are also commutation relations which involve both creation and annihilation operators. Consider the action of the operator $a_i^+ a_i$ on the determinantal wave-function $\Phi(n_1, \ldots, n_i, \ldots)$

$$a_i \Phi(n_1, \ldots, n_i, \ldots) = (-1)^{s_i} n_i \Phi(n_1, \ldots, n_i - 1, \ldots)$$

$$\therefore \quad a_i^+ a_i \Phi(n_1, \ldots, n_i, \ldots) = (-1)^{s_i} n_i (-1)^{s_i} \Phi(n_1, \ldots, n_i, \ldots)$$

$$\therefore \quad a_i^+ a_i \Phi(n_1, \ldots, n_i, \ldots) = n_i \Phi(n_1, \ldots, n_i, \ldots)$$

as $n_i = 0, 1$ for Fermi particles. The operator

$$N_i \equiv a_i^+ a_i \tag{11.63}$$

is the number operator for the state $\phi_{Q_i}(\mathbf{x})$. Its eigenvalues are the number of particles in this state. The total number operator is defined to be

$$N \equiv \sum_{i=1}^{\infty} N_i. \qquad (11.64)$$

Consider now the action of the operator $a_i a_i^+$

$$a_i^+\Phi(n_1, \ldots, n_i, \ldots) = (-1)^{s_i}(1 - n_i)\Phi(n_1, \ldots, n_i + 1, \ldots)$$

$$\therefore \quad a_i a_i^+\Phi(n_1, \ldots, n_i, \ldots) = (-1)^{s_i}(1 - n_i)(-1)^{s_i}(n_i + 1) \times \Phi(n_1, \ldots, n_i, \ldots)$$

i.e.

$$a_i a_i^+\Phi(n_1, \ldots, n_i, \ldots) = (1 - n_i)\Phi(n_1, \ldots, n_i, \ldots). \qquad (11.65)$$

From (11.62) and (11.65)

$$(a_i^+ a_i + a_i a_i^+)\Phi(n_1, \ldots, n_i, \ldots) = \Phi(n_1, \ldots, n_i, \ldots).$$

This is a special case of the commutation rule

$$a_i^+ a_i + a_i a_i^+ \equiv 1$$

or generally

$$[a_i^+, a_j]_+ \equiv \delta_{ij}. \qquad (11.66)$$

From the definition (11.63) and the commutation rules (11.59), (11.62) and (11.66), it is quite easy to show that the number of operators belonging to different states commute, i.e.

$$[N_i, N_j] \equiv 0. \qquad (11.67)$$

Consequently functions can be found that are simultaneously eigenfunctions of all these number operators. These are of course the determinantal wave-functions.

Finally it can be shown that

$$[N_i, a_j] \equiv -\delta_{ij} a_i \qquad (11.68)$$

and

$$[N_i, a_j^+] \equiv \delta_{ij} a_i^+. \qquad (11.69)$$

*Representation of observables*

A physical observable is represented in quantum mechanics by an operator. Let $\alpha(\mathbf{x})$ be the 'co-ordinate' operator that represents the observable $A$. As the set $\{\phi_{Q_i}(\mathbf{x})\}$ is complete the effect of $\alpha(\mathbf{x})$ on the single particle function $\phi_{Q_i}(\mathbf{x})$ is

$$\alpha\phi_{Q_i}(\mathbf{x}) = \sum_{j=1}^{\infty} \alpha_{ji}\phi_{Q_j}(\mathbf{x}) \tag{11.70}$$

with

$$\alpha_{ji} = \int \phi_{Q_j}^{*}\alpha\phi_{Q_i}\,\mathrm{d}\tau_x.$$

(The integration includes a sum over spin co-ordinates.) The right hand side of (11.70) can be written in terms of creation and annihilation operators.

$$\begin{aligned}\sum_{j=1}^{\infty} \alpha_{ji}a_j^{+}a_i\phi_{Q_i}(\mathbf{x}) &= \left(\sum_{j=1}^{\infty}\sum_{k=1}^{\infty} a_j^{+}\alpha_{jk}a_k\right)\phi_{Q_i}(\mathbf{x})\\ &= \left(\sum_{j=1}^{\infty}\sum_{k=1}^{\infty} a_j^{+}\alpha_{jk}a_k\right)(a_i^{+}\Phi_{\text{vac}}).\end{aligned} \tag{11.71}$$

From (11.70) and (11.71) it is seen that the operator $\alpha(\mathbf{x})$ which acts on functions of the co-ordinates $\mathbf{x}$ can be expressed as an operator acting on the occupation numbers $n_i$.

$$\alpha(\mathbf{x}) \equiv \sum_j\sum_k a_j^{+}\alpha_{jk}a_k. \tag{11.72}$$

This result has been shown for the simplest case of a one-electron function but it can be extended.

Let $\alpha(\mathbf{x}_1, \mathbf{x}_2, \ldots, \mathbf{x}_n)$ be an additive operator for a system of $n$ particles such that

$$\alpha(\mathbf{x}_1, \mathbf{x}_2, \ldots, \mathbf{x}_n) \equiv \alpha(\mathbf{x}_1) + \alpha(\mathbf{x}_2) + \ldots + \alpha(\mathbf{x}_n). \tag{11.73}$$

This operator is symmetrical with respect to all the particles and it can be shown that

$$\alpha(\mathbf{x}_1, \mathbf{x}_2, \ldots, \mathbf{x}_n) \equiv \sum_j\sum_k a_j^{+}\alpha_{jk}a_k \tag{11.74}$$

with $\alpha_{jk} = \int \phi_{Q_j}^{*}\alpha(\mathbf{x})\phi_{Q_k}\,\mathrm{d}\tau_x$ as before.

As an example consider a system of $n$ identical particles whose mutual interaction can be ignored. The system Hamiltonian is the additive operator

$$\mathscr{H} \equiv \sum_{r=1}^{n} \mathscr{H}_r \tag{11.75}$$

with

$$\mathscr{H}_r \equiv -\frac{\hbar^2}{2m}\nabla_r^2 + V(\mathbf{x}_r).$$

The simplest non-trivial case is a system of two particles. The wave-function for the system can be expressed as a linear sum of 2-particle determinantal wave-functions of the form

$$a_l^+ a_m^+ \Phi_{\text{vac}} = \frac{1}{\sqrt{2}} \left[\phi_{Q_l}(\mathbf{x}_1)\phi_{Q_m}(\mathbf{x}_2) - \phi_{Q_m}(\mathbf{x}_1)\phi_{Q_l}(\mathbf{x}_2)\right]$$

where the phase of the determinant has been sensibly chosen.

The effect of the system Hamiltonian on this determinant is

$$[\mathcal{H}_1 + \mathcal{H}_2]\frac{1}{\sqrt{2}} \left[\phi_{Q_l}(\mathbf{x}_1)\phi_{Q_m}(\mathbf{x}_2) - \phi_{Q_m}(\mathbf{x}_1)\phi_{Q_l}(\mathbf{x}_2)\right]$$

$$= \frac{1}{\sqrt{2}}\left[\phi_{Q_m}(\mathbf{x}_2)\sum_j \hbar_{jl}\phi_{Q_j}(\mathbf{x}_1) - \phi_{Q_l}(\mathbf{x}_2)\sum_j \hbar_{jm}\phi_{Q_j}(\mathbf{x}_1) + \phi_{Q_l}(\mathbf{x}_1)\sum_j \hbar_{jm}\phi_{Q_j}(\mathbf{x}_2) - \phi_{Q_m}(\mathbf{x}_1)\sum_j \hbar_{jl}\phi_{Q_j}(\mathbf{x}_2)\right] \quad (11.76)$$

as

$$\mathcal{H}_r\phi_{Q_l}(\mathbf{x}_r) = \sum_j \hbar_{jl}\phi_{Q_j}(\mathbf{x}_r) \qquad \text{for} \qquad r = 1, 2.$$

The right hand side of (11.76) is

$$\sum_j \hbar_{jl}\frac{1}{\sqrt{2}}\left[\phi_{Q_m}(\mathbf{x}_2)\phi_{Q_j}(\mathbf{x}_1) - \phi_{Q_m}(\mathbf{x}_1)\phi_{Q_j}(\mathbf{x}_2)\right]$$

$$-\sum_j \hbar_{jm}\frac{1}{\sqrt{2}}\left[-\phi_{Q_l}(\mathbf{x}_1)\phi_{Q_j}(\mathbf{x}_2) + \phi_{Q_l}(\mathbf{x}_2)\phi_{Q_j}(\mathbf{x}_1)\right].$$

In terms of the creation and annihilation operators this becomes

$$\sum_j [\hbar_{jl}a_j^+a_m^+\Phi_{\text{vac}} - \hbar_{jm}a_j^+a_l^+\Phi_{\text{vac}}]. \quad (11.77)$$

It is quite easy to show that (11.77) is equal to

$$\sum_j\sum_k a_j^+\hbar_{jk}a_k\,[a_l^+a_m^+\Phi_{\text{vac}}]$$

and hence deduce that

$$[\mathcal{H}_1 + \mathcal{H}_2] \equiv \sum_j\sum_k a_j^+\hbar_{jk}a_k$$

as required. For a system of $n$ non-interacting particles, the Hamiltonian in the occupation number representation is

$$\sum_j\sum_k a_j^+\hbar_{jk}a_k. \quad (11.78)$$

Suppose the expansion set $\{\phi_{Q_i}(\mathbf{x})\}$ is chosen to be the set of eigen-functions of the one-electron Hamiltonians so that

$$\mathscr{H}_r \phi_{Q_i}(\mathbf{x}_r) = E_{Q_i} \phi_{Q_i}(\mathbf{x}_r)$$

and

$$h_{jk} = E_{Q_j} \delta_{jk}. \tag{11.79}$$

The Hamiltonian (11.78) becomes

$$\sum_j E_{Q_j} a_j^+ a_j \equiv \sum_j E_{Q_j} N_j \tag{11.80}$$

and its eigenfunctions are also the simultaneous eigenfunctions of the number operator $N_j$. The eigenvalues are

$$\sum_j E_{Q_j} n_j \qquad n_j = 0, 1 \qquad \text{and} \qquad \sum_j n_j = n. \tag{11.81}$$

Unfortunately the operators in many particle systems are not always additive in the sense of (11.73). The particles often interact with each other and co-ordinate operators occur which can be written

$$\beta(\mathbf{x}_1, \mathbf{x}_2, \ldots, \mathbf{x}_n) \equiv \sum_{r>s}\sum \beta(\mathbf{x}_r, \mathbf{x}_s), \tag{11.82}$$

where $\beta(\mathbf{x}_r, \mathbf{x}_s)$ operates on functions involving $\mathbf{x}_r$ or $\mathbf{x}_s$. It can be shown that

$$\beta(\mathbf{x}_1, \mathbf{x}_2, \ldots, \mathbf{x}_n) \equiv \tfrac{1}{2} \sum_i \sum_j \sum_k \sum_l \beta_{kl}^{ij} a_i^+ a_j^+ a_l a_k, \tag{11.83}$$

with

$$\beta_{kl}^{ij} = \iint \phi_{Q_i}^*(\mathbf{x}_1) \phi_{Q_j}^*(\mathbf{x}_2) \beta(\mathbf{x}_1, \mathbf{x}_2) \phi_{Q_k}(\mathbf{x}_1) \phi_{Q_l}(\mathbf{x}_2)\, d\tau_{x_1}\, d\tau_{x_2}.$$

For example, the Coulomb interactions between electrons adds a term

$$\sum_{r>s}\sum V(\mathbf{r}_r - \mathbf{r}_s) \qquad \text{where} \qquad V(\mathbf{r}_r - \mathbf{r}_s) = \frac{e^2}{4\pi\epsilon_0 |\mathbf{r}_r - \mathbf{r}_s|}$$

to the $n$-particle Hamiltonian (11.75). $\mathbf{r}_r$ is the spatial part of $\mathbf{x}_r$. The Hamiltonian in the occupation number representation is then

$$\sum_j \sum_k a_j^+ h_{jk} a_k + \tfrac{1}{2} \sum_i \sum_j \sum_k \sum_l a_i^+ a_j^+ V_{kl}^{ij} a_l a_k \tag{11.84}$$

where $h_{jk}$ is defined as before and

$$V_{kl}^{ij} = \frac{e^2}{4\pi\epsilon_0} \iint \phi_{Q_i}^*(\mathbf{x}_1) \phi_{Q_j}^*(\mathbf{x}_2) \frac{1}{|\mathbf{r}_1 - \mathbf{r}_2|} \phi_{Q_k}(\mathbf{x}_1) \phi_{Q_l}(\mathbf{x}_2)\, d\tau_{x_1} d\tau_{x_2}.$$

*Electron field operators*

The rules for finding the form of the operators, representing observables, in the occupation number representation can be stated in terms of new 'electron field' operators $\Psi^+(\mathbf{x})$ and $\Psi(\mathbf{x})$ defined by

$$\Psi^+(\mathbf{x}) \equiv \sum_i \phi^*_{Q_i}(\mathbf{x}) a_i^+ \qquad \Psi(\mathbf{x}) \equiv \sum_i \phi_{Q_i}(\mathbf{x}) a_i. \tag{11.85}$$

The operator $\Psi^+(\mathbf{x}_0)$ creates a particle with the co-ordinates (including spin) $\mathbf{x}_0$.

It is quite easy to show that these operators satisfy the commutation relations

$$[\Psi(\mathbf{x}), \Psi(\mathbf{x}')]_+ \equiv 0$$

$$[\Psi(\mathbf{x}), \Psi^+(\mathbf{x}')]_+ \equiv \delta(\mathbf{x} - \mathbf{x}') \tag{11.86}$$

where

$$\delta(\mathbf{x} - \mathbf{x}') = \delta(x - x')\delta(y - y')\delta(z - z')\delta_{\sigma_z \sigma_z'}.$$

These operators are of interest because the equations (11.74) and (11.83) can be written

$$\alpha(\mathbf{x}_1, \mathbf{x}_2, \ldots, \mathbf{x}_n) \equiv \int \Psi^+(\mathbf{x})\alpha(\mathbf{x})\Psi(\mathbf{x})\, \mathrm{d}\tau_x \tag{11.87}$$

$$\beta(\mathbf{x}_1, \mathbf{x}_2, \ldots, \mathbf{x}_n) \equiv \tfrac{1}{2} \iint \Psi^+(\mathbf{x})\Psi^+(\mathbf{x}')\beta(\mathbf{x}, \mathbf{x}')\Psi(\mathbf{x}')\Psi(\mathbf{x})\, \mathrm{d}\tau_x \mathrm{d}\tau_x' \tag{11.88}$$

respectively. The reader is asked to confirm that the right hand sides of (11.87) and (11.88) agree with the right hand sides of (11.74) and (11.83) by substituting (11.85) for the electron field operators.

The electron field operators play a similar role to that of the wave-function in the Schrödinger theory. In the second quantization formalism wave-functions are in some sense replaced by operators.

*Bosons*

The work above has been specifically oriented towards Fermi particles. There are differences in the formalism when applied to bosons. The $n$-particle wave-functions $\Phi(n_1, n_2, \ldots)$ are now symmetric in interchanging any two particles and the occupation numbers can exceed unity.

Creation and annihilation operators are defined by

$$a_i^+\Phi(n_1, n_2, \ldots, n_i, \ldots) = \sqrt{n_i + 1}\; \Phi(n_1, n_2, \ldots, n_i + 1, \ldots)$$

$$a_i\Phi(n_1, n_2, \ldots, n_i, \ldots) = \sqrt{n_i}\; \Phi(n_1, n_2, \ldots, n_i - 1, \ldots)$$

respectively. The commutation relations are also different, e.g.

$$[a_i^+, a_j^+] \equiv 0. \tag{11.89}$$

This development is outside the scope of this text although it is certainly important in many problems, such as in a discussion of photons in an electromagnetic field or in the theory of phonons in lattice vibrations.

The interested reader is referred to *Introductory Quantum Electrodynamics,* by Power, and to *Concepts in Solids,* by Anderson. The latter text has an interesting derivation of the Hartree–Fock equations in second quantization formalism (p. 15).

## PROBLEMS

1 Use the vector coupling rule to show that

(a) $D^{(3)} \times D^{(1/2)} = D^{(7/2)} + D^{(5/2)}$

(b) $D^{(1)} \times D^{(1)} \times D^{(1)} = D^{(3)} + 2D^{(2)} + 3D^{(1)} + D^{(0)}$

(c) $D^{(1/2)} \times D^{(1/2)} \times D^{(1/2)} = D^{(3/2)} + 2D^{(1/2)}$.

2 Show that the inversion operator is Hermitian.

3 The electric dipole moment of a system of $n$ particles each of charge $e$ can be written

$$\mathbf{P} = e \sum_{k=1}^{n} \mathbf{r}_k$$

where $\mathbf{r}_k$ is the position vector of the $k$th particle. Explain why $\mathbf{P}$ changes sign under the action of the inversion operator.

Deduce that if the state function $\psi(\mathbf{r}_1, \mathbf{r}_2, \ldots, \mathbf{r}_n)$ has either even or odd parity then the expectation value of the dipole moment is zero, i.e.

$$\int \psi^* \mathbf{P} \psi \, d\tau_1 \ldots d\tau_n = 0.$$

This show that in the dipole approximation transitions are allowed only between even and odd states.

4 Consider a tightly bound collection of particles which behaves as a single particle at a point $\mathbf{r}$ with $S = 1$. The wave-functions may be written as a column matrix

$$\psi(\mathbf{r}, t) = \begin{bmatrix} \psi_+(\mathbf{r}, t) \\ \psi_0(\mathbf{r}, t) \\ \psi_-(\mathbf{r}, t) \end{bmatrix}$$

and $m_s$ can take the values $+1, 0, -1$.

Verify that the appropriate commutation rules are satisfied by the matrix representation

$$\mathbf{S}_x = \frac{\hbar}{\sqrt{2}}\begin{pmatrix} 0 & 1 & 0 \\ 1 & 0 & 1 \\ 0 & 1 & 0 \end{pmatrix} \qquad \mathbf{S}_y = \frac{\hbar}{\sqrt{2}}\begin{pmatrix} 0 & -i & 0 \\ i & 0 & -i \\ 0 & i & 0 \end{pmatrix}$$

$$\mathbf{S}_z = \hbar\begin{pmatrix} 1 & 0 & 0 \\ 0 & 0 & 0 \\ 0 & 0 & -1 \end{pmatrix}.$$

What are the eigenvectors corresponding to $m_s = 1, 0, -1$ respectively? Show that the matrix representing $S_{\rm op}^2$ is equal to $2\hbar^2$ times the 3 x 3 unit matrix as required.

5 Consider a system composed of three electrons. If the spin interaction terms are ignored, the system Hamiltonian is invariant under a separate rotation of each spin co-ordinate and the spin momentum of each electron is conserved. For each electron $s = \frac{1}{2}$. Using part (c) of problem (1), or otherwise, deduce that there are a quartet of system states with $S = \frac{3}{2}$ and $M_s = \frac{3}{2}, \frac{1}{2}, -\frac{1}{2}, -\frac{3}{2}$ and two separate doublets with $S = \frac{1}{2}$ and $M_s = \frac{1}{2}, -\frac{1}{2}$.

Obtain the values of $S$ and $M_s$ for each of the following spin functions and confirm that the quartet is

$$\alpha(\sigma_{1z})\alpha(\sigma_{2z})\alpha(\sigma_{3z})$$

$$\frac{1}{\sqrt{3}}\,[\alpha(\sigma_{1z})\alpha(\sigma_{2z})\beta(\sigma_{3z}) + \alpha(\sigma_{1z})\beta(\sigma_{2z})\alpha(\sigma_{3z}) + \beta(\sigma_{1z})\alpha(\sigma_{2z})\alpha(\sigma_{3z})]$$

$$\frac{1}{\sqrt{3}}\,[\beta(\sigma_{1z})\beta(\sigma_{2z})\alpha(\sigma_{3z}) + \beta(\sigma_{1z})\alpha(\sigma_{2z})\beta(\sigma_{3z}) + \alpha(\sigma_{1z})\beta(\sigma_{2z})\beta(\sigma_{3z})]$$

$$\beta(\sigma_{1z})\beta(\sigma_{2z})\beta(\sigma_{3z})$$

and that the two doublets may be written

$$\frac{1}{\sqrt{6}}\,[\alpha(\sigma_{1z})\alpha(\sigma_{2z})\beta(\sigma_{3z}) + \alpha(\sigma_{1z})\beta(\sigma_{2z})\alpha(\sigma_{3z}) - 2\beta(\sigma_{1z})\alpha(\sigma_{2z})\alpha(\sigma_{3z})]$$

$$\frac{1}{\sqrt{6}}\,[\beta(\sigma_{1z})\beta(\sigma_{2z})\alpha(\sigma_{3z}) + \beta(\sigma_{1z})\alpha(\sigma_{2z})\beta(\sigma_{3z}) - 2\alpha(\sigma_{1z})\beta(\sigma_{2z})\beta(\sigma_{3z})]$$

and

$$\frac{1}{\sqrt{2}}\left[\alpha(\sigma_{1z})\alpha(\sigma_{2z})\beta(\sigma_{3z}) - \alpha(\sigma_{1z})\beta(\sigma_{2z})\alpha(\sigma_{3z})\right]$$

$$\frac{1}{\sqrt{2}}\left[\beta(\sigma_{1z})\beta(\sigma_{2z})\alpha(\sigma_{3z}) - \beta(\sigma_{1z})\alpha(\sigma_{2z})\beta(\sigma_{3z})\right].$$

Observe that the first doublet is symmetric in the interchange of the second and third particles whereas the second doublet is antisymmetric in these particles.

6 Consider a linear box of unit length bounded by walls of infinite potential. If the origin is taken at one end of the box confirm that the wave-functions for an electron inside the box is

$$\psi_n(x) = \sqrt{2}\sin(n\pi x) \qquad n = 1, 2, \ldots.$$

Write down the spatial part of the determinantal wave-function $\psi(x_1, x_2)$ describing two electrons, with identical spin, in the two lowest energy states. If one electron is at the point $x = \frac{1}{4}$, the probability distribution of the other is

$$\psi^*(\tfrac{1}{4}, x_2)\psi(\tfrac{1}{4}, x_2).$$

Show that this is equal to

$$2\left[\sin \pi x_2 - \frac{1}{\sqrt{2}}\sin 2\pi x_2\right]^2.$$

Sketch this function and explain how it illustrates 'spin correlation'. Compare this with the distribution obtained from a simple product wave-function.

7 Show that the expectation value of the Hamiltonian (11.2) obtained using the Hartree product (11.24) is

$$\sum_k \epsilon_k - \sum_{j>k}\sum \iint |\phi_{Q_j}(\mathbf{r}_j)|^2 |\phi_{Q_k}(\mathbf{r}_k)|^2 \frac{e^2}{4\pi\epsilon_0 |\mathbf{r}_j - \mathbf{r}_k|}\, d\tau_j\, d\tau_k.$$

This shows that the energy of the system is less than the sum of the single particle energies.

8 In the approximation that the nuclei are fixed (Born–Oppenheimer) the Hamiltonian for the hydrogen molecule is, ignoring spin terms,

$$\mathscr{H} \equiv -\frac{\hbar^2}{2m}\nabla_1^2 - \frac{\hbar^2}{2m}\nabla_2^2 - \frac{1}{4\pi\epsilon_0}\left(\frac{e^2}{r_{1a}} + \frac{e^2}{r_{1b}} + \frac{e^2}{r_{2a}} + \frac{e^2}{r_{2b}}\right)$$

$$+ \frac{e^2}{4\pi\epsilon_0 r_{12}} + \frac{e^2}{4\pi\epsilon_0 R}$$

where $r_{1a}, r_{1b}$ are the distances of the first electron from the protons which are at '$a$' and '$b$' and similarly $r_{2a}, r_{2b}$ are the distances of the second electron from the protons. The separation of the electrons is $r_{12}$ and the protons are a distance $R$ apart.

Let $\phi_a(\mathbf{r})$, $\phi_b(\mathbf{r})$ be the real normalized hydrogen (1$s$) wave-functions centred on the protons at '$a$' and '$b$' respectively. A first approximation to the ground state wave-function for the molecule will be obtained by considering the four normalized determinantal wave-functions of the type

$$\psi(\mathbf{r}_1, \sigma_{1z}; \mathbf{r}_2, \sigma_{2z}) = \frac{1}{\sqrt{2(1-\Delta^2)}}\begin{vmatrix} \phi_a(\mathbf{r}_1)\alpha(\sigma_{1z}) & \phi_a(\mathbf{r}_2)\alpha(\sigma_{2z}) \\ \phi_b(\mathbf{r}_1)\alpha(\sigma_{1z}) & \phi_b(\mathbf{r}_2)\alpha(\sigma_{2z}) \end{vmatrix}$$

where

$$\Delta = \int \phi_a(\mathbf{r})\phi_b(\mathbf{r})\,\mathrm{d}\tau.$$

Show that from these it is possible to construct the singlet state with symmetric spatial part

$$\psi_s(\mathbf{r}_1, \mathbf{r}_2) = \frac{1}{\sqrt{2(1+\Delta^2)}}\,[\phi_a(\mathbf{r}_1)\phi_b(\mathbf{r}_2) + \phi_a(\mathbf{r}_2)\phi_b(\mathbf{r}_1)]$$

and a triplet state with antisymmetric spatial part

$$\psi_A(\mathbf{r}_1, \mathbf{r}_2) = \frac{1}{\sqrt{2(1-\Delta^2)}}\,[\phi_a(\mathbf{r}_1)\phi_b(\mathbf{r}_2) - \phi_a(\mathbf{r}_2)\phi_b(\mathbf{r}_1)].$$

Show that the energies of the system corresponding to the singlet and triplet states are respectively

$$E_s = 2E(1s) + \frac{J}{1+\Delta^2} + \frac{K}{1+\Delta^2}$$

$$E_A = 2E(1s) + \frac{J}{1-\Delta^2} - \frac{K}{1-\Delta^2}$$

where $E(1s)$ is the energy of the hydrogen $(1s)$ state and

$$J = \frac{e^2}{4\pi\epsilon_0} \int\int \phi_a(\mathbf{r}_1)\phi_b(\mathbf{r}_2) \left[ -\frac{1}{r_{1b}} - \frac{1}{r_{2a}} + \frac{1}{r_{12}} + \frac{1}{R} \right] \phi_a(\mathbf{r}_1)\phi_b(\mathbf{r}_2) \times d\tau_1 \, d\tau_2,$$

$$K = \frac{e^2}{4\pi\epsilon_0} \int\int \phi_a(\mathbf{r}_1)\phi_b(\mathbf{r}_2) \left[ -\frac{1}{r_{1b}} - \frac{1}{r_{2a}} + \frac{1}{r_{12}} + \frac{1}{R} \right] \phi_a(\mathbf{r}_2)\phi_b(\mathbf{r}_1) \times d\tau_1 \, d\tau_2.$$

Show that

$$J = -\frac{2e^2}{4\pi\epsilon_0} \int \frac{\phi_a^2(\mathbf{r}_1)}{r_{1b}} \, d\tau_1 + \frac{e^2}{4\pi\epsilon_0} \int\int \frac{\phi_a^2(\mathbf{r}_1)\phi_b^2(\mathbf{r}_2)}{r_{12}} \, d\tau_1 \, d\tau_2 + \frac{e^2}{4\pi\epsilon_0 R}$$

and that

$$K = -\frac{2e^2\Delta}{4\pi\epsilon_0} \int \frac{\phi_a(\mathbf{r}_1)\phi_b(\mathbf{r}_1)}{r_{1b}} \, d\tau_1 + \frac{e^2}{4\pi\epsilon_0} \int\int \frac{\phi_a(\mathbf{r}_1)\phi_b(\mathbf{r}_2)\phi_b(\mathbf{r}_1)\phi_a(\mathbf{r}_2)}{r_{12}} \times d\tau_1 \, d\tau_2 + \frac{e^2\Delta^2}{4\pi\epsilon_0 R}.$$

The first term in $K$ is dominant and so $K < 0$. Explain why this means that the singlet state lies deeper than the triplet state. (Compare this result with that obtained for the helium atom.)

9 Consider two particles each of mass $m$ at positions $\mathbf{r}_1, \mathbf{r}_2$ respectively which interact with one another through a potential $V(|\mathbf{r}_1 - \mathbf{r}_2|)$, which depends only upon their separation. The time-independent Schrödinger equation for the system is

$$-\frac{\hbar^2}{2m}(\nabla_1^2 + \nabla_2^2)\Omega + V(|\mathbf{r}_1 - \mathbf{r}_2|)\Omega = E\Omega$$

where $\Omega(\mathbf{r}_1, \mathbf{r}_2)$ is the spatial part of the state function. Following the argument from equations (4.45) to (4.51) show that the relative motion of the particles is described by the equation

$$\frac{1}{r^2}\frac{\partial}{\partial r}\left(r^2 \frac{\partial \psi}{\partial r}\right) + \frac{1}{r^2 \sin^2\theta}\frac{\partial^2 \psi}{\partial \phi^2} + \frac{1}{r^2 \sin\theta}\frac{\partial}{\partial \theta}\left(\sin\theta \frac{\partial \psi}{\partial \theta}\right) + \frac{2\mu}{\hbar^2}(E_s - V)\psi = 0$$

where $\mu = m/2$ and $r, \theta, \phi$ are the polar co-ordinates of the second particle relative to the first. (Note $r = |\mathbf{r}_1 - \mathbf{r}_2|$.) $E_s$ is the 'internal' energy of the system.

Show that if

$$\psi(\mathbf{r}, \theta, \phi) = R(r)\Theta(\theta)\Phi(\phi)$$

it is necessary that

$$\frac{\mathrm{d}^2\Phi}{\mathrm{d}\phi^2} = -m^2\Phi$$

$$\frac{1}{\sin\theta}\frac{\mathrm{d}}{\mathrm{d}\theta}\left(\sin\theta\frac{\mathrm{d}\Theta}{\mathrm{d}\theta}\right) - \frac{m^2}{\sin^2\theta}\Theta + \lambda\Theta = 0$$

$$\frac{1}{r^2}\frac{\mathrm{d}}{\mathrm{d}r}\left(r^2\frac{\mathrm{d}R}{\mathrm{d}r}\right) - \frac{\lambda}{r^2}R + \frac{2\mu}{\hbar^2}[E_s - V]R = 0$$

where $m, \lambda$ are separation constants.

Following the argument leading to equation (4.59) show that for a physically acceptable solution

$$\frac{1}{r^2}\frac{\mathrm{d}}{\mathrm{d}r}\left(r^2\frac{\mathrm{d}R}{\mathrm{d}r}\right) + \left[-\frac{l(l+1)}{r^2} + \frac{2\mu}{\hbar^2}(E_s - V)\right]R = 0 \qquad \text{(i)}$$

with

$$l \geqslant |m|$$

and $m$ is an integer.

10 In question 9 assume the 'Hooke's law' potential

$$V(r) = \frac{k}{2}(r - r_0)^2 + V_0$$

where $V_0, k, r_0$ are constants. In equation (i) substitute

$$R(r) = \frac{S(r)}{r} \qquad \text{and} \qquad \rho = r - r_0$$

and hence show that for those states with $l = 0$

$$-\frac{\hbar^2}{2\mu}\frac{\mathrm{d}^2 S}{\mathrm{d}\rho^2} + \frac{k}{2}\rho^2 S = (E_s - V_0)S.$$

Deduce that the allowed vibrational energies are

$$E_s = V_0 + (n + \tfrac{1}{2})\hbar\sqrt{\frac{k}{\mu}} \qquad n = 0, 1, 2, \ldots.$$

These energy levels are an approximation to the vibrational energy levels of a diatomic molecule. (The rotational levels are given by (4.60).) A better approximation would be obtained by assuming the Morse potential

$$V(r) = V_0\,[e^{-2(r-r_0)/a} - 2\,e^{-(r-r_0)/a}]$$

where $V_0, r_0, a$ are constants.

11 The spin functions $\alpha(\sigma_z)$ and $\beta(\sigma_z)$ may be both chosen to have even parity $p = +1$ (see Heine's book). Deduce that a configurational wave-function (including spin) has the parity $(-1)^{\Sigma l_i}$, where $l_i$ is the angular momentum of the $i$th particle. The complete atomic Hamiltonian including spin terms is invariant under the inversion operator. Use perturbation theory to deduce that configuration interaction only takes place between configurations with the same parity.

12 Show that the following configurations include only those terms given.

(a) $(2s)^2$ $\quad {}^1S$

(b) $(2p)^3$ $\quad {}^4S,\ {}^2D,\ {}^2P$

(c) $(3d)^2$ $\quad {}^3F,\ {}^3P,\ {}^1G,\ {}^1D,\ {}^1S$

13 Explain why the eigenfunctions belonging to the term $(L, S)$ can be characterized by the four quantum numbers $L, S, J, M_J$.

The spin-orbit coupling term in the many-electron Hamiltonian can be written (see Heine's book).

$$A(L, S)[\mathscr{L}_x S_x + \mathscr{L}_y S_y + \mathscr{L}_z S_z]$$

where $\mathscr{L}_x, S_x$ etc. are the system momentum operators. Deduce that the energy of the wave-function defined by $L, S, J, M_J$, is

$$E_0 + \tfrac{1}{2}A(L, S)[J(J+1) - L(L+1) - S(S+1)]\hbar^2$$

where $E_0$ is the term energy ignoring the spin-orbit coupling. (Hint: Compare the fine-structure theory leading to equation (10.31).)

Show that the energy difference between fine structure components with $\delta J = 1$ is equal to $A(L, S)J\hbar^2$. This is Landé's interval rule.

14 Confirm the commutation rules (11.67), (11.68) and (11.69).

15 Use the closure relation

$$\sum_i \phi^*_{Q_i}(\mathbf{x})\phi_{Q_i}(\mathbf{x}_0) \equiv \delta(\mathbf{x} - \mathbf{x}_0)$$

to show that the field operator $\Psi^+(\mathbf{x}_0)$ creates a particle with defined values of the co-ordinates (including spin) $\mathbf{x}_0$.

16 Show that the right hand sides of (11.87) and (11.88) agree with the right hand sides of (11.74) and (11.83) respectively.

17 Obtain the commutation rule (11.89) for Bosons.

## References

Adams, W. H. *J. Chem. Phys.* **34**, 89 (1961).
Anderson, P. W. *Concepts in Solids.* Benjamin, New York (1963).
Bates, D. R., *Quantum Theory II.* Academic, New York (1962).
Blatt, J. M. and Weisskopf, V. F. *Theoretical Nuclear Physics.* Wiley, New York (1952).
Condon, E. U. and Shortley, G. H. *The Theory of Atomic Spectra.* Cambridge University Press (1935).
Fock, V. *Z. Physik.* **61**, 126 (1930).
Hartree, D. R. *Proc. Camb. Phil. Soc.* **24**, 111 (1928).
Heine, V. *Group Theory in Quantum Mechanics.* Pergamon, Oxford (1964).
Landau, L. D. and Lifshitz, E. M. *Quantum Mechanics–Non-relativistic Theory.* Pergamon Press, London (1959).
Power, E. A. *Introductory Quantum Electrodynamics.* Longmans, London (1964).
Slater, J. C. *Phys. Rev.* **35**, 210 (1930).
Slater, J. C. *Phys. Rev.* **81**, 385 (1951).
Tinkham, M. *Group Theory and Quantum Mechanics*, McGraw-Hill, New York (1964).

CHAPTER 12

# The Dirac Equation

## 12.1 Introduction

The quantum mechanics described in the previous chapters is essentially non-relativistic and has been obtained by quantizing the appropriate classical expressions. It is necessary to develop the quantum theory to obtain a description that is valid for particles with speeds approaching that of light. The present chapter deals only with the relativistic theory for a single particle. A satisfactory many-particle theory requires quantum field operators and is beyond the scope of this book.

A useful feature of the relativistic equations is that particle spin is built into the theory and does not have to be added in the way that Pauli added the spin concept to Schrödinger's theory.

## 12.2 The Lorentz Transformation

According to the principle of special relativity, the velocity of light as measured by an observer is independent of the motion of the observer. Consider two inertial co-ordinate frames of reference $S$ and $S'$ such that $S'$ moves relatively to $S$ at the constant speed $v$ along the positive $x$-axis. The $x'$-axis coincides with the $x$-axis. $(x, y, z, t)$ and $(x', y', z', t)$ are the space and time co-ordinates referred to the reference frames $S$ and $S'$ respectively.

Suppose that at the instant $t = 0$ a spherical electromagnetic wave leaves the origin of $S$ which at that moment coincides with the origin of $S'$. The speed of propagation of the wave front is isotropic and equal to the same constant value $c$ in both reference frames and so

$$x^2 + y^2 + z^2 - c^2t^2 = 0 = x'^2 + y'^2 + z'^2 - c^2t'^2 \qquad (12.1)$$

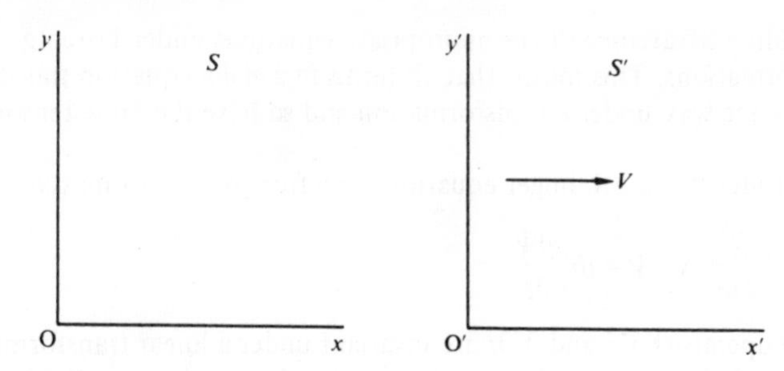


Figure 12.1

The transformation from $(x, y, z, t)$ to $(x', y', z', t')$ that satisfies (12.1) is

$$x' = \beta(x - vt), \qquad y' = y, \qquad z' = z, \qquad t' = \beta\left(t - \frac{v}{c^2}x\right) \quad (12.2)$$

with

$$\beta = 1 \bigg/ \sqrt{1 - \frac{v^2}{c^2}}\,.$$

Such a transformation is called a Lorentz transformation. Strictly this is a proper Lorentz transformation as it does not involve an inversion in space or time. When $v \ll c$, (12.2) reduces to the classical Galilean transformation

$$x' = x - vt, \qquad y' = y, \qquad z' = z, \qquad t' = t. \quad (12.3)$$

The transformation (12.2) can be inverted to give

$$x = \beta(x' + vt'), \qquad y = y', z = z', \qquad t = \beta\left(t' + \frac{v}{c^2}x'\right). \quad (12.4)$$

A rotation of axis in three-dimensional space leaves the quantity $x^2 + y^2 + z^2$ invariant. If a co-ordinate $T$ is defined by $T = ict$ then a Lorentz transformation is a linear transformation that leaves the quantity $x^2 + y^2 + z^2 + T^2$ invariant. A proper Lorentz transformation may be regarded as a rotation in four dimensional (Minkowski) space-time and the ordered 4-tuple $(x, y, z, T)$ is a vector the length of which remains unchanged under a Lorentz transformation. Clearly the ordinary rotations in $(x, y, z)$ space are a special type of Lorentz transformation.

Special relativity requires that the laws of nature take the same form in all inertial frames of reference. This can be expressed mathematically by demanding covariance of the appropriate equations under Lorentz transformations. This means that all terms in a given equation transform in the same way under a transformation and so have the same tensor rank.

Consider the Schrödinger equation for a free particle of mass $m$.

$$-\frac{\hbar^2}{2m}\nabla^2\Psi = i\hbar\frac{\partial\Psi}{\partial t}.$$

The operators $\nabla^2$ and $\partial/\partial t$ are invariant under a linear transformation of $(x, y, z)$ alone and so this equation is covariant under a spatial rotation. However it is clearly not covariant under a Lorentz transformation of the type (12.4). It may involve second derivatives involving the time after the transformation is applied. This result is not surprising as the Schrödinger equation is obtained from a non-relativistic, classical Hamiltonian.

## 12.3 The Klein-Gordon Equation

A relativistically invariant quantum theory must be based on a relativistic Hamiltonian.

Consider a free particle of rest mass $m_0$ moving relative to an observer with speed $v$. The ordered 4-tuple

$$\left(p_x, p_y, p_z, \frac{iH}{c}\right)$$

where the first three entries are the momentum components,

$$p_x = \beta m_0\dot{x} \qquad p_y = \beta m_0\dot{y} \qquad p_z = \beta m_0\dot{z}$$

$$\beta = 1\bigg/\sqrt{1-\frac{v^2}{c^2}}$$

and $H$ is the Hamiltonian, transforms as a vector in four-dimensional space-time and its length is invariant under a Lorentz transformation so that

$$p_x^2 + p_y^2 + p_z^2 - \frac{H^2}{c^2} = -m_0^2c^2. \tag{12.5}$$

The relativistic Hamiltonian is given by

$$H^2 = (p_x^2 + p_y^2 + p_z^2)c^2 + m_0^2c^4. \tag{12.6}$$

The quantum mechanical Hamiltonian is obtained by the usual substitution for the linear momentum components.

$$\mathscr{H}^2 \equiv -c^2\hbar^2\nabla^2 + m_0^2c^4. \tag{12.7}$$

The usual time-dependent Schrödinger equation (postulate 3, Chapter 3) is impossible to apply because of the square root in the Hamiltonian. However, since the operator $\mathscr{H}$ is equivalent to the operator

$$i\hbar \frac{\partial}{\partial t}$$

then a suitable equation of motion is

$$\mathscr{H}^2 \Psi = -\hbar^2 \frac{\partial^2 \Psi}{\partial t^2} \tag{12.8}$$

i.e.

$$\left[\nabla^2 \Psi - \frac{1}{c^2}\frac{\partial^2 \Psi}{\partial t^2}\right] = \frac{m_0^2 c^2}{\hbar^2}\,\Psi. \tag{12.9}$$

Note that this equation is second order in time. It is well-known that the operator

$$\square^2 \equiv \nabla^2 - \frac{1}{c^2}\frac{\partial^2}{\partial t^2}$$

is invariant under a Lorentz transformation and so equation (12.9) is Lorentz covariant as $m_0$ and $c$ are constant scalar quantities.

This equation was discovered by Schrödinger (1926) before he discovered the non-relativistic equation that bears his name. However as it did not give good agreement with the known hydrogen spectrum he discarded it. Klein and Gordon (1926) studied equation (12.9) and it carries their names.

*Equation of continuity*

In section 5.5 it was shown that in the non-relativistic theory, the probability density $P$ and the probability current density $\mathbf{S}$ satisfy the equation of continuity at each point of space, i.e.

$$\frac{\partial P}{\partial t} + \operatorname{div} \mathbf{S} = 0 \tag{12.10}$$

where

$$P = \Psi^* \Psi$$

and

$$\mathbf{S} = \frac{i\hbar}{2m_0}\left[\Psi \operatorname{grad} \Psi^* - \Psi^* \operatorname{grad} \Psi\right]. \tag{12.11}$$

In the same way by multiplying the Klein-Gordon equation (12.9) by $\Psi^*$ and subtracting from this result its complex conjugate, the equation of continuity (12.10) is again obtained. However in this case the probability density is

$$P = \frac{\hbar}{2im_0c^2}\left(\Psi\frac{\partial\Psi^*}{\partial t} - \Psi^*\frac{\partial\Psi}{\partial t}\right), \tag{12.12}$$

with the probability current density taking the Schrödinger form (12.11).

The Klein-Gordon probability density (12.12) introduces a difficulty that does not arise in the Schrödinger case. The Schrödinger $P$ (12.11) is nowhere negative as expected for a probability density. Unfortunately this is not the case for the case for the Klein-Gordon probability density (12.12). This expression can be positive, zero or negative. This arises because the Klein-Gordon equation, unlike the Schrödinger equation, is second order in time and the values of $\Psi$ and $\partial\Psi/\partial t$ at a given instant can be arbitrary. Even if these values were chosen to make $\mathbf{P} \geqslant 0$ at all points of space at some initial time it would not follow that $P$ would not become negative at a later time. It was some time before this curious result was interpreted. The quantitites $eP$ and $e\mathbf{S}$ can be regarded as the electric charge density and current density respectively and it is reasonable for both to be positive or negative if both signs for charge occur. Also, at high energies there is the possibility of creation and annihilation of particle pairs so that the number of particles is not conserved. The charge density $eP$ depends on the relative number of positive and negative charges and so may be positive or negative. It can be shown from the equation of continuity that the total charge is conserved, i.e.

$$\int eP \, \mathrm{d}\tau = \text{constant}.$$

The integral is over all space co-ordinates.

In explaining the change in sign of $eP$ and $e\mathbf{S}$ it has been necessary to allow the presence of both positive and negative (and neutral) particles and this is essentially a many-particle theory. Also, it can be shown that in the non-relativistic limit of low energies the Klein-Gordon equation reduces to the Schrödinger equation without spin. The solutions of the Klein-Gordon equation describe particles with zero-spin. Pions are such particles. However the Klein-Gordon equation certainly cannot describe particles with spin $\frac{1}{2}$ and so cannot describe the motion of electrons. Dirac discovered the correct equation for electrons.

## 12.4 The Dirac Equation

By an ingenious method Dirac in 1928 succeeded in obtaining a linear relativistic equation which is first order in time. For a free particle, the relativistic Hamiltonian is given by

$$\mathscr{H}^2 \equiv (\hat{p}_x^2 + \hat{p}_y^2 + \hat{p}_y^2)c^2 + m_0^2 c^4, \tag{12.13}$$

where $\hat{p}_x$, $\hat{p}_y$, $\hat{p}_z$ represent the usual quantum operators.

The basic postulate of Dirac was to express the right side of (12.13) as a perfect square so that the relativistic quantum Hamiltonian is

$$\mathscr{H} \equiv c(\alpha_x \hat{p}_x + \alpha_y \hat{p}_y + \alpha_z \hat{p}_z + \beta m_0 c) \tag{12.14}$$

with $\alpha_x, \alpha_y, \alpha_z, \beta$ independent of position, momentum and time. The negative root could be taken for $\mathscr{H}$ but this choice does not lead to any extra solutions. From (12.13) and (12.14) it is clearly necessary for

$$c^2(\alpha_x \hat{p}_x + \alpha_y \hat{p}_y + \alpha_z \hat{p}_z + \beta m_0 c)^2 \equiv (\hat{p}_x^2 + \hat{p}_y^2 + \hat{p}_z^2)c^2 + m_0^2 c^4.$$

It follows immediately that

$$\beta^2 = \alpha_x^2 = \alpha_y^2 = \alpha_z^2 = 1 \tag{12.15a}$$

and that any pair of $\alpha_x, \alpha_y, \alpha_z, \beta$ anticommute, i.e.

$$\alpha_x \alpha_y + \alpha_y \alpha_x = 0$$

$$\alpha_x \beta + \beta \alpha_x = 0, \quad \text{etc.} \tag{12.15b}$$

It is clear from (12.15) that the $\alpha$'s and $\beta$ cannot be simply numbers as they do not commute. They are in fact operators and can be represented by square matrices $\boldsymbol{\alpha}_x, \boldsymbol{\alpha}_y, \boldsymbol{\alpha}_z$ and $\boldsymbol{\beta}$. The Hamiltonian is

$$\mathscr{H} \equiv c\,\frac{\hbar}{i}\,\boldsymbol{\alpha}\,.\,\text{grad} + \boldsymbol{\beta} m_0 c^2 \tag{12.16}$$

where

$$\boldsymbol{\alpha}\,.\,\text{grad} \equiv \boldsymbol{\alpha}_x \frac{\partial}{\partial x} + \boldsymbol{\alpha}_y \frac{\partial}{\partial y} + \boldsymbol{\alpha}_z \frac{\partial}{\partial z}.$$

The time-dependent equation (3.10) becomes

$$\left(c\,\frac{\hbar}{i}\,\boldsymbol{\alpha}\,.\,\text{grad} + \boldsymbol{\beta} m_0 c^2\right)\boldsymbol{\Phi} = i\hbar\,\frac{\partial \boldsymbol{\Phi}}{\partial t}. \tag{12.17}$$

Equation (12.17), which is first order in both space and time co-ordinates, is relativistically covariant and is Dirac's equation for a free particle. It must be understood that $\boldsymbol{\Phi}$ is a column matrix representing a many-component wave-function.

As the Hamiltonian must be Hermitian then the $\boldsymbol{\alpha}$ and $\boldsymbol{\beta}$ matrices must also be Hermitian. Hermitian matrices that satisfy the commutation relations (12.15) can only be found if they have at least 4 rows and columns. One possible choice is to diagonalize $\boldsymbol{\beta}$ with

$$\boldsymbol{\beta} = \begin{pmatrix} 1 & 0 & 0 & 0 \\ 0 & 1 & 0 & 0 \\ 0 & 0 & -1 & 0 \\ 0 & 0 & 0 & -1 \end{pmatrix} = \begin{pmatrix} \mathbf{I} & \mathbf{0} \\ \mathbf{0} & -\mathbf{I} \end{pmatrix}$$

and then

$$\boldsymbol{\alpha}_x = \begin{pmatrix} 0 & 0 & 0 & 1 \\ 0 & 0 & 1 & 0 \\ 0 & 1 & 0 & 0 \\ 1 & 0 & 0 & 0 \end{pmatrix} = \begin{pmatrix} \mathbf{0} & \boldsymbol{\sigma}_x \\ \boldsymbol{\sigma}_x & \mathbf{0} \end{pmatrix}$$

$$\boldsymbol{\alpha}_y = \begin{pmatrix} 0 & 0 & 0 & -i \\ 0 & 0 & i & 0 \\ 0 & -i & 0 & 0 \\ i & 0 & 0 & 0 \end{pmatrix} = \begin{pmatrix} \mathbf{0} & \boldsymbol{\sigma}_y \\ \boldsymbol{\sigma}_y & \mathbf{0} \end{pmatrix} \tag{12.18}$$

$$\boldsymbol{\alpha}_z = \begin{pmatrix} 0 & 0 & 1 & 0 \\ 0 & 0 & 0 & -1 \\ 1 & 0 & 0 & 0 \\ 0 & -1 & 0 & 0 \end{pmatrix} = \begin{pmatrix} \mathbf{0} & \boldsymbol{\sigma}_z \\ \boldsymbol{\sigma}_z & \mathbf{0} \end{pmatrix}$$

where $\boldsymbol{\sigma}_x, \boldsymbol{\sigma}_y, \boldsymbol{\sigma}_z$ are the Pauli spin matrices. This choice is certainly not unique. Any new set of matrices obtained by premultiplying the above by an arbitrary unitary matrix $\mathbf{S}$ and post multiplying by $\mathbf{S}^{-1}$ will also satisfy the necessary conditions. The choice (12.18) is called the Dirac representation.

These matrices operate on a four-dimensional 'spinor space' which is spanned by a set of four basis functions. The Dirac wave-function satisfying (12.17) is represented in this space by a column matrix.

$$\boldsymbol{\Phi}(\mathbf{r}, t) = \begin{pmatrix} \Psi_1(\mathbf{r}, t) \\ \Psi_2(\mathbf{r}, t) \\ \Psi_3(\mathbf{r}, t) \\ \Psi_4(\mathbf{r}, t) \end{pmatrix}. \tag{12.19}$$

The $\Psi_i(\mathbf{r}, t)$ are the components of the wave-function in the 'spinor-space' and the Dirac equation (12.17) is equivalent to four simultaneous first order equations that are linear and homogeneous in $\Psi_1, \Psi_2, \Psi_3$ and $\Psi_4$. Each of these component wave-functions separately satisfy the Klein–Gordon equation (12.9).

For stationary states, the time dependence can be separated from the spatial part by writing

$$\mathbf{\Phi}(\mathbf{r}, t) = \boldsymbol{\phi}(\mathbf{r})e^{-iEt/\hbar} \tag{12.20}$$

where $\mathbf{\Phi}(\mathbf{r}, t)$ and $\boldsymbol{\phi}(\mathbf{r})$ are both four-component column matrices and from (12.17)

$$\left(c\frac{\hbar}{i}\boldsymbol{\alpha}.\text{grad} + \boldsymbol{\beta}m_0c^2\right)\boldsymbol{\phi}(\mathbf{r}) = E\boldsymbol{\phi}(\mathbf{r}). \tag{12.21}$$

*Probability and current density*

The Dirac equation, unlike the Klein–Gordon equation, leads to a satisfactory equation of continuity with a non-negative probability density. The time-dependent Dirac equation for a free particle is

$$i\hbar\frac{\partial\mathbf{\Phi}}{\partial t} + ic\hbar\boldsymbol{\alpha}.\text{grad}\,\mathbf{\Phi} - \boldsymbol{\beta}m_0c^2\mathbf{\Phi} = 0. \tag{12.17}$$

This is a matrix equation and its Hermitian conjugate is (take the complex conjugate and transpose and note $(\boldsymbol{\alpha}_y^T)^* = \boldsymbol{\alpha}_y$, etc.).

$$-ih\frac{\partial\mathbf{\Phi}^+}{\partial t} - ich\,\text{grad}\,\mathbf{\Phi}^+.\boldsymbol{\alpha} - m_0c^2\mathbf{\Phi}^+\boldsymbol{\beta} = 0. \tag{12.22}$$

$\mathbf{\Phi}^+$ is the row matrix Hermitian conjugate to $\mathbf{\Phi}$, i.e.

$$\mathbf{\Phi}^+(\mathbf{r}, t) = (\Psi_1^*(\mathbf{r}, t), \Psi_2^*(\mathbf{r}, t), \Psi_3^*(\mathbf{r}, t), \Psi_4^*(\mathbf{r}, t)).$$

Observe that in obtaining (12.22), the order of matrices in the second and third terms of (12.17) has been reversed.

The equation of continuity is obtained in the usual way. Premultiply (12.17) by $\mathbf{\Phi}^+$ and postmultiply (12.22) by $\mathbf{\Phi}$ and subtract to obtain

$$i\hbar\left(\mathbf{\Phi}^+\frac{\partial\mathbf{\Phi}}{\partial t} + \frac{\partial\mathbf{\Phi}^+}{\partial t}\mathbf{\Phi}\right) + ic\hbar(\mathbf{\Phi}^+\boldsymbol{\alpha}.\,\text{grad}\,\mathbf{\Phi} + \text{grad}\,\mathbf{\Phi}^+.\boldsymbol{\alpha}\mathbf{\Phi}) = 0$$

i.e.

$$\frac{\partial}{\partial t}(\mathbf{\Phi}^+\mathbf{\Phi}) + c\,\text{div}\,(\mathbf{\Phi}^+\boldsymbol{\alpha}\mathbf{\Phi}) = 0. \tag{12.23}$$

This is the equation of continuity with probability density

$$p = \mathbf{\Phi}^{+}\mathbf{\Phi} = |\Psi_1|^2 + |\Psi_2|^2 + |\Psi_3|^2 + |\Psi_4|^2 \tag{12.24}$$

and probability current density

$$\mathbf{S} = c\mathbf{\Phi}^{+}\boldsymbol{\alpha}\mathbf{\Phi}. \tag{12.25}$$

Equation (12.24) ensures that the probability density is nowhere negative and is the expected result.

The expression representing the probability current density is very interesting as the eigenvalues of $\boldsymbol{\alpha}$ are $\pm 1$. The operator representing the velocity components of the particle is

$$\dot{\mathbf{r}} \equiv \frac{i}{\hbar}[\mathscr{H}, \mathbf{r}] \equiv c\boldsymbol{\alpha} \tag{12.26}$$

and so the eigenvalues of $\dot{\mathbf{r}}$ are $\pm c$. An exact measurement of a velocity component must give $\pm c$. This represents a 'shuddering' of the particle with speed $c$. Trigg (*Quantum Mechanics*), examines this result in more detail and suggests that it is meaningless to specify the position of the particle to an accuracy of less than its Compton wave-length $h/m_0c$. In this case the uncertainty principle would require that the uncertainty in the energy should be of the order of $2m_0c^2$ and this would be sufficient to produce a pair of particles (see last section).

The equation of continuity (12.23), the probability density (12.24) and the probability current density (12.25) remain valid in the presence of an electromagnetic field.

*Covariance of the Dirac equation*

The Dirac equation can be shown to be covariant under a Lorentz transformation. The reader is referred to *Group Theory and Quantum Mechanics*, by Heine, and *Quantum Mechanics*, by Davydov, for a discussion.

*Plane wave solutions*

It is expected that the free-particle Dirac equation (12.17) will have plane wave solutions of the type

$$\Psi_j(\mathbf{r}, t) = a_j \exp i(\mathbf{k} . \mathbf{r} - Et/\hbar) \qquad j = 1, 2, 3, 4 \tag{12.27}$$

with

$$\mathbf{k} . \mathbf{r} = xk_x + yk_y + zk_z \tag{12.28}$$

and $a_j$ are constants. Substitution of (12.27) into (12.17) gives the four simultaneous equations

$$(m_0c^2 - E)a_1 + c\hbar k_z a_3 + c\hbar(k_x - ik_y)a_4 = 0$$
$$(m_0c^2 - E)a_2 + c\hbar(k_x + ik_y)a_3 - c\hbar k_z a_4 = 0$$
$$c\hbar k_z a_1 + c\hbar(k_x - ik_y)a_2 - (m_0c^2 + E)a_3 = 0$$
$$c\hbar(k_x + ik_y)a_1 - c\hbar k_z a_2 - (m_0c^2 + E)a_4 = 0. \quad (12.29)$$

For a non-trivial solution, the determinant of the coefficients of the $a_j$ must vanish, i.e.

$$(E^2 - m_0^2c^4 - c^2\hbar^2k^2)^2 = 0 \quad (12.30)$$

where

$$k^2 = k_x^2 + k_y^2 + k_z^2. \quad (12.31)$$

The energy can take the two values (both double roots)

$$E_+ = +(m_0^2c^4 + c^2\hbar^2k^2)^{1/2}$$
$$E_- = -(m_0^2c^4 + c^2\hbar^2k^2)^{1/2} \quad (12.32)$$

There are no allowed energy states between $\pm m_0c^2$.

There are two linearly independent solutions corresponding to the positive energy $E_+$. In a non-normalized form, they can be chosen to be

$$a_1 = 1 \qquad a_2 = 0 \qquad a_3 = \frac{c\hbar k_z}{m_0c^2 + E_+} \qquad a_4 = \frac{c\hbar(k_x + ik_y)}{m_0c^2 + E_+} \quad (12.33)$$

and

$$a_1 = 0 \qquad a_2 = 1 \qquad a_3 = \frac{c\hbar(k_x - ik_y)}{m_0c^2 + E_+} \qquad a_4 = \frac{-c\hbar k_z}{m_0c^2 + E_+}. \quad (12.34)$$

Similarly, for the negative energy $E_-$, there are two linearly independent solutions

$$a_1 = \frac{-c\hbar k_z}{m_0c^2 - E_-} \qquad a_2 = \frac{-c\hbar(k_x + ik_y)}{m_0c^2 - E_-} \qquad a_3 = 1 \qquad a_4 = 0$$

$$a_1 = \frac{-c\hbar(k_x - ik_y)}{m_0c^2 - E_-} \qquad a_2 = \frac{c\hbar k_z}{m_0c^2 - E_-} \qquad a_3 = 0 \qquad a_4 = 1. \quad (12.35)$$

In the non-relativistic limit $E_+ \sim m_0c^2$ and the positive energy solutions (12.33) have two 'large' components $a_1$, $a_2$ and two 'small' components of order $v/2c$. The negative energy solutions then have two 'small' components $a_1$, $a_2$ and two 'large' components $a_3$, $a_4$.

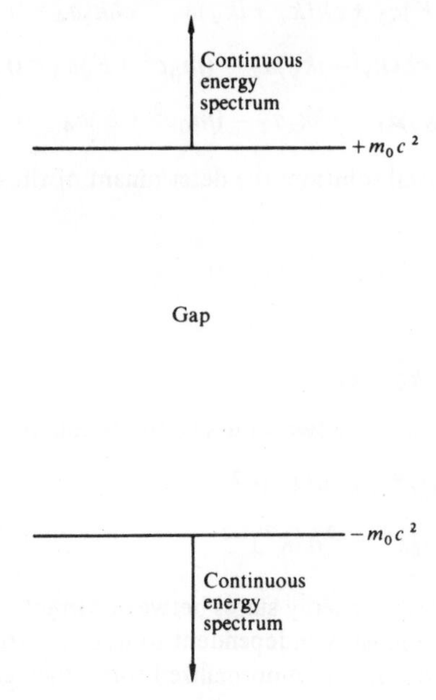


Figure 12.2 Continuous energy spectrum.

The energy gap between $\pm m_0c^2$ occurs in 'classical' relativistic theory (12.6). However these negative energies can be ignored in classical theory as energies can only change continuously and a particle whose energy is positive at some instant will remain so. This is not the case in quantum theory as the particle energy can change discontinuously. The relation between positive and negative energies will be discussed later.

*Introduction of spin*

For a free particle it is expected that angular momentum will be conserved. However it is simple to show that orbital angular momentum is not a constant of the motion.

In the matrix representation the operator representing the $z$-component of orbital angular momentum is

$$\mathscr{L}_z \equiv \frac{h}{i}\left(x\frac{\partial}{\partial y} - y\frac{\partial}{\partial x}\right)\mathbf{I} \tag{12.36}$$

where **I** is the 4 × 4 unit matrix. The free particle Dirac Hamiltonian is

$$\mathscr{H} \equiv \left( c\frac{\hbar}{i}\boldsymbol{\alpha}\,.\,\text{grad} + \boldsymbol{\beta} m_0 c^2 \right). \tag{12.37}$$

The second term in the Hamiltonian is diagonal and so commutes with $\mathscr{L}_z$. It is simple to show that the commutator

$$\begin{aligned}[\mathscr{H}, \mathscr{L}_z] &\equiv \left[ c\frac{\hbar}{i}\boldsymbol{\alpha}\,.\,\text{grad},\ \frac{\hbar}{i}\left( x\frac{\partial}{\partial y} - y\frac{\partial}{\partial x} \right)\mathbf{I} \right] \\ &\equiv -\hbar^2 c\left( \boldsymbol{\alpha}_x \frac{\partial}{\partial y} - \boldsymbol{\alpha}_y \frac{\partial}{\partial x} \right) \\ &\equiv -\hbar^2 c(\boldsymbol{\alpha} \wedge \text{grad})_z .\end{aligned} \tag{12.38}$$

As $\mathscr{L}_z$ and $\mathscr{H}$ do not commute then the $z$-component of orbital angular momentum is not conserved. The generalization of (12.38) to include the two other components gives

$$[\mathscr{H}, \mathscr{L}] \equiv -\hbar^2 c(\boldsymbol{\alpha} \wedge \text{grad}). \tag{12.39}$$

Now consider the matrix operator

$$\boldsymbol{\sigma}_z' = \begin{pmatrix} 1 & 0 & 0 & 0 \\ 0 & -1 & 0 & 0 \\ 0 & 0 & 1 & 0 \\ 0 & 0 & 0 & -1 \end{pmatrix} = \begin{pmatrix} \boldsymbol{\sigma}_z & \mathbf{0} \\ \mathbf{0} & \boldsymbol{\sigma}_z \end{pmatrix}. \tag{12.40}$$

It is not difficult to show that the commutator

$$\left[ \mathscr{H}, \frac{\hbar}{2}\boldsymbol{\sigma}_z' \right] = +\hbar^2 c(\boldsymbol{\alpha} \wedge \text{grad})_z . \tag{12.41}$$

By defining

$$\boldsymbol{\sigma}_x' = \begin{pmatrix} \boldsymbol{\sigma}_x & \mathbf{0} \\ \mathbf{0} & \boldsymbol{\sigma}_x \end{pmatrix} \qquad \boldsymbol{\sigma}_y' = \begin{pmatrix} \boldsymbol{\sigma}_y & \mathbf{0} \\ \mathbf{0} & \boldsymbol{\sigma}_y \end{pmatrix} \tag{12.42}$$

it can similarly be shown that

$$\left[ \mathscr{H}, \frac{\hbar}{2}\boldsymbol{\sigma}' \right] = +\hbar^2 c(\boldsymbol{\alpha} \wedge \text{grad}) \tag{12.43}$$

where $\boldsymbol{\sigma}'$ has the three components $\boldsymbol{\sigma}_x'$, $\boldsymbol{\sigma}_y'$ and $\boldsymbol{\sigma}_z'$.

By combining (12.38) and (12.43) then

$$\left[\mathscr{H}, \mathscr{L} + \frac{\hbar}{2}\boldsymbol{\sigma}'\right] = 0 \tag{12.44}$$

and

$$\mathscr{L} + \frac{\hbar}{2}\boldsymbol{\sigma}'$$

is a constant of the motion.

By analogy with the Pauli spin theory a spin operator $\mathbf{S}'$ is defined by

$$\mathbf{S}' = \frac{\hbar}{2}\boldsymbol{\sigma}' \tag{12.45}$$

and the total angular momentum $\mathbf{J}$ is defined by

$$\mathbf{J} = \mathscr{L} + \mathbf{S}'. \tag{12.46}$$

The components of both $\mathbf{S}'$ and $\mathscr{L}$ satisfy the necessary commutation relations for angular momentum operators (10.4). From (12.44), total angular momentum is conserved for a free particle.

The operators $\boldsymbol{\sigma}_x', \boldsymbol{\sigma}_y', \boldsymbol{\sigma}_z'$ have eigenvalues $\pm 1$ corresponding to eigenvalues $\pm \hbar/2$ for the spin operators $\mathbf{S}_x', \mathbf{S}_y', \mathbf{S}_z'$. The Dirac equations describes particles whose spin along any specified direction can only take the two values $\pm \hbar/2$. An electron is such a particle with spin $\frac{1}{2}$. (Other such particles are protons, neutrons, etc.)

Consider the plane wave energy functions obtained in the previous section. In the non-relativistic limit, putting the small components equal to zero, the column vectors representing the two solutions with energy $E_+$ are, from (12.33)

$$\begin{pmatrix} 1 \\ 0 \\ 0 \\ 0 \end{pmatrix} \quad \text{and} \quad \begin{pmatrix} 0 \\ 1 \\ 0 \\ 0 \end{pmatrix}$$

These are simultaneously eigenvectors of $\mathbf{S}_z'$ with eigenvalues $+\hbar/2$ and $-\hbar/2$ respectively. Of course these vectors are not eigenvectors of $\mathbf{S}_x'$ or $\mathbf{S}_y'$ as these operators do not commute with $\mathbf{S}_z'$.

Similarly, the 'non-relativistic' eigenvectors with energy $E_-$ are, from (12.35)

$$\begin{pmatrix} 0 \\ 0 \\ 1 \\ 0 \end{pmatrix} \quad \text{and} \quad \begin{pmatrix} 0 \\ 0 \\ 0 \\ 1 \end{pmatrix}$$

and are simultaneously eigenvectors of $\mathbf{S}_z'$ with eigenvalues $+\hbar/2$ and $-\hbar/2$ respectively.

## 12.5 The Dirac Particle in an E.M. Field

An electromagnetic field is described completely when the vector potential $\mathbf{A}$ and the scalar potential $\phi$ are given. It can be shown (*Classical Electricity and Magnetism*, by Panofsky and Phillips), that the ordered 4-tuple

$$\left(A_x, A_y, A_z, \frac{i\phi}{c}\right)$$

is a vector in Minkowski space and its length is invariant under a Lorentz transformation.

The momentum $p_x$ conjugate to $x$ for a particle (electron) with charge $-e$ in an electromagnetic field is given by ((1.43) holds in relativistic mechanics)

$$p_x = \beta m_0 \dot{x} - eA_x \tag{12.47}$$

and similarly for the other two components. The 'classical' relativistic Hamiltonian for the particle in the field is obtained from the free particle Hamiltonian in the usual way by substituting for the linear momentum in terms of the new conjugate momentum (12.47) and also replacing the Hamiltonian by $H + e\phi$. From (12.6)

$$(H + e\phi)^2 = c^2 [(p_x + eA_x)^2 + (p_y + eA_y)^2 + (p_z + eA_z)^2] + m_0^2 c^4. \tag{12.48}$$

A quantum-mechanical wave equation is obtained by the substitutions

$$H \to i\hbar \frac{\partial}{\partial t}$$

and

$$p_x \to \frac{\hbar}{i}\frac{\partial}{\partial x} \text{ etc., i.e.}$$

$$\left(i\hbar\frac{\partial}{\partial t} + e\phi\right)^2 \Psi = c^2 \left[\left(\frac{\hbar}{i}\frac{\partial}{\partial x} + eA_x\right)^2 + \left(\frac{\hbar}{i}\frac{\partial}{\partial y} + eA_y\right)^2 + \left(\frac{\hbar}{i}\frac{\partial}{\partial z} + eA_z\right)^2 + m_0^2 c^2\right]\Psi. \tag{12.49}$$

This relativistically covariant equation is the Klein-Gordon equation for a particle of charge $-e$ in an electromagnetic field and will not be considered further.

The Dirac equation for a particle of charge $-e$ in the field is obtained in a similar manner. From (12.17)

$$\left[i\hbar\mathbf{I}\frac{\partial}{\partial t}+\mathbf{I}\,e\phi-c\boldsymbol{\alpha}.\left(\frac{\hbar}{i}\,\mathrm{grad}+e\mathbf{A}\right)-\boldsymbol{\beta}m_0c^2\right]\boldsymbol{\Phi}=0. \tag{12.50}$$

This equation is also relativistically covariant but its solutions are not solutions of the Klein-Gordon equation unlike the field-free case. To see this operate on (12.50) using

$$\left[i\hbar\mathbf{I}\frac{\partial}{\partial t}+\mathbf{I}\,e\phi+c\boldsymbol{\alpha}.\left(\frac{\hbar}{i}\,\mathrm{grad}+e\mathbf{A}\right)+\boldsymbol{\beta}m_0c^2\right]$$

to obtain a similar but different equation to (12.49).

To find the stationary states of the Dirac equation put

$$\boldsymbol{\Phi}(\mathbf{r},t)=\boldsymbol{\phi}(\mathbf{r})\,e^{-iEt/\hbar}$$

in (12.50) and then

$$\left[\mathbf{I}E+\mathbf{I}\,e\phi-c\boldsymbol{\alpha}.\left(\frac{\hbar}{c}\,\mathrm{grad}+e\mathbf{A}\right)-\boldsymbol{\beta}m_0c^2\right]\boldsymbol{\phi}(\mathbf{r})=0. \tag{12.51}$$

It has already been demonstrated that in the non-relativistic limit the four-component Dirac wave-function can be divided into large and small components. With this limiting case in mind put

$$\boldsymbol{\phi}(\mathbf{r})=\begin{pmatrix}\boldsymbol{\chi}_l\\ \boldsymbol{\chi}_s\end{pmatrix}$$

where $\boldsymbol{\chi}_l$ and $\boldsymbol{\chi}_s$ are two-component column vectors. With this substituion equation (12.51) becomes

$$\mathbf{I}[E+e\phi-m_0c^2]\boldsymbol{\chi}_l-c\boldsymbol{\sigma}.\left(\frac{\hbar}{i}\,\mathrm{grad}+e\mathbf{A}\right)\boldsymbol{\chi}_s=0$$

$$\mathbf{I}[E+e\phi+m_0c^2]\boldsymbol{\chi}_s-c\boldsymbol{\sigma}.\left(\frac{\hbar}{i}\,\mathrm{grad}+e\mathbf{A}\right)\boldsymbol{\chi}_l=0 \tag{12.52}$$

with

$$\boldsymbol{\sigma}.\left(\frac{\hbar}{i}\,\mathrm{grad}+e\mathbf{A}\right)=\sigma_x\left(\frac{\hbar}{i}\frac{\partial}{\partial x}+e\mathbf{A}_x\right)+\sigma_y\left(\frac{\hbar}{i}\frac{\partial}{\partial y}+eA_y\right)$$

$$+\sigma_z\left(\frac{\hbar}{i}\frac{\partial}{\partial z}+eA_z\right).$$

Consider a positive energy solution with

$$E=m_0c^2+E'.$$

Eliminating $\chi_s$ from (12.52)

$$\left\{c^2\frac{\left[\boldsymbol{\sigma}.\left(\frac{\hbar}{i}\,\text{grad}+e\mathbf{A}\right)\right]^2}{2m_0c^2+E'+e\phi}-\mathbf{I}\,e\phi\right\}\chi_l=\mathbf{I}E'\chi_l. \qquad (12.53)$$

In the non-relativistic limit of small kinetic energy and if the field is weak then $E'\ll m_0c^2$ and $e\phi\ll m_0c^2$ and so

$$\frac{1}{(2m_0c^2+E'+e\phi)}=\frac{1}{2m_0c^2}\left[1-\frac{E'+e\phi}{2m_0c^2}+\ldots\right].$$

Keeping the first term only of this binomial expansion (12.53) becomes

$$\left\{\frac{1}{2m_0}\left[\boldsymbol{\sigma}.\left(\frac{\hbar}{i}\,\text{grad}+e\mathbf{A}\right)\right]^2-\mathbf{I}\,e\phi\right\}\chi_l=\mathbf{I}E'\chi_l. \qquad (12.54)$$

This is a two-component equation and gives the large components $\chi_l$.

In this non-relativistic limit the magnitude of the momentum is given by $m_0V$ where $V$ is the electron speed and the second of the equations (12.52) becomes

$$\chi_s\simeq\frac{1}{2}\frac{V}{c}\chi_l.$$

This result confirms that $\chi_s$ is 'small' compared with $\chi_l$. As the operator

$$\left(\frac{\hbar}{i}\,\text{grad}+e\mathbf{A}\right)$$

commutes with the components of $\boldsymbol{\sigma}$ then it can be shown that (problem 6)

$$\left[\boldsymbol{\sigma}.\left(\frac{\hbar}{i}\,\text{grad}+e\mathbf{A}\right)\right]^2=\mathbf{I}\left(\frac{\hbar}{i}\,\text{grad}+e\mathbf{A}\right)^2+i\boldsymbol{\sigma}.\left[\left(\frac{\hbar}{i}\,\text{grad}+e\mathbf{A}\right)\right.$$

$$\left.\wedge\left(\frac{\hbar}{i}\,\text{grad}+e\mathbf{A}\right)\right]=\mathbf{I}\left(\frac{\hbar}{i}\,\text{grad}+e\mathbf{A}\right)^2+e\hbar\boldsymbol{\sigma}.\,\text{curl}\,\mathbf{A}. \qquad (12.55)$$

The magnetic induction

$$\mathbf{B}=\text{curl}\,\mathbf{A}$$

and so in the non-relativistic limit, the two large components of the Dirac equation satisfy

$$\left[\frac{\mathbf{I}}{2m_0}\left(\frac{\hbar}{i}\,\text{grad}+e\mathbf{A}\right)^2-\mathbf{I}e\phi+\frac{e\hbar}{2m_0}\,\sigma.\mathbf{B}\right]\chi_l=\mathbf{I}E'\chi_l. \qquad (12.56)$$

This equation was suggested by Pauli for the electron. It is of course simply the equation obtained by combining the Schrödinger non-relativistic theory and the Pauli electron spin theory of Chapter 10. (c.f. (8.24), (10.24), (10.25)). The Pauli theory is obtained in the non-relativistic limit (i.e. small velocity) from the Dirac theory. The final term on the left hand side of (12.56) implies that the particle has an intrinsic magnetic moment of

$$\frac{-e\hbar}{2m_0}\boldsymbol{\sigma}$$

in agreement with the value assumed in Chapter 10 for the electron. Clearly the particles described by the Dirac equation are electrons.

*Coulomb field*

A Coulomb field due to a charge of $+Ze$ can be described by

$$\mathbf{A} = 0 \qquad \phi = \frac{+Ze}{4\pi\epsilon_0 r}.$$

The Dirac Hamiltonian is then

$$\mathscr{H} \equiv c\frac{\hbar}{i}\boldsymbol{\alpha}\,.\,\mathrm{grad} + \boldsymbol{\beta} m_0 c^2 - \frac{Ze^2}{4\pi\epsilon_0 r}.$$

The solutions of the corresponding equation is dealt with in detail by Davydov (*Quantum Mechanics*, p. 279). It is sufficient here to point out that the total angular momentum as defined by (12.46) is conserved for a particle in a central field. The reader is asked to verify this himself.

## 12.6 The Positron

It has been shown that the Dirac equation has both positive and negative energy states and this leads to a difficulty in interpretation. These negative energy states correspond to the negative square root of the classical 'relativistic' expression (12.6). However, in classical theory a particle energy changes continuously and so a particle in a state with positive energy will remain so and the negative energy states can be ignored. This is not the case in quantum mechanics as a particle can make discontinuous jumps in energy and so cross the 'forbidden' gap $2m_0c^2$. There are an infinite number of negative energy states. In this case it would seem reasonable to expect a Dirac particle to fall to lower energy states radiating continuously. However electrons are not observed to do this and some other interpretation of the negative energy states must be forthcoming.

In 1930 Dirac proposed that in the normal state the negative energy states are filled and assuming the particles obey the Pauli principle a particle with positive energy cannot jump to a state with negative energy. The charge density due to the filled negative energy states must be assumed to be unobservable. A vacuum is then a space in which all negative energy states are filled and all positive energy states are empty. Only deviations from the vacuum state are observed.

A positive energy solution of the Dirac equation (12.50) corresponds to an ordinary negatively charged particle (electron). A negative energy state is only observed when it is empty. Now a filled negative energy state corresponds to a particle with both negative charge and negative mass. Consequently the observed empty negative energy state appears as a particle with equal positive charge and positive mass. This 'anti-particle' appears as a hole in the 'sea' of filled negative energy states. The energy of the anti-particle is positive as it corresponds to the energy required to lift a negatively charged particle out of the negative energy state.

It is quite easy to relate the behaviour of an electron to its anti-particle. The Dirac equation for an electron of charge $-e$ and rest mass $m_0$ in an electromagnetic field is

$$\left[i\hbar \mathbf{I}\frac{\partial}{\partial t} + \mathbf{I}e\phi - c\boldsymbol{\alpha}.\left(\frac{\hbar}{i}\,\text{grad} + e\mathbf{A}\right) - \boldsymbol{\beta}m_0c^2\right]\boldsymbol{\Phi} = 0. \tag{12.57}$$

It is useful to choose a representation in which the $\boldsymbol{\alpha}$ matrices are all real and $\boldsymbol{\beta}$ is imaginary. A choice satisfying the conditions (12.15) is to take $\boldsymbol{\alpha}_x$ and $\boldsymbol{\alpha}_z$ as before (12.18) and interchange $\boldsymbol{\alpha}_y$ and $\boldsymbol{\beta}$ to obtain

$$\boldsymbol{\alpha}_y = \begin{pmatrix} 1 & 0 & 0 & 0 \\ 0 & 1 & 0 & 0 \\ 0 & 0 & -1 & 0 \\ 0 & 0 & 0 & -1 \end{pmatrix} \qquad \boldsymbol{\beta} = \begin{pmatrix} 0 & 0 & 0 & -i \\ 0 & 0 & i & 0 \\ 0 & -i & 0 & 0 \\ i & 0 & 0 & 0 \end{pmatrix}. \tag{12.58}$$

With this choice take the complex conjugate of equation (12.57).

$$-\left[i\hbar \mathbf{I}\frac{\partial}{\partial t} + \mathbf{I}(-e)\phi - c\boldsymbol{\alpha}.\left(\frac{\hbar}{i}\,\text{grad} + (-e)\mathbf{A}\right) - \boldsymbol{\beta}m_0c^2\right]\boldsymbol{\Phi}^* = 0. \tag{12.59}$$

If in (12.57) $\boldsymbol{\Phi}$ describes an electron with charge $-e$, then $\boldsymbol{\Phi}^*$ describes a particle with equal mass but opposite charge $+e$.

Consider now the stationary states. The Dirac equation for an electron with energy $E$ is

$$\left[\mathbf{I}E + \mathbf{I}e\phi - c\boldsymbol{\alpha}.\left(\frac{\hbar}{i}\,\text{grad} + e\mathbf{A}\right) - \boldsymbol{\beta}m_0c^2\right]\boldsymbol{\Phi} = 0. \tag{12.60}$$

Take the complex conjugate of this equation.

$$-\left[\mathbf{I}(-E) + \mathbf{I}(-e)\phi - c\boldsymbol{\alpha}.\left(\frac{\hbar}{i}\,\text{grad} + (-e)\mathbf{A}\right) - \boldsymbol{\beta}m_0c^2\right]\boldsymbol{\Phi}^* = 0. \tag{12.61}$$

If in (12.60) $\boldsymbol{\Phi}$ describes an electron with negative energy ($E < 0$) then $\boldsymbol{\Phi}^*$ describes a positively charged particle with positive energy. The electron and its antiparticle appear symmetrically in the theory. The pair of equations (12.60) and (12.61) should be considered together and the negative energy solutions of one correspond to the positive energy solutions of the other.

From this discussion it is seen that the electron and its antiparticle possess the same mass, spin and differ only in the sign of their charge. The electron antiparticle was discovered in 1932 and is called the positron.

It is observed that a positron always appears as a pair together with an electron and that an energy greater than $2m_0c^2$ is absorbed in the pair creation. This is explained as in pair creation an electron must be excited from a negative energy state through the forbidden gap of $2m_0c^2$ to a positive energy state. The empty 'hole' manifests itself as the positron and the electron in the positive energy state behaves as an ordinary electron.

The reverse process takes place when an electron falls into an unoccupied negative energy state. This occurs as electron-positron annihilation and the energy is emitted as photons.

To explain the negative energy state it is necessary to extend the single particle Dirac theory to a type of many-particle theory. This is not very satisfactory. The difficulties can be overcome using the method of quantized fields but this development is outside the scope of this text. The second quantization formalism mentioned in the previous chapter is necessary to cope with the annihilation and creation of particles.

## PROBLEMS

1 Equation (12.9) is the Klein–Gordon equation for a free particle. Consider the transformation

$$\Psi(\mathbf{r}, t) = \Psi_0(\mathbf{r}, t)\exp(-im_0c^2t/\hbar).$$

In the non-relativistic limit $E = E' + m_0c^2$ with $E' \ll m_0c^2$. Also in general

$$\left|\frac{\partial \Psi_0}{\partial t}\right| \sim \frac{E'}{\hbar}\Psi_0.$$

Hence show that

$$\frac{\partial \Psi}{\partial t} \simeq \frac{-im_0c^2}{\hbar}\Psi_0 \exp(-im_0c^2t/\hbar)$$

and obtain a similar expression for $\partial^2\Psi/\partial t^2$.

Show further that in this approximation of $E' \ll m_0c^2$ the Klein-Gordon equation becomes

$$-\frac{\hbar^2}{2m_0}\nabla^2\Psi_0 = i\hbar\frac{\partial \Psi_0}{\partial t}.$$

This is Schrödinger's non-relativistic equation for a free particle.

2 Using the same transformation as in the above question and in the same non-relativistic limit, show that the Klein–Gordon probability density (12.12) reduces to the Schrödinger probability density (12.11).

3 Show that conditions (12.15) are necessary for the square of (12.14) to be equal to (12.13).

4 Show that the Dirac matrices $\boldsymbol{\alpha}_x$, $\boldsymbol{\alpha}_y$, $\boldsymbol{\alpha}_z$ and $\boldsymbol{\beta}$ are both Hermitian and unitary. Show that all four matrices have eigenvalues of $\pm 1$ only.

5 Confirm that the total angular momentum operator (12.46) commutes with the Hamiltonian for a Dirac particle in a central field.

6 If **A** and **B** are two vector matrix operators that commute with the Pauli spin matrices show that

$$(\boldsymbol{\sigma}.\mathbf{A})(\boldsymbol{\sigma}.\mathbf{B}) = (\mathbf{A}.\mathbf{B}) + i[\boldsymbol{\sigma}.(\mathbf{A}\wedge\mathbf{B})]$$

where $\boldsymbol{\sigma}.\mathbf{A} = \boldsymbol{\sigma}_x\mathbf{A}_x + \boldsymbol{\sigma}_y\mathbf{A}_y + \boldsymbol{\sigma}_z\mathbf{A}_z$, etc. Hence prove equation (12.55).

## References

Davydov, A. S. *Quantum Mechanics.* Pergamon, Oxford (2nd edition, 1976).
Dirac, P. A. M. *Proc. Roy. Soc.* **A117**, 610 (1928).
Dirac, P. A. M. *Quantum Mechanics.* O.U.P., Oxford (1947).
Gordon, W. *Z. Phys.* **40**, 117 (1926).
Heine, V. *Group Theory in Quantum Mechanics.* Pergamon, Oxford (1960).
Klein, O. *Z. Phys.* **37**, 895 (1926).

Panofsky, W. K. H. and Phillips, *Classical Electricity and Magnetism*, Addison-Wesley, Cambridge, Mass. (2nd edition, 1962).
Schrödinger, E. *Ann. d. Physik* **81**, 109 (1926).
Trigg. G. L. *Quantum Mechanics.* Van Nostrand, London (1964).

## Suggested Further Reading

Bethe, H. A. *Intermediate Quantum Mechanics.* Benjamin, New York (1964).
Gottfried, K. *Quantum Mechanics,* Vol. I. Benjamin, New York (1966).
Mackey, G. W. *The Mathematical Foundations of Quantum Mechanics.* Benjamin, New York (1963).
Merzbacher, E. *Quantum Mechanics.* John Wiley, New York (1961).
Messiah, A. *Quantum Mechanics,* Vols. I and II. North-Holland, Amsterdam (1961; 1963).
Trigg, G. L. *Quantum Mechanics.* Van Nostrand, Princeton (1964).

CHAPTER 13

# Electrons in Solids

## 13.1 Introduction

The fundamental property of crystalline solids is their degree of order, whereas gases and liquids are in a state of disorder. An ideal crystal has a periodic structure. The deviations in real crystals, both at the surface, where the periodicity clearly fails, and in the interior due to dislocations, impurities and other imperfections, can often be considered as perturbations of an ideal crystal.

The lattice periodicity has important consequences for the dynamics of the electrons since it leads to the concept of allowed bands of electron energies and forbidden gaps.

A solid may be classified as a metal, a semiconductor or an insulator by the value of its resistivity. At room temperature metals have a resistivity of about $10^{-8}\,\Omega$ m, semiconductors have values in the range $10^{-4}-10^{6}\,\Omega$ m, and insulators have resistivities greater than about $10^{8}\,\Omega$ m. The chief factors that decide this classification for a solid are the valencies of the constituent atoms and the magnitudes of the energy gaps between the bands.

In a crystal, the great multiplicity of particles, both electrons and nuclei, makes solution of the complete Schrödinger equation impossible. Sweeping approximations have to be made.

*The Born-Oppenheimer approximation*

The Born-Oppenheimer approximation attempts to separate the ionic motion from that of the valence electrons.

The stationary states of a crystal are the solution of the Schrödinger equation

$$\begin{aligned}&\mathscr{H}(\mathbf{R}_1,\mathbf{R}_2,\ldots;\mathbf{r}_1,\mathbf{r}_2,\ldots)\,\Psi(\mathbf{R}_1,\mathbf{R}_2,\ldots;\mathbf{r}_1,\mathbf{r}_2,\ldots)\\&\quad=E_{\mathrm{c}}\,\Psi(\mathbf{R}_1,\mathbf{R}_2,\ldots;\mathbf{r}_1,\mathbf{r}_2,\ldots)\end{aligned}\qquad(13.1)$$

where the state function $\Psi$ and the Hamiltonian are functions of all the nuclear and electronic co-ordinates $\mathbf{R}_i$ and $\mathbf{r}_i$ respectively. The number of variables is reduced if the atomic electrons are separated somewhat arbitrarily into core and valence electrons. The core electrons have atomic wave-functions that do not appreciably overlap at the observed interatomic spacing in the crystal and remain unaffected when the atoms combine to form the crystal. The valence electronic wave-functions do overlap and are affected by crystal formation.

The non-relativistic Hamiltonian of (13.1) contains all the interactions of an assembly of ionic cores and valence electrons.

$$\text{i.e. } \mathscr{H}(\mathbf{R}_1, \mathbf{R}_2, \ldots; \mathbf{r}_1, \mathbf{r}_2, \ldots) \equiv \mathscr{H}_{\text{ion}}(\mathbf{R}_1, \mathbf{R}_2, \ldots) + \mathscr{H}_{\text{electron}}(\mathbf{r}_1, \mathbf{r}_2, \ldots) + \mathscr{H}_{\substack{\text{electron}\\ \text{-ion}}}(\mathbf{R}_1, \mathbf{R}_2, \ldots; \mathbf{r}_1, \mathbf{r}_2, \ldots)$$

with

$$\mathscr{H}_{\text{ion}}(\mathbf{R}_1, \mathbf{R}_2, \ldots) \equiv -\sum_i \frac{\hbar^2}{2M} \nabla^2_{\mathbf{R}_i} + \tfrac{1}{2} \sum_{i \neq j} \sum \mathrm{W}(\mathbf{R}_i - \mathbf{R}_j) \qquad (13.2)$$

$$\mathscr{H}_{\text{electron}}(\mathbf{r}_1, \mathbf{r}_2, \ldots) \equiv -\sum_i \frac{\hbar^2}{2m} \nabla^2_{\mathbf{r}_i} + \tfrac{1}{2} \sum_{i \neq j} \sum \frac{e^2}{4\pi\epsilon_0 |\mathbf{r}_i - \mathbf{r}_j|}$$

$$\mathscr{H}_{\substack{\text{electron}\\ \text{-ion}}}(\mathbf{R}_1, \mathbf{R}_2, \ldots; \mathbf{r}_1, \mathbf{r}_2, \ldots) \equiv \sum_i \sum_j \mathrm{V}(\mathbf{r}_i - \mathbf{R}_j)$$

where $M$ is the ionic mass, $m$ is the electron mass, $\mathrm{W}(\mathbf{R}_i - \mathbf{R}_j)$ is the potential energy of interaction of two ions at $\mathbf{R}_i$ and $\mathbf{R}_j$ respectively and $\mathrm{V}(\mathbf{r}_i - \mathbf{R}_j)$ is the interaction potential between an electron at $\mathbf{r}_i$ and an ion at $\mathbf{R}_j$.

If $\mathscr{H}_{\substack{\text{electron}\\ \text{-ion}}}$ could be neglected the separation of the ionic and electronic motions would be exact. Even though this cannot be assumed, since $m \ll M$ the electrons move much more rapidly than the ions and the ionic motion can approximately be separated from that of the electrons. In the Born-Oppenheimer approximation (1927), the Schrödinger equation is solved for all the electrons in the field of the ions in an assumed fixed configuration. The electronic wave-function $\psi(\mathbf{R}_1, \mathbf{R}_2, \ldots; \mathbf{r}_1, \mathbf{r}_2, \ldots)$ depends upon the variable electronic positions $\mathbf{r}_i$ and also on the nuclear co-ordinates $\mathbf{R}_i$ regarded as parameters. The full crystal wave-function is written

$$\Psi(\mathbf{R}_1, \mathbf{R}_2, \ldots; \mathbf{r}_1, \mathbf{r}_2, \ldots) = \Phi(\mathbf{R}_1, \mathbf{R}_2, \ldots)\, \psi(\mathbf{R}_1, \mathbf{R}_2, \ldots; \mathbf{r}_1, \mathbf{r}_2, \ldots) \qquad (13.3)$$

where $\Phi$ is the wave-function describing the ionic motion and is a function of the nuclear co-ordinates $\mathbf{R}_i$ only.

Assume $\psi$ satisfies the Schrödinger equation

$$(\mathscr{H}_{\text{electron}} + \mathscr{H}_{\substack{\text{electron}\\\text{-ion}}})\, \psi\, (\mathbf{R}_1, \mathbf{R}_2, \ldots; \mathbf{r}_1, \mathbf{r}_2, \ldots)$$
$$= E(\mathbf{R}_1, \mathbf{R}_2, \ldots)\, \psi\, (\mathbf{R}_1, \mathbf{R}_2, \ldots; \mathbf{r}_1, \mathbf{r}_2, \ldots) \tag{13.4}$$

The electronic energy eigenvalues $E\,(\mathbf{R}_1, \mathbf{R}_2, \ldots)$ depend upon the particular ionic configuration assumed.

This electronic energy then is part of the effective potential for the approximate Schrödinger equation describing the ionic motion

$$(\mathscr{H}_{\text{ion}} + E\,(\mathbf{R}_1, \mathbf{R}_2, \ldots))\, \Phi\, (\mathbf{R}_1, \mathbf{R}_2, \ldots) = E_{\text{c}}\, \Phi\, (\mathbf{R}_1, \mathbf{R}_2, \ldots) \tag{13.5}$$

The error introduced by the Born-Oppenheimer approximation is small when the characteristic vibration frequencies of the ions are much smaller than the characteristic electronic frequencies. Generally this is the case, since the ionic mass is much greater than the electronic mass. However, if the electronic excitation energy is small, implying a small electronic frequency, the approximation breaks down. This is the case for metals and the ionic function $\Phi$ depends on the electronic co-ordinates as well as the ionic co-ordinates.

Equation (13.4) describes the electronic motion in a static lattice and is solved assuming both the crystal symmetry and the lattice constant. Essentially the Hamiltonian of (13.4) is a function of the electronic co-ordinates only, the nuclear co-ordinates being given their observed mean values, and so

$$\psi = \psi\, (\mathbf{r}_1, \mathbf{r}_2, \ldots) \tag{13.6}$$

*One-electron approximation*

It is very difficult to solve problems involving more than one electron because of the electron-electron interaction terms. In Section 11.4 approximate methods for reducing the many electron problem to a one-electron self-consistent field problem were outlined.

In a crystal, the Hartree-Fock one-electron Hamiltonian (11.32b) takes the form

$$\mathscr{H}_k \equiv -\frac{\hbar^2}{2m} \nabla_1^2 + \sum_i \mathrm{V}(\mathbf{r}_1 - \mathbf{R}_i) + \frac{e^2}{4\pi\epsilon_0} \sum_j \int \frac{|\phi_{Q_j}(\mathbf{r}_2)|^2}{|\mathbf{r}_1 - \mathbf{r}_2|}\, \mathrm{d}\tau_2$$
$$- \frac{e^2}{4\pi\epsilon_0} \sum_j^{\|} \int \frac{\phi^*_{Q_j}(\mathbf{r}_2)\, \phi^*_{Q_k}(\mathbf{r}_1)\, \phi_{Q_k}(\mathbf{r}_2)\, \phi_{Q_j}(\mathbf{r}_1)}{\phi^*_{Q_k}(\mathbf{r}_1)\, \phi_{Q_k}(\mathbf{r}_1)\, |\mathbf{r}_1 - \mathbf{r}_2|}\, \mathrm{d}\tau_2 \tag{13.7}$$

The Hartree-Fock energy $E_{\text{H.F.}}$ for a crystal is the expectation value of the Hamiltonian for a crystal with fixed ions obtained using a determinantal wave-function. This is the expectation value of the Hamiltonian of (13.4) (with the ions fixed) together with the inter-ionic repulsion terms

$$\tfrac{1}{2} \sum_{i \neq j} \sum \mathbf{W}(\mathbf{R}_i - \mathbf{R}_j)$$

Although the interaction between two electrons may be expected to be of the same magnitude as the interaction between an electron and an ion the same distance apart, the one-electron approximation has proved to be useful. Bohm and Pines (1953) have shown why the one-electron approximation gives such good results. They have shown that the screening effect of the mobile electrons limits the long-range Coulomb interactions between them.

The one-electron approximation is used for the rest of this chapter but is not sufficient to fully explain such phenomena as cohesive energy, ferromagnetism or superconductivity.

## 13.2 The Sommerfeld Model

In the one-electron approximation,the Hartree-Fock potential takes account of the interactions of the electron with the ionic cores and with the averaged charge of the electrons. Due to the positive ions at each atom site, the potential must possess singularities and the potential plot along a section through a line of atoms is illustrated in Fig. 13.1.

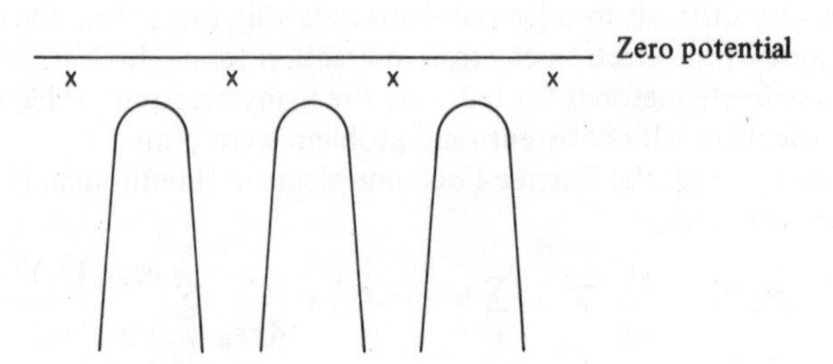


Figure 13.1 Section through a line of atoms.

In 1928 Sommerfeld suggested a model for a metal in which the potential is taken as a constant inside the metal. This implies the force acting on the electron is zero. The metal is replaced by a box containing a non-interacting gas of electrons; the free-electron gas. The electrons are not assumed altogether independent since they are taken to obey the Pauli principle. The free-electron gas does bear some relation to the state of the valence electrons in the alkali metals.

Consider a cube of side '*a*' with the origin of the co-ordinate axes at a corner of the box. Inside the box the potential energy of an electron is zero and outside the potential is infinite. This problem is solved in Section 4.3 and the eigenfunctions are

$$A \sin k_x x \sin k_y y \sin k_z z \tag{13.8}$$

with

$$k_x = \frac{\ell\pi}{a}, \; k_y = \frac{m\pi}{a}, \; k_z = \frac{n\pi}{a}$$

where $\ell, m, n$ are integers. The electronic energy takes the values

$$E_k = \frac{\hbar^2}{2m}(k_x^2 + k_y^2 + k_z^2) = \frac{h^2}{8ma^2}(\ell^2 + m^2 + n^2) \tag{13.9}$$

From (13.8) all the linearly independent solutions are obtained if $\ell$, $m$, $n$ are restricted to positive values only. Since '*a*' has macroscopic dimensions (e.g. $a \sim 10^{-2}$ m) the energy levels lie very close together ($\sim 10^{-15}\,eV$ apart) and for most purposes may be taken as continuous; the energy levels are said to be quasi-continuous.

The number of electronic states (neglecting spin) with kinetic energies less than some given energy $E_k$ is the number of integral points within the positive octant of the sphere of radius $r$ where

$$E_k = \frac{h^2}{8ma^2}\, r^2 \tag{13.10}$$

For large $r$, the number of points is the octant volume

$$\frac{1}{8}\left(\frac{4}{3}\pi r^3\right)$$

The number of electronic states with energy less than $E_k$ is

$$M(E_k) = \Omega \frac{\pi}{6}\left(\frac{8m}{h^2}\right)^{3/2} E_k^{3/2} \tag{13.11}$$

where $\Omega = a^3$ is the volume of the box. The number of states per unit energy range is called the density of states $N(E)$ and is related to $M(E)$ by

$$\int_0^{E_k} N(E)\,\mathrm{d}E = M(E_k) \tag{13.12}$$

$$\therefore N(E_k) = \left(\frac{\mathrm{d}M}{\mathrm{d}E}\right)_{E_k} \tag{13.13}$$

Hence

$$N(E_k) = \Omega \frac{\pi}{4} \left(\frac{8m}{h^2}\right)^{3/2} E_k^{1/2} \tag{13.14}$$

The number of states per unit energy range per unit volume is independent of the size of the box. When spin is included the values of $M(E_k)$ and $N(E_k)$ given by (13.11) and (13.14) must be doubled.

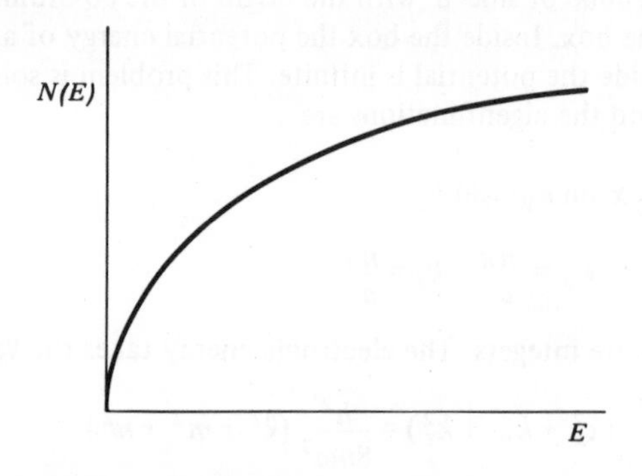


Figure 13.2 Density of state function for free electrons.

*Periodic boundary conditions*

The solution (13.8), (13.9) is a consequence of the boundary conditions. Suppose that instead of a box in which the electrons are trapped there is an infinity of such boxes stacked together and all identical. The appropriate boundary conditions are the cyclic or Born von Kármán conditions.

$$\psi (x + a, y, z) = \psi (x, y + a, z) = \psi (x, y, z + a) = \psi (x, y, z) \tag{13.15}$$

The normalized solutions of the wave equation (4.34) are now

$$\psi (x, y, z) = \frac{1}{\sqrt{\Omega}} \exp [i(k_x x + k_y y + k_z z)] \tag{13.16}$$

with

$$k_x = \ell \left(\frac{2\pi}{a}\right), k_y = m \left(\frac{2\pi}{a}\right), \; k_z = \left(\frac{2\pi}{a}\right)$$

where $\ell$, $m$, $n$ are integers. Here the integers take all values, including positive and negative, since the corresponding solutions are linearly independent.

The electronic energy is (cf. 13.9)

$$E_k = \frac{h^2}{2ma^2} (\ell^2 + m^2 + n^2) \tag{13.17}$$

The physically significant formulae (13.11) and (13.14) still apply. (Remember in calculating $M(E_k)$ all eight octants of a sphere now need to be taken into account.)

*Fermi-Dirac statistics*

Consider a gas of $N$ non-interacting electrons contained in the box at absolute zero temperature. The electrons are Fermi particles and the Pauli principle prevents more than two electrons occupying a state with the same set of quantum numbers ($\ell$, $m$, $n$). At absolute zero the electrons occupy energy levels from zero to a maximum value $E_m$ defined by

$$\frac{N}{2} = \Omega \frac{\pi}{6} \left(\frac{8m}{h^2}\right)^{3/2} E_m^{3/2} \tag{13.18}$$

i.e. $$E_m = \left(\frac{h^2}{8m}\right)\left(\frac{3}{\pi}\right)^{\frac{2}{3}}\left(\frac{N}{\Omega}\right)^{\frac{2}{3}} \tag{13.19}$$

The value of $E_m$ corresponding to the valence electron density in metals is of the order of a few electron volts.

At absolute zero temperature there is a cut-off energy below which all the electron states are filled and above which all are unoccupied. As the temperature is increased some electrons are excited into higher states and the boundary between occupied and unoccupied states becomes blurred. In this case it is appropriate to talk about the probability of a state being filled.

The fundamental result of statistical mechanics is that the probability of the system being in an allowed state with energy $E$ is proportional to $\exp(-E/kT)$ where $k$ is the Boltzmann constant and $T$ is the absolute temperature. This is only useful in quantum mechanics if the many-electron eigenstates are known.

The Fermi-Dirac distribution law gives the probability that a single-electron state is occupied, remembering that the many electron wave-function must be antisymmetric with respect to interchange of electrons. The Fermi-Dirac law is not as fundamental as the Boltzmann law since it depends on a Hartree-Fock approximate wave-function, but it is very important in electron band theory.

The Fermi-Dirac law states that the probability $f(E_i)$ that a single-electron state with energy $E_i$ is occupied at temperature $T$ is given by

$$f(E_i) = \frac{1}{e^{(E_i - E_F)/kT} + 1} \tag{13.20}$$

where the Fermi energy $E_F$ is the chemical potential per electron. At $T = 0, f = 1$ for $E_i < E_F$ and $f = 0$ for $E_i > E_F$. All states with energies less than $E_F$ are filled whereas all states with energies greater than $E_F$ are empty.

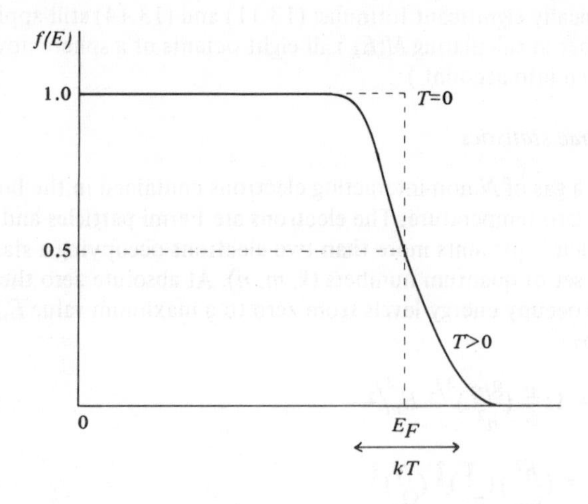


Figure 13.3 Fermi-Dirac distribution.

As the temperature is increased from absolute zero the distribution function departs from this step character. Some electrons in the states immediately below $E_F$ are excited into levels just above. The only region where $f$ is appreciably different from one or zero is in the neighbourhood of $E_F$ over a range of energies of the order of $kT$.

The criterion for quantum statistics to be necessary is that the Pauli principle be important. This occurs when there are 'too many' electrons for the available energy states. If electrons obey classical statistics then almost all electrons will have energies up to $\sim kT$ and at room temperature this is $\sim 1/40$ eV. Silver is a 'typical' monovalent metal where one electron per atom is donated to a sea of loosely bound conduction electrons. In one mole of silver, the volume of which is $\sim 10^{-5}$ m$^3$, the number of electronic states less than $1/40$ eV is $\sim 10^{20}$. But one mole contains $10^{23}$ conduction electrons. There are far more electrons than states available and quantum statistics must be used. This is the case for metals where to accommodate all electrons, states up to about 5 eV are filled. Quantum statistics is necessary when $kT \ll E_F$ and then the electron gas is said to be degenerate. A degeneracy temperature may be defined by $kT_0 = E_F$ and the condition for degeneracy is $T \ll T_0$. For metals $T_0 \sim 10^4$ K.

Not withstanding its obvious crudity the Sommerfeld model has been applied with some success to the problems of electron specific heat and weak spin paramagnetism in metals. In particular the linear dependence with temperature of the small specific heat of the conduction electrons in metals is readily explained. This is swamped by the much larger lattice contribution except at very low temperatures.

## 13.3 The Bravais Lattice

Since crystals possess translational symmetry, it must be possible to define a minimum periodic displacement in any direction. Any two points separated by such a displacement must be equivalent in every respect. For a set of identical points placed at equal intervals along a straight line, the distance between any two consecutive points is the minimum repeating distance. In the case of a multi-dimensional periodic array of points, the minimum displacement is very long for most directions. For a three-dimensional array of points it is possible to choose three basis vectors $\mathbf{a}_1, \mathbf{a}_2, \mathbf{a}_3$ so that all equivalent points in the array are given by

$$\mathbf{R}_n = n_1\mathbf{a}_1 + n_2\mathbf{a}_2 + n_3\mathbf{a}_3 \tag{13.21}$$

where $n_1, n_2, n_3$ are integers. The array is called a Bravais lattice. The basis vectors are not unique but they must satisfy the requirement that all equivalent points in the lattice are given by (13.21) with suitable choice for integers, $n_i$. In addition, all integral choices for $n_i$ must represent a lattice point.

The three basis vectors define a parallelepiped which is called a primitive cell. The primitive cell has a volume given by the triple product $\mathbf{a}_1 \cdot (\mathbf{a}_2 \wedge \mathbf{a}_3)$ and is the smallest unit of volume which will build up the lattice by periodic repetition.

Of course a Bravais lattice has other symmetry properties apart from translational symmetry such as a centre of inversion or mirror plane or rotation axis.

*The reciprocal lattice*

Each Bravais lattice has an associated reciprocal lattice defined by the three basis vectors $\mathbf{b}_1, \mathbf{b}_2, \mathbf{b}_3$ with

$$\mathbf{b}_j \cdot \mathbf{a}_i = 2\pi\,\delta_{ij} \tag{13.22}$$

Vectors $\mathbf{b}_i$ that satisfy these relations are

$$\mathbf{b}_1 = \frac{2\pi}{V_d}(\mathbf{a}_2 \wedge \mathbf{a}_3),\ \mathbf{b}_2 = \frac{2\pi}{V_d}(\mathbf{a}_3 \wedge \mathbf{a}_1),\ \mathbf{b}_3 = \frac{2\pi}{V_d}(\mathbf{a}_1 \wedge \mathbf{a}_2) \tag{13.23}$$

where $V_d = \mathbf{a}_1 \cdot (\mathbf{a}_2 \wedge \mathbf{a}_3)$ is the primitive cell volume of the direct lattice.

The reciprocal lattice is composed of the end points of all the vectors

$$\mathbf{K}_m = m_1\mathbf{b}_1 + m_2\mathbf{b}_2 + m_3\mathbf{b}_3 \tag{13.24}$$

The primitive cell has a volume

$$V_R = \mathbf{b}_1 \cdot (\mathbf{b}_2 \wedge \mathbf{b}_3) \tag{13.25}$$

and it can be shown that

$$V_R = \frac{(2\pi)^3}{V_d} \tag{13.26}$$

An important result is that the scalar product

$$\mathbf{K}_m \cdot \mathbf{R}_n = 2\pi\,(m_1 n_1 + m_2 n_2 + m_3 n_3) = 2\pi \ell \tag{13.27}$$

where $\ell$ is an integer.

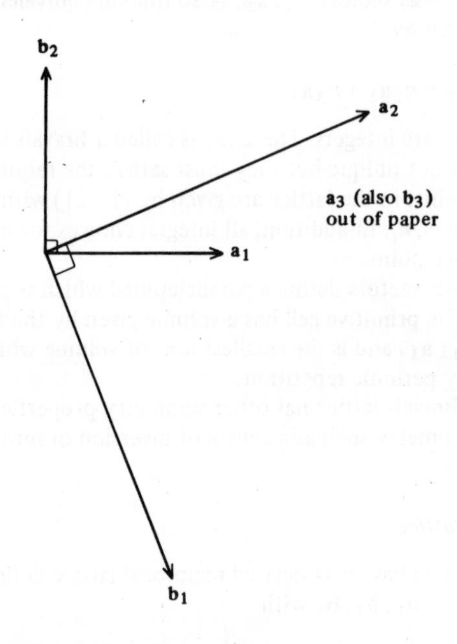


Figure 13.4 Direct and reciprocal lattice vectors.

## 13.4 The Bloch Theorem

The central theorem of electron band theory is the Bloch Theorem. The single-electron Schrödinger Hamiltonian is

$$\mathscr{H} \equiv -\frac{\hbar^2}{2m}\nabla^2 + V(\mathbf{r}) \tag{13.28}$$

The crystal potential has the lattice periodicity

$$V(\mathbf{r}-\mathbf{R}_n) = V(\mathbf{r}) \tag{13.29}$$

with $\mathbf{R}_n = n_1\mathbf{a}_1 + n_2\mathbf{a}_2 + n_3\mathbf{a}_3$ $n_1, n_2, n_3$ integers.

The substitutional operator representing the translation $\mathbf{R}_n$ is written

$$\{\epsilon \,|\mathbf{R}_n\}$$

and

$$\{\epsilon \,|\, \mathbf{R}_n\}\; V(\mathbf{r}) = V(\mathbf{r}-\mathbf{R}_n) \tag{13.30}$$

Strictly, a finite crystal is not invariant under a translation. For an infinite crystal the number of translations is infinite. To keep the number of distinct translations finite it is usual to use periodic boundary conditions. The infinite crystal is considered to be composed of identical microcrystals of size $N\mathbf{a}_1$ by $N\mathbf{a}_2$ by $N\mathbf{a}_3$ fitted together. This is a mathematical convenience and will suffice within the crystal for suitably large $N$ but cannot be expected to include surface effects. The three translations $N\mathbf{a}_1, N\mathbf{a}_2, N\mathbf{a}_3$ produce no change at all.

$$\{\epsilon \,|\mathbf{O}\} \equiv \{\epsilon \,|\, N\mathbf{a}_1\} \equiv \{\epsilon \,|\, N\mathbf{a}_2\} \equiv \{\epsilon \,|\, N\mathbf{a}_3\} \tag{13.31}$$

The translation operators form a group of $N^3$ elements and as all translations commute this group is Abelian. Each element in an Abelian group forms a class in itself and the number of irreducible representations is equal to the number of elements $N^3$. All these representations are one-dimensional.

The operator $\nabla^2$ is invariant under a translation of the co-ordinate system and so the Hamiltonian is invariant under a crystal translation. Consequently the energy eigenfunctions can be chosen to transform according to one of the irreducible representations of the group. An alternative view is to note that the Hamiltonian and translation operators commute with each other so wave-functions may be chosen to be simultaneously eigenfunctions of $\mathcal{H}$ and all the operators.

If $\psi$ is the basis for one of the irreducible representations

$$\{\epsilon \,|\, \mathbf{a}_i\}\;\; \psi(\mathbf{r}) = \lambda_i\, \psi(\mathbf{r}) \qquad i = 1, 2, 3 \tag{13.32}$$

where $\lambda_i$ are constants which may be complex. The periodic boundary conditions require

$$\lambda_1^{\,N} = \lambda_2^{\,N} = \lambda_3^{\,N} = 1 \tag{13.33}$$

and so each $\lambda_i$ must be one of the $N$ $N$th roots of unity.

$$\lambda_i = \exp\left[-i\frac{2\pi r_i}{N}\right] \qquad r_i = 1,2,3 \ldots, N \tag{13.34}$$

For a general translation $\{\epsilon \,|\, n_1\mathbf{a}_1 + n_2\mathbf{a}_2 + n_3\mathbf{a}_3\} \equiv \{\epsilon \,|\, \mathbf{R}_n\}$

$$\lambda = \exp\left[-i\frac{2\pi}{N}(r_1 n_1 + r_2 n_2 + r_3 n_3)\right] \tag{13.35}$$

If a real quasi-continuous 'wave-vector' **k** is defined by

$$\mathbf{k} = \frac{1}{N}(r_1\mathbf{b}_1 + r_2\mathbf{b}_2 + r_3\mathbf{b}_3) \qquad r_i = 1,2,3,\ldots,N \tag{13.36}$$

where $\mathbf{b}_1, \mathbf{b}_2, \mathbf{b}_3$ are the reciprocal lattice basis vectors, then the irreducible representations and hence the energy eigenfunctions $\psi_\mathbf{k}(\mathbf{r})$ may be classified by **k** with

$$\{\epsilon \mid \mathbf{R}_n\}\ \psi_\mathbf{k}(\mathbf{r}) = e^{-i\mathbf{k}\cdot\mathbf{R}_n}\ \psi_\mathbf{k}(\mathbf{r}) \tag{13.37}$$

This is Bloch's theorem and a wave-function satisfying this condition is

$$\psi_\mathbf{k}(\mathbf{r}) = e^{i\mathbf{k}\cdot\mathbf{r}}\ U_\mathbf{k}(\mathbf{r}) \tag{13.38}$$

where $U_\mathbf{k}(\mathbf{r})$ has the lattice periodicity.

The **k**-vector that defines an eigenstate (13.37) is not unique since **k** and $\mathbf{k}+\mathbf{K}_m$ are equivalent from (13.27)

$$e^{-i(\mathbf{k}+\mathbf{K}_m)\cdot\mathbf{R}_n} = e^{-i\mathbf{k}\cdot\mathbf{R}_n} \quad \text{for all } \mathbf{R}_n \tag{13.39}$$

All non-equivalent **k**-values are obtained if **k** is restricted to a primitive cell of the reciprocal lattice, this value of **k** being the 'reduced wave vector'. The symmetric primitive cell obtained by bisecting the lines joining $\mathbf{k} = \mathbf{0}$ to the nearest reciprocal lattice points by planes is called 'the Brillouin zone'. For a three-dimensional lattice this cell is a polyhedron.

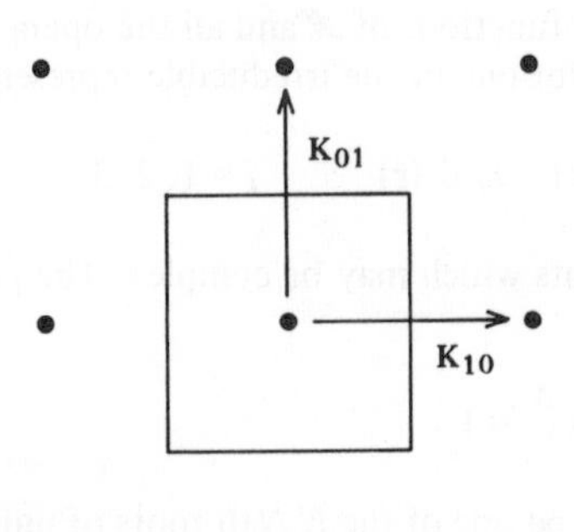


Figure 13.5 Brillouin zone for the square reciprocal net.

Each point outside the Brillouin zone is equivalent to some interior point and a point on the surface of the zone is equivalent to at least one other point on the surface.

There are $N^3$ **k**-vectors of the type (13.36) inside the zone and this gives the total number of distinct irreducible representations of the translation group.

Since $\psi_{\mathbf{k}}(\mathbf{r})$ is an energy eigenfunction, the energy may be regarded as a function of the 'quasi-continuous' **k**-vector.

$$\mathscr{H}\psi_{\mathbf{k}}(\mathbf{r}) = E(\mathbf{k})\,\psi_{\mathbf{k}}(\mathbf{r}) \tag{13.40}$$

In the spin-free case each **k**-state may be occupied by two electrons. In the reduced zone scheme this function is multivalued and may be written $E_n(\mathbf{k})$, $n$ denoting the branch.

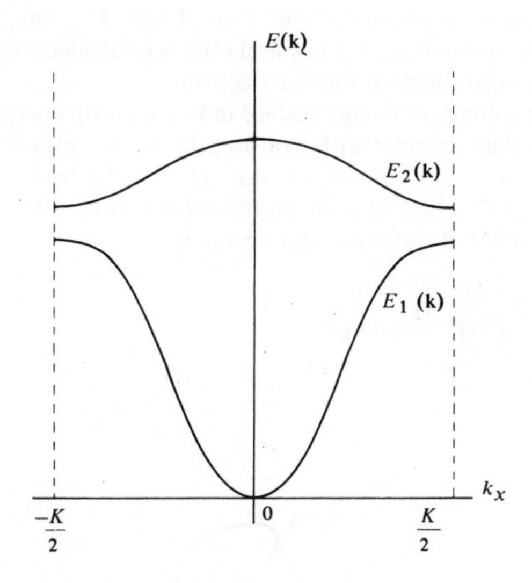


Figure 13.6 Energy band structure.

Within the Brillouin zone, the energy is a continuous function of **k** with a continuous derivative. A single continuous branch is called an energy band. Discontinuities in the energy occur only on the zone surfaces.

The energy surfaces in **k**-space have inversion symmetry

$$\text{i.e.}\quad E(-\mathbf{k}) = E(\mathbf{k}) \tag{13.41}$$

This follows from (13.40) since if spin is ignored $\mathscr{H}$ is real and $\psi_{\mathbf{k}}^*$ and $\psi_{\mathbf{k}}$ both satisfy (13.40) with the same energy. But, from (13.38) $\psi_{\mathbf{k}}^*$

transforms according to $\psi_{-\mathbf{k}}$.

In general, the energy surfaces satisfy other symmetry relations. The collection of rotations and reflections associated with a crystal forms the crystal point group. The wave vectors generated from $k$ by the crystal point group form a set called the star of the k-vector and each member of the set has the same energy since the elements of the crystal point group commute with the crystal Hamiltonian.

If there is sufficient symmetry so that the two opposite sides of a Brillouin zone face are symmetrically equivalent then in general the smoothness of the energy ensures that the normal derivative of $E(\mathbf{k})$ vanishes at a zone boundary.

*Density of states*

The density of state function $N(E)$ is defined so that $N(E)\mathrm{d}E$ is the number of states in a band between $E$ and $E$+d$E$. This function is very important in determining the physical effects of an electron band and (13.14) gives this function for free electrons.

Consider a constant energy surface in **k**-space with energy $E$. Construct a right cylinder with base area d$S$ on this surface and with length d$k$. The number of states in this cylinder (13.36) is $\Omega\, \mathrm{d}k\, \mathrm{d}S\, /\, (2\pi)^3$ where $\Omega$ is the crystal volume. Since $\mathrm{d}E = |\mathrm{grad}_{\mathbf{k}}E| \mathrm{x} |\mathrm{d}\mathbf{k}|$ then the number of states per unit energy is

$$N(E) = \frac{\Omega}{(2\pi)^3} \iint \frac{1}{|\mathrm{grad}_{\mathbf{k}}E|}\, \mathrm{d}S \qquad (13.42)$$

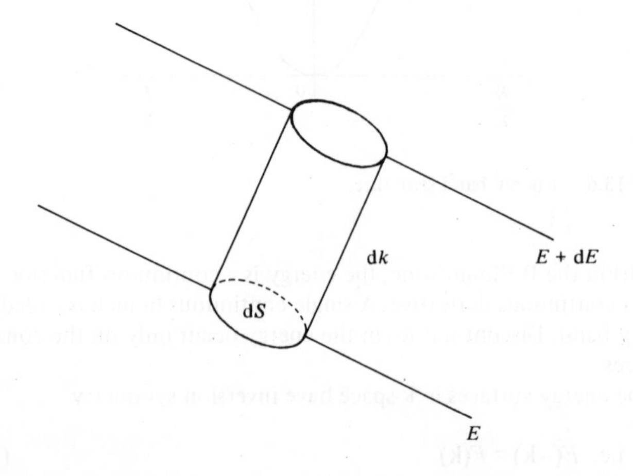


Figure 13.7 Two neighbouring surfaces in k-space.

where the integration is over the surface of constant energy. Each of these states may be occupied by two electrons.

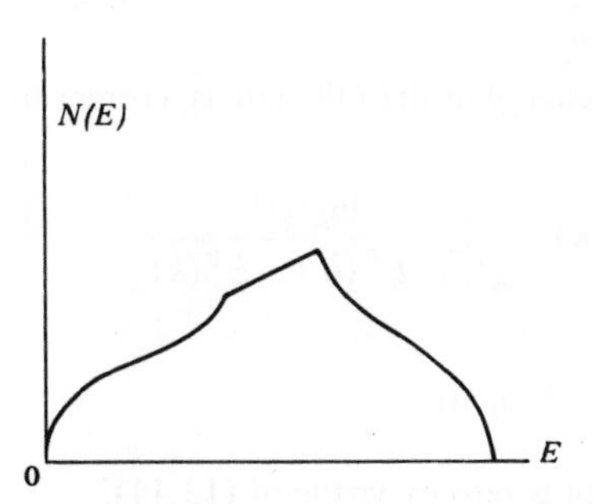


Figure 13.8 'Typical' density of state function.

## 13.5 Nearly Free-Electron Model

This is a crystal model in which the potential $V(\mathbf{r})$ is assumed to be very weak compared with the electronic kinetic energy. This is not the case for valence electrons since the variations of potential energy are generally comparable to the kinetic energy. However, the method is worth describing since it is one of the limiting cases (the other is tight-binding) and is the basis of more sophisticated approaches. This approximation will apply to the motion of a beam of fast electrons injected into a crystal.

The one-dimensional case is discussed below. The Schrödinger equation is

$$-\frac{\hbar^2}{2m}\frac{d^2\psi}{dx^2} + V\psi = E\psi \tag{13.43}$$

with

$$V(x-a) = V(x)$$

where '$a$' is the lattice period. It is convenient to choose the zero of energy so that the mean value of the potential function is zero.

$$\text{i.e.}\ \int_0^a V(x)\,dx = 0 \tag{13.44}$$

Since $V$ is assumed to be small compared with the kinetic energy, perturbation theory is employed.

The unperturbed wave-functions, corresponding to $V$=0 are the free-electron plane waves

$$\phi_k(x) = \frac{1}{\sqrt{Na}}\ e^{ikx} \tag{13.45}$$

normalized over a microcrystal containing $N$ cells. The unperturbed electronic energies are

$$E^0(k) = \frac{\hbar^2 k^2}{2m} \tag{13.46}$$

The perturbed energy of the $k^{\text{th}}$ state is, correct to second-order terms (7.59)

$$E(k) = E^0(k) - \sum_{k' \neq k} \frac{|h_{k'k}|^2}{E^0(k') - E^0(k)} \tag{13.47}$$

where

$$h_{k'k} = \int \phi_{k'}^* \, V \, \phi_k \, dx \tag{13.48}$$

The first-order term is zero by virtue of (13.44).

The real periodic potential may be expressed as a Fourier series

$$V(x) = \sum_{n \neq 0} V_n \, e^{-in \frac{2\pi}{a} x}, \quad V_n^* = V_{-n} \tag{13.49}$$

Clearly the integral

$$\begin{aligned} h_{k'k} &= V_n \quad \text{if } k - k' = n\frac{2\pi}{a} \\ &= 0 \quad \text{otherwise.} \end{aligned} \tag{13.50}$$

The state $k$ is mixed only with the states $k - n\frac{2\pi}{a}$ by the periodic perturbation and the perturbed energy is

$$E(k) = E^0(k) - \sum_{n \neq 0} \frac{|V_n|^2}{E^0(k - n\frac{2\pi}{a}) - E^0(k)} \tag{13.51}$$

This expansion for the energy is satisfactory if the Fourier components $V_n$ tend to zero rapidly as $n$ increases and if there are no degeneracies among the plane waves such that

$$E^0(k - n\frac{2\pi}{a}) = E^0(k) \tag{13.52}$$

From (13.46) degeneracies of this type occur when

$$|k - n\frac{2\pi}{a}| = |k| \tag{13.53}$$

Hence, near a Brillouin zone boundary with $k = n\frac{\pi}{a}$ the energy expansion (13.51) does not apply.

As in (7.63) it is necessary to write the wave-function

$$\psi_k = C_o \, \phi_k + C_n \, \phi_{k - n\frac{2\pi}{a}} \tag{13.54}$$

with incident and diffracted rays on an equal footing. This wave-function is substituted into (13.43) and on premultiplying the equation by $\phi_k^*$ and $\phi^*_{k-n\frac{2\pi}{a}}$ in turn and then integrating over the microcrystal, the following simultaneous equations are obtained.

$$C_o\,[E(k) - E^0(k)] - C_n V_n^* = 0$$

$$-C_o V_n + C_n\,[E(k) - E^0(k - n\tfrac{2\pi}{a})] = 0 \qquad (13.55)$$

The determinant of coefficients must vanish for a non-trivial solution and the resulting quadratic in $E(k)$ has the two solutions

$$E(k) = \tfrac{1}{2}\,\{E^0(k) + E^0\,(k - n\tfrac{2\pi}{a}) \pm [(E^0(k) - E^0(k - n\tfrac{2\pi}{a}))^2 + 4\,|V_n|^2]^{1/2}\} \qquad (13.56)$$

At the zone boundary

$$E^+\,(\tfrac{n\pi}{a}) = E^0\,(\tfrac{n\pi}{a}) + |V_n|$$

$$E^-\,(\tfrac{n\pi}{a}) = E^0\,(\tfrac{n\pi}{a}) - |V_n| \qquad (13.57)$$

The energy function has a discontinuity at the zone boundary and the energy gap is

$$\Delta E = 2|V_n| \qquad (13.58)$$

The quasi-continuous free-electron energy parabola is broken into bands.

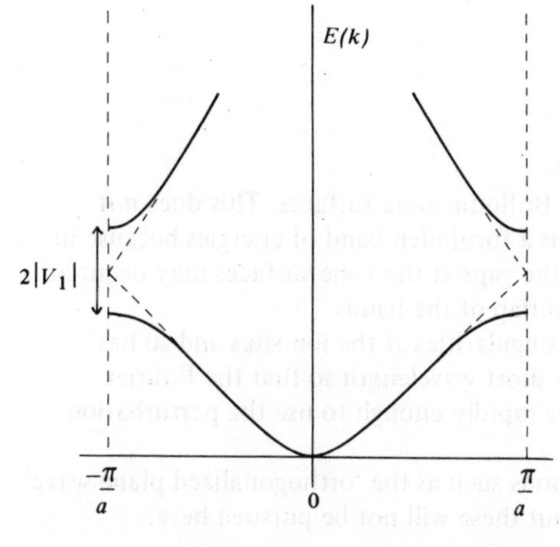


Dotted line indicates free-electron energy parabola.

Figure 13.9 Reduced zone scheme. $k \equiv k + (2\pi/a)$.

An external electron beam incident on the crystal with energy lying in one of the gaps will be totally reflected.

*Three-dimensional case*

In the three-dimensional case, energy gaps occur when

$$E^0(\mathbf{k}) = E^0(\mathbf{k}-\mathbf{K})$$

with **K** a reciprocal lattice vector.

$$\text{i.e.} \quad |\mathbf{k}| = |\mathbf{k}-\mathbf{K}| \tag{13.59}$$

This condition implies that **k** lies on the perpendicular bisector of the reciprocal lattice vector **K** (Bragg condition).

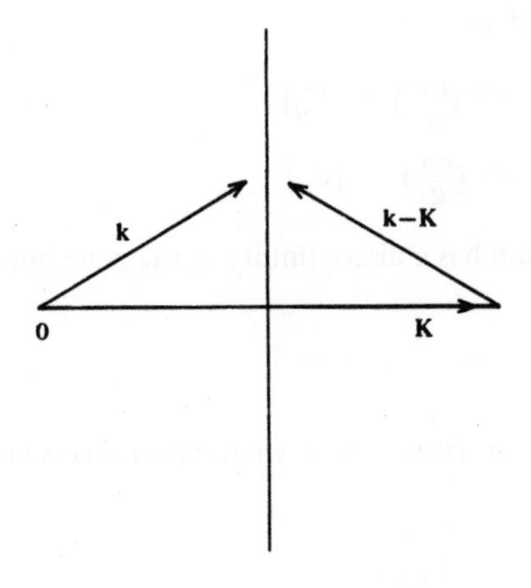


Figure 13.10

Energy gaps occur at the Brillouin zone surfaces. This does not necessarily mean that there is a forbidden band of energies because in the three-dimensional case, the gaps at the zone surfaces may occur at different energies causing overlap of the bands.

In a real crystal $V(\mathbf{r})$ has singularities at the ion sites and so has Fourier components of very short wavelength so that the Fourier components do not converge rapidly enough to use the perturbation expansion.

There are improved methods such as the 'orthogonalized plane wave' technique (Herring, 1940) but these will not be pursued here.

## 13.6 Tight-Binding Model

The tight-binding model is the extreme case that can be considered to apply when the lattice constant is so large that the wave-functions of adjacent atoms do not overlap to any appreciable extent. This model may apply to the inner core electrons bound to the ions.

In 1928 Bloch showed that the wave-function

$$\psi_{\mathbf{k}}(\mathbf{r}) = A(\mathbf{k}) \sum_{\mathbf{R}_n} e^{i\mathbf{k}\cdot\mathbf{R}_n} \phi(\mathbf{r}-\mathbf{R}_n) \tag{13.60}$$

where $\phi(\mathbf{r}-\mathbf{R}_n)$ is an atomic orbital centred on the $\mathbf{R}_n$th site, satisfies the Bloch condition (13.37). Near an ion site this function resembles an atomic wave-function. The constant $A(\mathbf{k})$ may be chosen to normalize $\psi_{\mathbf{k}}$ over the microcrystal containing $N^3$ cells.

The expectation value of the energy for this wave-function is

$$E(\mathbf{k}) = \frac{\int \psi_{\mathbf{k}}^* \left\{-\frac{\hbar^2}{2m} \nabla^2 + V(\mathbf{r})\right\} \psi_{\mathbf{k}} \, d\tau}{\int \psi_{\mathbf{k}}^* \psi_{\mathbf{k}} \, d\tau} \tag{13.61}$$

giving

$$E(\mathbf{k}) = |A|^2 N^3 \sum_{\mathbf{R}_\ell} e^{-i\mathbf{k}\cdot\mathbf{R}_\ell} \epsilon(\mathbf{R}_\ell) \tag{13.62}$$

where

$$\epsilon(\mathbf{R}_\ell) = \int \phi^*(\mathbf{r}-\mathbf{R}_\ell) \left\{-\frac{\hbar^2}{2m} \nabla^2 + V(\mathbf{r})\right\} \phi(\mathbf{r}) \, d\tau \tag{13.63}$$

The energy function (13.62) is periodic in **k**-space and the Fourier components $\epsilon(\mathbf{R}_\ell)$ involve three-centre integrals since the crystal potential is a superposition of atomic potentials.

If $V_a(\mathbf{r})$ is the free-atom potential then $\phi(\mathbf{r})$ satisfies

$$\left[-\frac{\hbar^2}{2m} \nabla^2 + V_a(\mathbf{r})\right] \phi(\mathbf{r}) = E_a \phi(\mathbf{r})$$

where $E_a$ is the atomic eigenvalue. Hence

$$\left[-\frac{\hbar^2}{2m} \nabla^2 + V(\mathbf{r})\right] \phi(\mathbf{r}) = \left[E_a + (V(\mathbf{r}) - V_a(\mathbf{r}))\right] \phi(\mathbf{r}) \tag{13.64}$$

Neglecting the overlap between atomic functions on different ions $|A|^2 = 1/N^3$ and so

$$\epsilon(\mathbf{o}) = E_a + \alpha$$

with

$$\alpha = \int \phi^*(\mathbf{r}) \left[V(\mathbf{r}) - V_a(\mathbf{r})\right] \phi(\mathbf{r}) \, d\tau \tag{13.65}$$

In fact some overlap must be assumed for $\alpha \neq 0$. If the atomic functions are *s*-type, all nearest neighbours $\mathbf{R}_\ell$ give

$$\epsilon(\mathbf{R}_\ell) = \int \phi^*(\mathbf{r}-\mathbf{R}_\ell) \left[V(\mathbf{r}) - V_a(\mathbf{r})\right] \phi(\mathbf{r}) \, d\tau = \beta \tag{13.66}$$

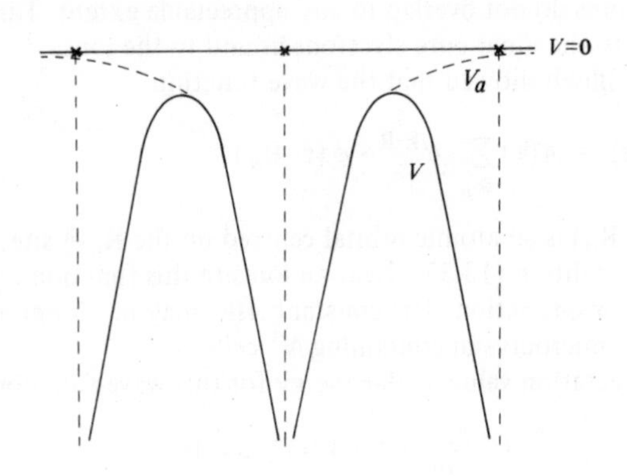


Figure 13.11 Atomic and crystal potentials.

The interaction integrals will be smaller for second and further neighbours and in the 'nearest neighbour' approximation are neglected. Then

$$E(\mathbf{k}) = E_a + \alpha + \beta \sum_{\mathbf{R}\ell} e^{-i\mathbf{k}\cdot\mathbf{R}_\ell} \tag{13.67}$$

Since $V - V_a < 0$ then both the interaction integrals $\alpha, \beta$ are negative and may be of the order of a few electronvolts.

If the lattice is simple cubic with nearest neighbours at the six sites $(a, 0, 0)$ etc.,

$$E(\mathbf{k}) = E_a + \alpha + 2\beta\,[\cos a\,k_x + \cos a\,k_y + \cos a\,k_z] \tag{13.68}$$

The minimum energy is at the centre of the zone

$$E(\mathbf{o}) = E_a + \alpha + 6\beta \tag{13.69}$$

and the maximum energy is at a zone corner

$$E\left(\frac{\pi}{a}, \frac{\pi}{a}, \frac{\pi}{a}\right) = E_a + \alpha - 6\beta \tag{13.70}$$

so that the band width is $12|\beta|$

Each atomic level of the free atom splits into a band of energies as the atoms are brought together to form a crystal. The $N^3$ atoms, at infinite separation, contain $N^3$ degenerate levels. The crystal band covers a range of energies and contains $N^3$ different states. For the deeply lying inner core electrons, the atomic function overlap is very small so that $\beta$ is small and consequently the band width may be negligible.

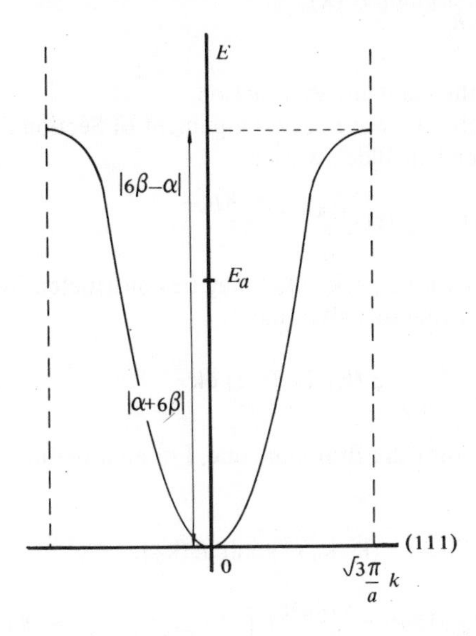


Figure 13.12 Energy plot along the (111)-direction. Simple cubic lattice.

If two atomic levels broaden so much that their bands almost intersect, it is necessary to assume the wave-function (13.60) is a linear combination of atomic orbitals (L.C.A.O.) involving more than one atomic function on each site.

Energy band calculations involving the tight-binding method have certain disadvantages. The bound atomic states do not form a complete set and also the overlap integrals between atomic orbitals on different sites are not negligible. Slater and Koster (1954) used the L.C.A.O. method in an interpolation scheme fitting the interaction integrals to values either observed empirically or calculated by other methods.

## 13.7 Dynamics of Band Electrons

Both the crystal models discussed above imply that there exist allowed bands of energies with the possibility of forbidden energy gaps. This band structure is of prime importance in the understanding of the division of materials into insulators, semiconductors and conductors.

In general each Bloch wave is complex and describes a state with a mean velocity. From (5.70)

$$\mathbf{V}(\mathbf{k}) \equiv \frac{d}{dx} \langle x \rangle = \frac{1}{m} \int \psi_{\mathbf{k}}^* \left(\frac{\hbar}{i} \frac{\partial \psi_k}{\partial x}\right) d\tau \tag{13.71}$$

The integral can be calculated to give

$$\mathbf{V}(\mathbf{k}) = \frac{1}{\hbar}\,\text{grad}_{\mathbf{k}}\, E(\mathbf{k}) \tag{13.72}$$

where

$E(\mathbf{k})$ is the band energy function.

Alternatively, the wave-packet argument of Section 5.4 can be used. The time-dependent Bloch wave is

$$\Psi_{\mathbf{k}}(\mathbf{r}, t) = U_{\mathbf{k}}(\mathbf{r})\, e^{i[\mathbf{k}\cdot\mathbf{r} - \frac{E(\mathbf{k})t}{\hbar}]} \tag{13.73}$$

A wave-packet with a peak at $\mathbf{k}_0$ may be constructed from Bloch functions belonging to a given band.

$$\Psi(\mathbf{r}, t) = \int_{\text{B.Z.}} a(\mathbf{k})\, \Psi_{\mathbf{k}}(\mathbf{r}, t)\, d\mathbf{k} \tag{13.74}$$

The integral is over the Brillouin zone. For all **k**-vectors in the wave-packet

$$E(\mathbf{k}) \simeq E(\mathbf{k}_0) + (\mathbf{k}-\mathbf{k}_0)\cdot \text{grad}_{\mathbf{k}}\, E(\mathbf{k}_0) \tag{13.75}$$

and

$$\Psi(\mathbf{r}, t) = e^{i[\mathbf{k}_0\cdot\mathbf{r} - \frac{E(\mathbf{k}_0)t}{\hbar}]} \int_{\text{B.Z.}} a(\mathbf{k})\, U_{\mathbf{k}}\; e^{i(\mathbf{k}-\mathbf{k}_0)(\mathbf{r} - \frac{t}{\hbar}\text{grad}_{\mathbf{k}}E)}\, d\mathbf{k}$$

The probability density $\Psi^*\Psi$ depends only on the integral and is essentially constant along a path for which

$$\mathbf{r} = \frac{t}{h}\,\text{grad}_{\mathbf{k}}E$$

and so the velocity of the wave-packet is

$$\mathbf{V}(\mathbf{k}) = \frac{1}{\hbar}\,\text{grad}_{\mathbf{k}}E \tag{13.76}$$

*Static electric field*

When a static electric field is applied to a crystal the electrons are accelerated and gain kinetic energy although total energy remains constant. The Schrödinger equation has an additional potential

$$U(\mathbf{r}) = e\,\mathscr{E}\cdot\mathbf{r}$$

where $\mathscr{E}$ is the electric field strength. Perturbation theory cannot be used since this term becomes arbitrarily large for large **r**. $U(\mathbf{r})$ does not have the lattice periodicity so the Hamiltonian of a crystal in an external field no longer commutes with the translation operators; **k** is no longer a constant of the motion and the field induces transitions from one **k**-state to another within a band. There is a possibility for a transition from one band to another but for 'normal' fields ($<10^6$ $Vm^{-1}$) such band transitions are improbable.

The change in band energy of the electron in state **k** in unit time is

$$\dot{\mathbf{k}} \cdot \mathrm{grad}_{\mathbf{k}} E$$

The work done on the electron by the field in unit time is

$$-e\,\mathscr{E} \cdot \mathbf{V}(\mathbf{k}) = -\frac{e}{\hbar}\,\mathscr{E} \cdot \mathrm{grad}_{\mathbf{k}} E$$

Equating these then

$$\hbar\,\dot{\mathbf{k}} = -e\,\mathscr{E} \tag{13.77}$$

is a solution.

For the mean **k**-vector of a wave-packet to have significance it is necessary that the packet be spread over several crystal primitive cells. So, the argument above is only valid if the field is sufficiently slow-varying.

## 13.8 Classification of Solids

Each occupied Bloch state represents a charge $-e$ moving with a mean velocity **V(k)** (13.76). The total current carried by a complete band is proportional to the integral

$$-e \int_{\mathrm{B.Z.}} f(\mathbf{k})\,\mathbf{V}(\mathbf{k})\,d\mathbf{k} \tag{13.78}$$

where $f(\mathbf{k})$ is the probability the **k**-state is occupied. Since the band energy function has inversion symmetry (13.41) the velocity satisfies, (13.76)

$$\mathbf{V}(-\mathbf{k}) = -\mathbf{V}(\mathbf{k})$$

For a full band $f(\mathbf{k}) = 1$ for all states and the current is zero. Even for a partially-filled band, if it is symmetrically filled, the current will be zero.

However, the effect of an electric field on a partially-filled band is to displace the distribution into an asymmetric state representing a net current. When no field is applied, the **k**-states are filled symmetrically about **k=o** up to the Fermi energy surface. This is only strictly true at absolute zero temperature. When the electric field is applied, the electronic distribution will move in **k**-space against the field according to (13.77). In time $\delta t$ the centre will move a distance $e\mathscr{E}\delta t/\hbar$.

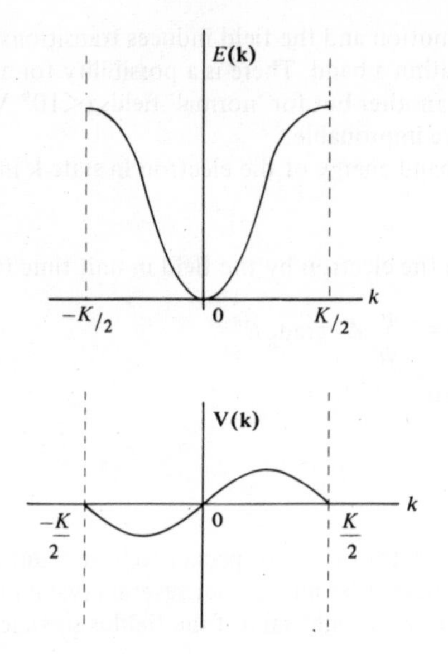


Figure 13.13

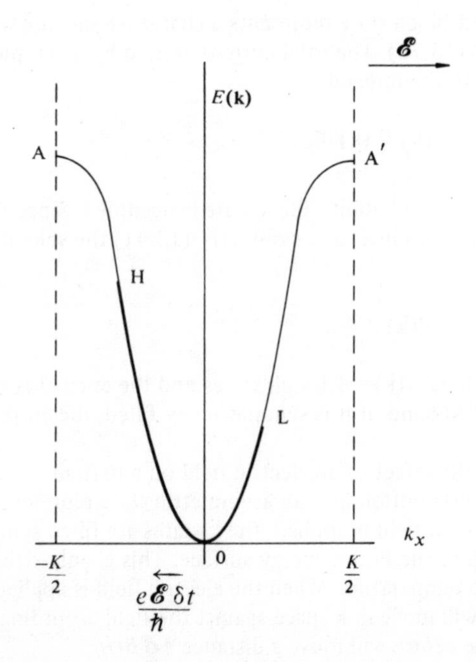


Figure 13.14 Asymmetric partially-filled band.

In a perfect crystal this process would continue indefinitely, the electrons moving towards A and reappearing at A′ corresponding to an oscillation in real space. However, in a real crystal this process is interrupted by the scattering of the electrons from the regions of high kinetic energy H to the regions of low energy L, by interaction with imperfections and the thermal vibrations of the lattice. If $\tau$ is the average lifetime before a collision takes place, the electronic distribution undergoes a shift of $e\mathscr{E}\tau/\hbar$ in **k**-space. This shift represents a net current. At room temperature $\tau \sim 10^{-12}_{\rm s}$ in metals.

Materials with partially-filled energy bands are conductors. For conductors the conductivity is high but will decrease as the temperature increases due to the greater scattering by the lattice. A material with a full valence band separated by a large gap (~ few eV) from an empty band cannot carry a current. Such materials are insulators.

If a material has a full valence band separated from an empty conduction band by a small gap then for temperatures above absolute zero some electrons will be thermally excited into the upper band, leaving some unoccupied states (holes) in the lower band. Both bands are then capable of carrying current and the conductivity, though small, increases with temperature as more electrons are excited. Such materials are semiconductors.

A complete band contains as many different **k**-states as there are primitive cells in the microcrystal. Each state may contain up to two electrons of opposite spin. For a crystal with $N^3$ cells, each energy band can contain $2N^3$ electrons. If a solid has an odd number of valence electrons per cell, there must be partially-filled bands and the material must be a conductor. For example, sodium is monovalent with one atom per cell and so is a conductor with a half-filled band.

If a solid has an even number of valence electrons per cell, there may be full bands and so the material would then be an insulator or perhaps a semiconductor. Diamond has two quadrivalent carbon atoms per cell (i.e. 8 electrons) and is an insulator. Silicon also has two quadrivalent atoms per cell but is a semiconductor. However, an even number of electrons per cell does not necessarily imply full bands, since bands may overlap in energy and parts of the lower band may be unfilled and some electrons may be in the upper band. The divalent metals are the result of this effect.

### *Metal-insulator transition*

In the crystal model described above, a monovalent material such as potassium would be a metal no matter how large the interatomic separation. Clearly, this cannot be correct. An array of widely separated atoms cannot be a conductor. The single-electron band model is not sufficient since the detailed interaction between electrons has been ignored. A more detailed analysis indicates that as the monovalent atoms are brought together, a sudden 'Mott transition' will take place and the array will change from an insulator to a conductor.

A similar type of argument explains why not all liquids and glasses are conductors. The simple band model depends upon the existence of translation symmetry. In an amorphous material there would be no regular band structure and no energy gaps. This would apparently imply that all liquids and glasses should be conductors. This is not the case. The reader is referred to a review paper by Mott on 'Electrons in glass' and to the text *Electronic Processes in Non-Crystalline Materials* by Mott and Davis.

## 13.9 The Effective Mass

When studying the charge carriers in semiconductors (also metals) it is very useful to introduce the concept of effective mass. Assuming the energy band function $E(\mathbf{k})$ is analytic it can be expanded in a Taylor series. At the top or bottom of an energy band, the first derivatives of $E(\mathbf{k})$ are zero and in the neighbourhood of such a stationary point $\mathbf{k}_0$, with a suitable choice of axes,

$$E(\mathbf{k}) = E(\mathbf{k}_0) + \frac{\hbar^2}{2}\left(\frac{k_1'^2}{m_1^*} + \frac{k_2'^2}{m_2^*} + \frac{k_3'^2}{m_3^*}\right) + \ldots \tag{13.79}$$

where

$$\mathbf{k}' = \mathbf{k} - \mathbf{k}_0$$

and

$$m_i^* = \hbar^2 \Big/ \left(\frac{\partial^2 E}{\partial k_i^2}\right)_{\mathbf{k}_0} \tag{13.80}$$

$m_i^*$ are the components of the effective mass. Equation 13.79 has similarities to the energy function for free electrons (13.9) although the effective mass in general has different components in the three axes directions.

Near an energy minimum, the effective mass components are positive. From (13.76)

$$V_i(\mathbf{k}) = \frac{\hbar k_i'}{m_i^*} \tag{13.81}$$

and then using (13.77)

$$m_i^* \dot{V}_i(\mathbf{k}) = -e\mathscr{E}_i \tag{13.82}$$

This equation is analogous to Newton's law and the effective mass description can be used to describe the motion of the electrons excited into the conduction band of a semiconductor, if they are close to the energy minimum of the band. The electron-crystal interaction enters (13.82) through the effective mass components and a reasonable value is $m^* \sim 0.1$ (electron mass) for the conduction band in silicon.

A nearly full valence band in a semiconductor can conveniently be described in terms of the few unoccupied **k**-states (i.e. holes) rather than in terms of the many occupied states. Since a full band carries no current, a band with one electron missing must carry the negative of the current which that single electron could carry. If the hole description is used to describe the band, then a hole must carry the negative of the current of an electron in that state. The real space velocities of a hole and of an electron are identical and are both given by (13.76). Consequently a hole must have a positive charge. Near an energy maximum (13.79) and (13.82) still apply although $m_i^*$ is now negative. Since it is necessary for a hole to have a positive charge (13.82) can be rewritten

$$m_{ih}^* \dot{V}_i(\mathbf{k}) = + e\mathscr{E}_i \tag{13.83}$$

where

$$m_{ih}^* = -m_i^* = -\hbar^2 / \left(\frac{\partial^2 E}{\partial k_i^2}\right)_{\mathbf{k}_0}$$

Then $m_{ih}^*$ is positive. The nearly full valence band can be described by a few positive charge carriers with positive masses and the current carried by the band is proportional to

$$+e \int_{\text{B.Z.}} [1 - f(\mathbf{k})]\, \mathbf{V}(\mathbf{k})\, d\mathbf{k}$$

*Carrier densities in semiconductors*

Suppose the Fermi energy lies within the band gap so that for all conduction band energies

$$E - E_F >> kT$$

In this case the Fermi function approximates to

$$f(E) \simeq e^{-(E-E_F)/kT} \tag{13.84}$$

Let $N_c(E)$ be the density of state function for the conduction band excluding spin, then $N_c(E)/\Omega$ is the density of states per unit volume where $\Omega$ is the crystal volume and so the number of electrons thermally excited across the gap into the conduction band is, per unit volume

$$n = \frac{2}{\Omega} \int_{\substack{\text{conduction}\\ \text{band}}} N_c(E) f(E)\, dE \tag{13.85}$$

Near the bottom of the conduction band $E_c$, if the density of state function can be taken to be free-electron like

$$N_c(E) = \Omega \frac{\pi}{4} \left(\frac{8m_c^*}{h^2}\right)^{3/2} (E - E_c)^{1/2}$$

where $m_c^*$ is the conduction band effective mass. Then

$$n = n_c\, e^{(E_F - E_c)/kT}, \quad E_c - E_F >> kT \tag{13.86}$$

with

$$n_c = 2\left(\frac{m_c^* kT}{2\pi\hbar^2}\right)^{3/2}$$

since

$$\int_0^\infty e^{-x/kT}\, x^{1/2}\, dx = \frac{\sqrt{\pi}}{2}\,(kT)^{3/2}$$

The number of holes in the valence band per unit volume produced by the thermal excitation of the electrons is

$$p = \frac{2}{\Omega} \int_{\substack{\text{valence}\\ \text{band}}} N_v(E)\, [1 - f(E)]\, dE$$

where $N_v(E)$ is the valence band density of state function, For holes near the top of the valence band $E_v$,

$$p = n_h\, e^{(E_v - E_F)/kT} \quad \text{if } E_F - E_v >> kT \tag{13.87}$$

with

$$n_h = 2\left(\frac{m_h^* kT}{2\pi\hbar^2}\right)^{3/2}$$

where $m_h^* > 0$ is the valence band hole effective mass.

The product of the carrier densities is

$$np = n_c\, n_h\, e^{-(E_c - E_v)/kT} \tag{13.88}$$

where $E_c - E_v \equiv E_g$ is the energy gap.

In a pure semiconductor crystal the number of negative carriers must equal the number of positive carriers and so $n = p$. Equate (13.86) and (13.87) and take logarithms to obtain

$$E_F = \tfrac{1}{2}(E_v + E_c) + \tfrac{3}{4} kT \ln\left[\frac{m_h^*}{m_c^*}\right]$$

Since $E_g >> kT$ then to a good approximation $E_F$ is in the middle of the band gap.

In pure germanium $n, p \sim 10^{19}$ m$^{-3}$ at room temperatures.

*Impurities*

The presence of impurities can significantly alter the conduction properties of a semiconductor. If in a pure germanium semiconductor a quadrivalent germanium atom is replaced by a pentavalent impurity atom such as arsenic, it can be shown that bound states that can accommodate the extra electron are produced in the energy gap just below the bottom of the conduction band. A very small ionization energy (~ 0.005eV) will excite this electron into the conduction band. Such an impurity that readily donates an electron to the conduction band is called a donor.

Similarly, if a germanium atom is replaced by a trivalent impurity atom such as aluminium, unoccupied bound states are produced in the energy gap just above the top of the valence band. These states can accept electrons that are excited from the valence band and consequently produce holes in the valence band. Such impurities are called acceptors.

If impurities are present then in general $n \neq p$ but (13.88) still holds. If there are more donor atoms than acceptors $n > p$ and the material is $n$-type. Conversely if acceptor atoms predominate $p > n$ and the material is $p$-type.

Since the impurity levels are excited so readily the extrinsic conductivity of an impure semiconductor is much greater than the intrinsic conductivity of a pure crystal.

**Problems**

1. Assume that $\Psi(\mathbf{R}_1, \mathbf{R}_2, \ldots; \mathbf{r}_1, \mathbf{r}_2, \ldots)$ satisfies the complete crystal equation (13.1) and that the electronic wave-function $\psi_m(\mathbf{R}_1, \mathbf{R}_2, \ldots; \mathbf{r}_1, \mathbf{r}_2, \ldots)$ satisfies (13.4) with energy $E_m(\mathbf{R}_1, \mathbf{R}_2, \ldots)$. Expand

$$\Psi(\mathbf{R}_1, \mathbf{R}_2, \ldots; \mathbf{r}_1, \mathbf{r}_2, \ldots) = \sum_m \Phi_m(\mathbf{R}_1, \mathbf{R}_2, \ldots)$$
$$\psi_m(\mathbf{R}_1, \mathbf{R}_2, \ldots; \mathbf{r}_1, \mathbf{r}_2, \ldots)$$

and substitute into (13.1). Premultiply the resulting equation by $\psi_n^*(\mathbf{R}_1, \mathbf{R}_2, \ldots; \mathbf{r}_1, \mathbf{r}_2, \ldots)$ and integrate over all the electronic co-ordinates to obtain (assuming the $\psi_m$ are orthonormal)

$$(\mathscr{H}_{\text{ion}} + E_n)\, \Phi_n(\mathbf{R}_1, \mathbf{R}_2, \ldots) = E_c\, \Phi_n(\mathbf{R}_1, \mathbf{R}_2, \ldots) + S$$

where

$$S = \frac{\hbar^2}{2M} \sum_m \Phi_m(\mathbf{R}_1, \mathbf{R}_2, \ldots) \sum_j \int \psi_n^*(\mathbf{R}_1, \mathbf{R}_2, \ldots; \mathbf{r}_1, \mathbf{r}_2, \ldots)$$
$$\nabla^2_{\mathbf{R}_j} \psi_m(\mathbf{R}_1, \mathbf{R}_2, \ldots; \mathbf{r}_1, \mathbf{r}_2, \ldots)\, d\mathbf{r}$$
$$+ \frac{\hbar^2}{M} \sum_m \sum_j \int \psi_n^*(\mathbf{R}_1, \mathbf{R}_2, \ldots; \mathbf{r}_1, \mathbf{r}_2, \ldots)\, \text{grad}_{\mathbf{R}_j} \Phi_m(\mathbf{R}_1, \mathbf{R}_2, \ldots).$$
$$\text{grad}_{\mathbf{R}_j} \psi_m(\mathbf{R}_1, \mathbf{R}_2, \ldots; \mathbf{r}_1, \mathbf{r}_2, \ldots)\, d\mathbf{r}$$

Equation (13.5) now follows if $S$ can be neglected and this will be the case if $\psi_m$ is effectively independent of $\mathbf{R}_j$.

You may assume

$$\nabla^2 (fg) \equiv \text{div}\,(\text{grad}\, fg) = f\,\nabla^2 g + g\,\nabla^2 f + 2\,\text{grad}\, f \cdot \text{grad}\, g$$

where $f$, $g$ are scalar functions of position.

2. Using the periodic boundary condition solution (13.16), (13.17) for the free-electron problem, obtain the density of state function (13.14).

3. Use the formula

$$\overline{E}_k = \int_0^{E_m} E_k\, N(E_k)\, \mathrm{d}E \Big/ \int_0^{E_m} N(E_k)\, \mathrm{d}E$$

to show that the average energy of a free-electron gas at absolute zero is $\overline{E}_k = \frac{3}{5} E_m$. (That $\overline{E}_k$ is less than the free atom energy is the main cause of binding in metals).

4. Use equation (13.9) for free-electrons to confirm that the general density of state expression (13.42) gives (13.14).

5. The simple cubic lattice can be defined by the basis vectors

$$\mathbf{a}_1 = a\,(1, 0, 0),\ \mathbf{a}_2 = a\,(0, 1, 0),\ \mathbf{a}_3 = a\,(0, 0, 1)$$

Confirm that the reciprocal lattice is also simple cubic.

6. The body-centred cubic lattice can be defined by the basis vectors

$$\mathbf{a}_1 = a\,(1, 1, 1)\,,\ \mathbf{a}_2 = a\,(1, -1, 1)\,,\ \mathbf{a}_3 = a\,(1, 1, -1)$$

Confirm that the reciprocal lattice basis vectors are

$$\mathbf{b}_1 = \frac{\pi}{a}(0, 1, 1),\ \mathbf{b}_2 = \frac{\pi}{a}(1, -1, 0),\ \mathbf{b}_3 = \frac{\pi}{a}(1, 0, -1)$$

Verify the reciprocal lattice is face-centred cubic.

7. Use (13.27) to show that the sum

$$\sum_{\mathbf{K}_m} A_m\, \mathrm{e}^{i\,\mathbf{K}_m \cdot \mathbf{r}}$$

represents a function $f(\mathbf{r})$ that is periodic with the period of the direct lattice.

8. Substitute the energies (13.57) into (13.55) to show that the wave-functions at the zone boundary $k = \frac{\pi}{a}$ are

$$\psi^+ = i\sqrt{\frac{2}{Na}}\ \sin\frac{\pi x}{a}$$

$$\psi^{-} = \sqrt{\frac{2}{Na}} \cos \frac{\pi x}{a}$$

if $V_1 < 0$. Find where the charge densities of these wave-functions are maximum.

9. The face-centred cubic lattice has 12 neighbours at

$$a\,(0, \pm 1, \pm 1),\ a\,(\pm 1, 0, \pm 1),\ a\,(\pm 1, \pm 1, 0)$$

Use the tight-binding method to show the energy band constructed from an atomic *s*-state is, up to nearest neighbours, given by

$$E(\mathbf{k}) = E_a + \alpha + 4\beta\,[\cos a\,k_y \cos a\,k_z + \cos a\,k_z \cos a\,k_x$$

$$+ \cos a\,k_x \cos a\,k_y]$$

where $\alpha, \beta$ are appropriate interaction integrals and $E_a$ is the free atom energy.

10. In a certain cubic crystal the energy band function can be written locally as

$$E(\mathbf{k}) = \frac{\hbar^2}{2}\left[\frac{(k_x - k_0)^2}{m_1^*} + \frac{k_y^2 + k_z^2}{m_2^*}\right]$$

where $k_0, m_1^*, m_2^*$ are constants. Find 6 points in the Brillouin zone where the electron velocity is zero.

11. The free-electron wave-function $e^{i\mathbf{k}\cdot\mathbf{r}}$ has momentum $\mathbf{p} = \hbar\,\mathbf{k}$. When an electron field $\mathscr{E}$ is applied a potential $e\,\mathscr{E}\cdot\mathbf{r}$ is introduced. Use (5.72) to show that a wave-packet with mean momentum $\hbar\,\mathbf{k}$ undergoes a change of momentum according to

$$\hbar\,\dot{\mathbf{k}} = -e\,\mathscr{E}$$

12. The basis vectors

$$\mathbf{a}_1 = \frac{a}{2}\,(1, 1, 0)$$

$$\mathbf{a}_2 = \frac{a}{2}\,(1, 0, 1)$$

$$\mathbf{a}_3 = \frac{a}{2}\,(0, 1, 1)$$

define a Bravais lattice. Identify the lattice and obtain the basis vectors for the reciprocal lattice.

Show that if a sphere is inscribed in the Brillouin zone its radius is $\sqrt{3}\,\pi/a$ and also that the Fermi sphere will just touch the zone boundary for a conduction electron concentration of 1.36 electrons per atom. (This result is important in the Hume-Rothery theory of alloys.)

13. (a) Show that in the presence of an electric field $\mathscr{E}$ and a magnetic induction field **B** a band electron satisfies

$$\hbar \dot{\mathbf{k}} = -e\,(\mathscr{E} + \mathbf{V} \wedge \mathbf{B})$$

where $-e$ is the electron charge.

(b) By considering two neighbouring constant energy orbits in **k**-space show that in the presence of a magnetic induction field **B** alone, an electron undergoes a periodic motion with period

$$T = \frac{\hbar^2}{e\,B}\,\frac{\mathrm{d}A}{\mathrm{d}E}$$

where $A\,(E)$ is the area in **k**-space of an orbit of energy $E$. If the energy surfaces are spheres of the form

$$E\,(\mathbf{k}) = \frac{\hbar^2}{2m^*}\,(k_x^2 + k_y^2 + k_z^2)$$

show that $T = 2\pi\, m^*/e\,B$
(This cyclotron resonance effect may be observed at very low temperatures when collisions are less frequent).

14. The hexagonal lattice can be defined by the basis vectors

$$\mathbf{a}_1 = a\,(\tfrac{1}{2}, \tfrac{\sqrt{3}}{2}, 0) \quad \mathbf{a}_2 = a\,(-\tfrac{1}{2}, \tfrac{\sqrt{3}}{2}, 0) \quad \mathbf{a}_3 = b\,(0, 0, 1)$$

$b > a$ constants

(a) Confirm that there are six nearest neighbours and an axis of 6-fold symmetry.

(b) Suppose two atoms are associated with each lattice point, those associated with the origin being at

$$(0, 0, 0) \qquad \text{and} \qquad (0, \frac{a}{\sqrt{3}}, \frac{b}{2})$$

Show that the condition that each atom has 12 nearest neighbours each at distance $a$ is that

$$b = \sqrt{\frac{8}{3}}\,a$$

(This is the hexagonal close-packed structure, e.g. Mg, Ca .)

## Recommended References and Further Reading

Bohm, D. and Pines, D. *Phys. Rev.* **92**, 609 (1953).
Born, M. and Oppenheimer, R. *Ann. Physik* **87**, 457 (1927).
Callaway, J. *Quantum Theory of the Solid State*, Parts A and B. Academic Press, New York (1974).

Clark, H. *Solid State Physics – An introduction to its theory.* Macmillan, London (1968).

Herring, C. *Phys. Rev.* **57**, 1169 (1940).

Mott, N.F. *Reviews of Modern Physics* **50**, No. 2, 203 (1978).

Mott, N.F. and Davis, E.A. *Electronic Processes in Non-crystalline Materials.* **O.U.P., New York (1978).**

Slater, J.C. 'The Electronic Structure of Solids', *Encyclopaedia of Physics.* Springer-Verlag, Berlin (1956).

Slater, J.C. and Koster, G.F. *Phys. Rev.* **94**, 1498 (1954).

Smith, R.A. *Wave Mechanics of Crystalline Solids.* Chapman & Hall, London (1961).

# Index